# THE WATER PROFESSIONAL'S GUIDE TO INFECTIOUS DISEASE OUTBREAKS

WEF Special Publication

2023

Water Environment Federation
601 Wythe Street
Alexandria, VA 22314-1994 USA
http://www.wef.org

ISBN 978-1-57278-412-3



## *About WEF*

The Water Environment Federation (WEF) is a not-for-profit technical and educational organization of 30,000 individual members and 75 affiliated Member Associations representing water quality professionals around the world. Since 1928, WEF and its members have protected public health and the environment. As a global water sector leader, our mission is to connect water professionals; enrich the expertise of water professionals; increase the awareness of the impact and value of water; and provide a platform for water sector innovation. To learn more, visit www.wef.org.

Prepared by **The Water Professional's Guide to Infectious Disease Outbreaks** Task Force of the **Water Environment Federation**


Rasha Maal-Bared, PhD, *Chair*

Nija Ali
Tiong Gim Aw, PhD
Akin Babatola
Adam Baron
Philip Block, PhD
Kari Brisolara
Tom Burgess, MS, CSP, CIH
Stephanie Corso
Andrew F. DeGraca, PE
Shea Dunifon
Tricia Garrison
Lee P. Gary Jr.
Chuck Gerba
David Gill
Maureen Holman
Krystin Kadonsky
Alexandra Lampson
Bo Li
Jen Loudon
Dusti Lowndes
Mia Mattiolli
Anna Mehrotra
Bina Nayak, PhD
Kevin O'Brien
Keerthi Palanisamy
Sehul Patel, PE
Leilani Perriatt
Scott Perry
Aditya Ramamurthy
Bob Reimers
Jenny Reina
Albert Rubin
Prakit Saingam
Samendra Sherchan
Stan Shmia
Nelson Sims
Mark Sobsey
Jay Swift
Dienye Tolofari, PhD
Mark Wilson
Yue Xu, PhD


Under the Direction of the **Disinfection and Public Health Subcommittee** of the **Technical Practice Committee**

*Special Publications of the Water Environment Federation*

The WEF Technical Practice Committee (formerly the Committee on Sewage and Industrial Wastes Practice of the Federation of Sewage and Industrial Wastes Associations) was created by the Federation Board of Control on October 11, 1941. The primary function of the Committee is to originate and produce, through appropriate subcommittees, publications dealing with technical aspects of the broad interests of the Federation. These publications are intended to provide background information through a review of technical practices and detailed procedures that research and experience have shown to be functional and practical.

# Contents

# List of Figures

# List of Tables

# Preface

It is undeniable that our mission as water professionals is to protect our workers, public health, and the environment in which we operate. However, when infectious disease outbreaks hit, our ability to do so is potentially compromised and we are inundated with information of various levels of reliability and relevance that we are assumed to understand and incorporate into decision-making.

In April 2019, nine members of the Water Environment Federation's Disinfection and Public Health Community met to discuss the need for a team focused on infectious disease outbreaks. We agreed that the 2014–2016 Ebola outbreak in West Africa was a wake-up call. The Water Sector, as a community, needed a trusted resource we could depend on during times of need. That was how the Waterborne Infectious Disease Outbreak Control (WIDOC) Working Group was formed. Little did we know that our WIDOC team would grow and produce tens of guidances, papers, presentations, and graphics to support WEF and the community we love and respect, as COVID-19 became the new normal.

It was during that time in exchanges with water professionals across the United States and Canada that we realized that when infectious disease outbreaks happen, we all have the same concerns regardless of utility size, geography or the pathogens involved. We ask ourselves and the experts around us: (1) Is our process effective against these pathogens? (2) Are our workers safe? and (3) Do we need to do anything differently to ensure we meet our permits or approvals to operate? Continuing on the path of responsiveness to water professionals' needs, we decided to tackle these complex questions and invite more experts to the table that can help us do these questions justice. Thirty-six coauthors, many meetings, and 2 years later, we produced the book in your hands.

The authors hope that this book equips you—the reader—with the basic literacy to consume and evaluate the vast information produced during outbreaks. It explains concepts involving pathogens, treatment practices, risk assessment, health and safety evaluation approaches, adaptations in facility management, communications, and wastewater surveillance. In an ideal world, it will even empower you to make the best decisions or recommendations possible to protect our workers, public health, and the environment in a One Water context. Who can ask for anything more?

Authors' and reviewers' efforts were supported by the following organizations:

District of Columbia Water and Sewer Authority
Rogue Water
Sanitary Consortium, Inc.
Tulane University, International Development Studies
Tulane University, School of Public Health and Tropical Medicine, Department of Environmental Health Services

# Chapter 1

# What Are Infectious Disease Outbreaks?

*Rasha Maal-Bared, PhD; Kari Brisolara; Keerthi Palanisamy; Leilani Perriatt; & Albert Rubin*



## 1.0 INFECTIOUS DISEASES AND ZOONOSES

### 1.1 History of Infectious Agents and Disease Outbreaks

In December of 2019, a small cluster of viral pneumonia cases diagnosed in Wuhan City in China was reported to the World Health Organization (WHO). When the Chinese authorities sequenced the genome of the causative agent, the virus was determined to be new (i.e., never having infected humans in a sustained manner previously) but closely related to both Severe Acute Respiratory Syndrome Coronavirus 1 (SARS-CoV-1) and Middle Eastern Respiratory Syndrome Coronavirus (MERS-CoV), making this the third documented spillover of an animal coronavirus to humans in only two decades (Morens & Fauci, 2020). By January 30, 2020, the number of global infections had increased sufficiently to warrant declaring this outbreak a public health emergency of international concern (WHO, 2020a). On March 11, 2020, the WHO declared coronavirus disease (COVID-19) a pandemic (WHO, 2020b). Since then, more than 648 million COVID-19 cases have been confirmed, resulting in 6.65 million deaths worldwide (Johns Hopkins University, 2023).

Without a doubt, COVID-19 is unprecedented in its impacts on the global community, but humans have been facing infectious disease outbreaks for millennia. In 1918 and 1919, the world saw the onslaught of the Spanish Influenza, which resulted in the deaths of an estimated 50 to 100 million people. The HIV/AIDS epidemic has killed more than 30 million individuals over the last few decades. Cholera continues to plague many developing countries and most recently resulted in the Haiti Cholera Outbreak of 2010. SARS-CoV-1 and the 2009 Influenza A (H1N1) epidemics followed. Although low-resource settings and developing countries continue to struggle with outbreaks related to water treatment and sanitation (e.g., Dengue fever, resistant typhoid fever, polio), high-income countries are seeing a shift toward waterborne infectious disease outbreaks from environmental pathogens that can grow in drinking water distribution systems, plumbing systems, recreational water venues, and industrial water systems (Collier et al., 2021).

In 2014, biofilm-associated pathogens in the drinking water distributions system—such as nontuberculous mycobacteria, *Pseudomonas*, and *Legionella*—were estimated to cause the most hospitalizations and deaths in the United States and cost the health care system approximately US $2.39 billion (Collier et al., 2021). In 2016, cases of Legionnaires' disease in Genesee County in Flint, Michigan, had spiked during the period of Flint River usage, causing 45 cases with 5 deaths from 2014 to 2015 and 46 cases with 7 deaths in 2015 (Schwake et al., 2016). Similarly, 32 Legionnaires' disease outbreaks were reported in New York City from 2003 to 2012 and five outbreaks of Legionella within a 4-month period in 2016 in the Sydney area of Australia (MacIntyre et al., 2018). Although it is becoming increasingly difficult to predict what is next and who will be affected, the only thing we know with certainty is that outbreaks are inevitable. Figure 1.1 shows a brief history of epidemics starting in the Neolithic Age (Morens & Fauci, 2020; Spyrou et al., 2019).

With the unprecedented mobility of people, products, and food, and the fact that disease-causing microorganisms—called pathogens—are

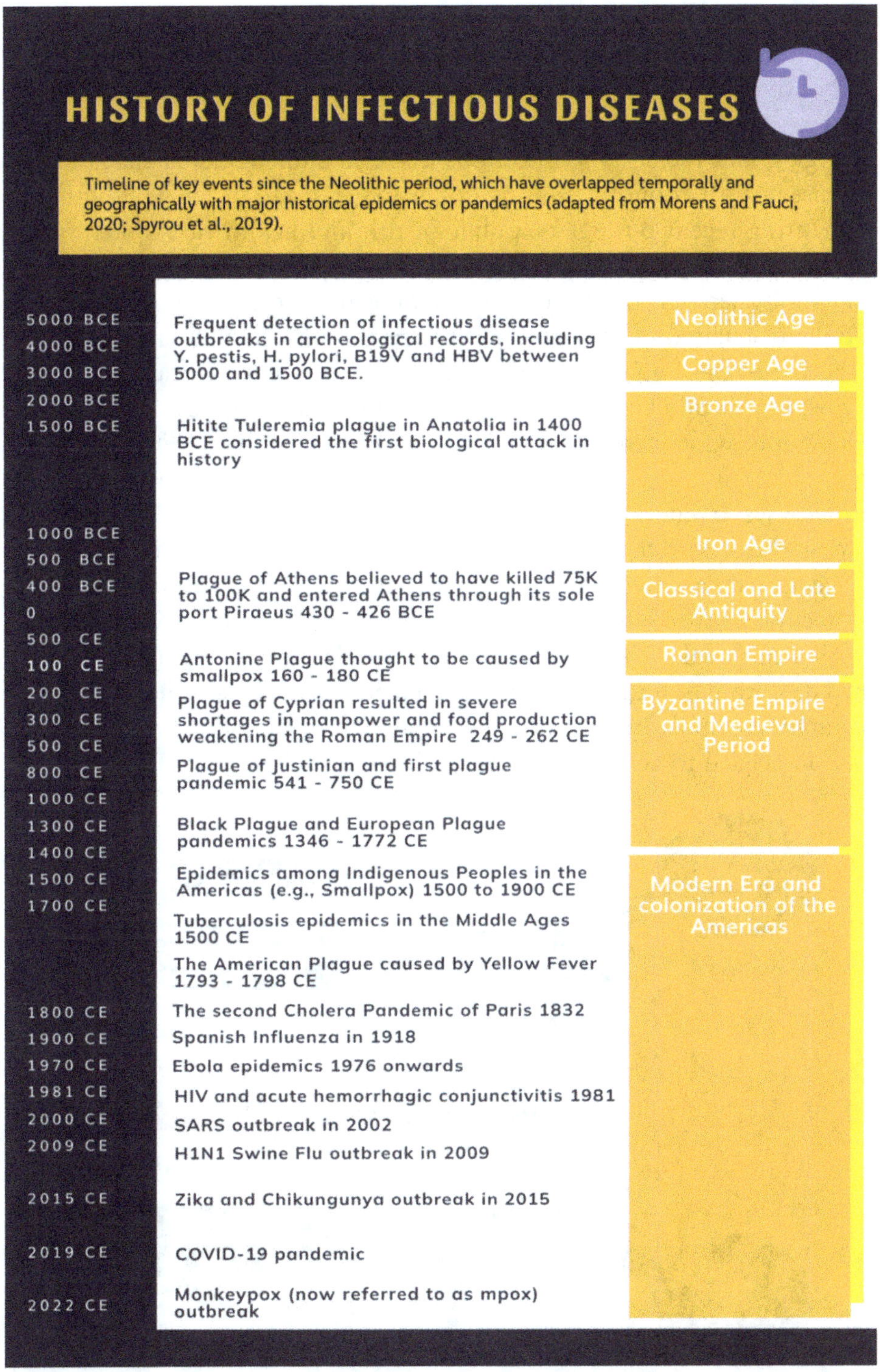


**FIGURE 1.1** History of Infectious Disease Outbreaks (Adapted from Morens & Fauci, 2020; Spyrou et al., 2019)

increasingly treatment resistant and mobile, this is no longer an issue that impacts only developing countries. No nation is immune to the growing global threat posed by an isolated infectious disease outbreak in a seemingly remote part of the world. Morens and Fauci (2013) discussed newly emerging and reemerging infectious diseases and suggested that the best defense against zoonotic transmission is to get ahead of the next outbreak.

During recent outbreaks, public health and health care systems have played an undisputed role in provision of treatment services, surveillance, and public health protection against pathogens on the frontlines. Less commonly articulated in the press and the public imagination is the key contribution of the water sector. Water professionals play a critical role in the protection of public, occupational, and environmental health. Collection system and wastewater treatment workers often put themselves at risk to keep water flowing and waste treated. The objective of this book is to provide water and wastewater professionals, as well as utility managers and workers, with the critical knowledge needed to be an active, informed, and engaged participant in public, occupational, and environmental health protection during infectious disease outbreaks, as outlined in Figure 1.2. During outbreaks, the water professional's day job is expanded to often include data-based risk assessment, evidence-based risk management, and communication and outreach. The chapters in this book provide the information needed to accomplish these tasks effectively.

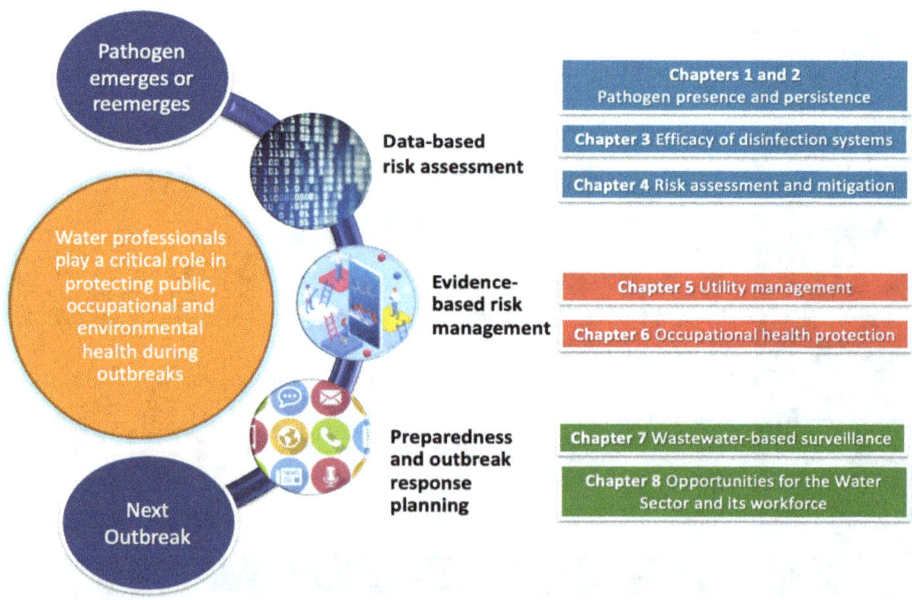


**FIGURE 1.2** Chapters in the *Water Professional's Guide to Infectious Disease Outbreaks* and How They Relate to the Water Professional's Possible Roles During Infectious Disease Outbreaks

## 1.2 What Are Zoonoses?

When a pathogen (e.g., bacterial, viral, protozoan, helminths, fungal) makes the leap from an animal host to a human host, that transition is referred to as "zoonotic." A zoonosis (plural zoonoses) is any disease or infection that is naturally transmissible from vertebrate animals to humans (e.g., rabies). There are more than 200 zoonoses known to man, many of which are viral and pose significant health consequences to humans (O'Brien & Xagoraraki, 2019). Viruses historically have been the most relevant pathogens, resulting in significant outbreaks, infections, and mortality (Morens & Fauci, 2020).

Emerging infectious diseases (EIDs) remain a significant concern for public health officials worldwide. Emerging infectious diseases have plagued human populations for millennia; in fact, the plagues we read about from ancient Rome (Justinian's Plague) and the Middle Ages (the Bubonic Plague) were diseases of zoonotic origin (see Figure 1.1). The sciences of medicine and epidemiology were not well developed during these plagues, and the causative agents were not discovered in time to abate these public health crises. The cause of these plagues was the bacteria *Yersinia pestis*—a bacterium carried by rats and fleas. The loss of life attributed to these plagues was significant: historians estimate that Justinian's Plague resulted in loss of 20% of the population within the empire, and the Bubonic Plague of the 14th century is reportedly the single most fatal pandemic reported in Western history, resulting in the loss of between 100 and 200 million people in Europe, Asia, and North Africa (Bernstein et al., 2022; Morens & Fauci, 2020).

More recently, the 1993 cryptosporidiosis outbreak in Milwaukee, Wisconsin, was caused by *Cryptosporidium parvum* and attributed to fecal contamination of the watersheds supplying the drinking water system. Although the source of the fecal contamination was not definitively identified, suspected sources were livestock and wildlife. The pathogenic organism responsible for the outbreak is chlorine resistant, and the water treatment system was ineffective in controlling source water contamination. That outbreak impacted more than 400 000 individuals and resulted in 70 reported deaths (Corso et al., 2003). *Cryptosporidium* is widespread throughout the world and is responsible for millions of cases of disease annually (Rose, 1997).

The National Institutes of Health estimates that 75% or more of the EID burden today is associated with zoonotic transmission (Ogden et al., 2017). To cope with the increasing threat levels associated with zoonotic disease, the U.S. Centers for Disease Control and Prevention (CDC) developed the *One Health Zoonotic Disease Prevention Tool* to help governmental and nongovernmental entities identify and control emerging infectious diseases of concern to human populations. The CDC suggests that emerging and endemic zoonotic diseases pose significant health risks to both human and animal populations, potentially resulting in localized transmission, epidemics, and even pandemics, as seen in Figure 1.3. The CDC estimated that zoonotic EIDs are responsible for 2.5 billion cases of illness and 2.7 million deaths annually worldwide (Nabarro & Wannous, 2014).

Zoonoses originate from close interactions between humans and animals and are often linked to agriculture and animal husbandry practices. Emerging infectious diseases not only impact human health but are also increasingly associated with sound animal husbandry, wildlife protection, and animal welfare. Transmission of health threats from humans to animal populations (called anthroponoses) can dramatically reduce the viability of livestock production operations.

Zoonotic diseases are transmitted via animal and insect bites, aerosol transmission of fine droplets containing infectious agents, water contamination from fecal material, and other routes. The types of organisms typically associated with disease transmission are viruses, bacteria, parasitic protozoans and microorganisms, helminths or worms, plant-based organisms like fungi, and others. The organisms associated as carriers of these disease-producing organisms include insects, rodents, poultry and other avian species, swine, cattle, deer, elk, and others. Infectious diseases may not always be the causative agent associated with casualties. Immunocompromised individuals exposed to these infectious agents may develop disease stresses that lead to serious health consequences and death. Opportunistic infections may result

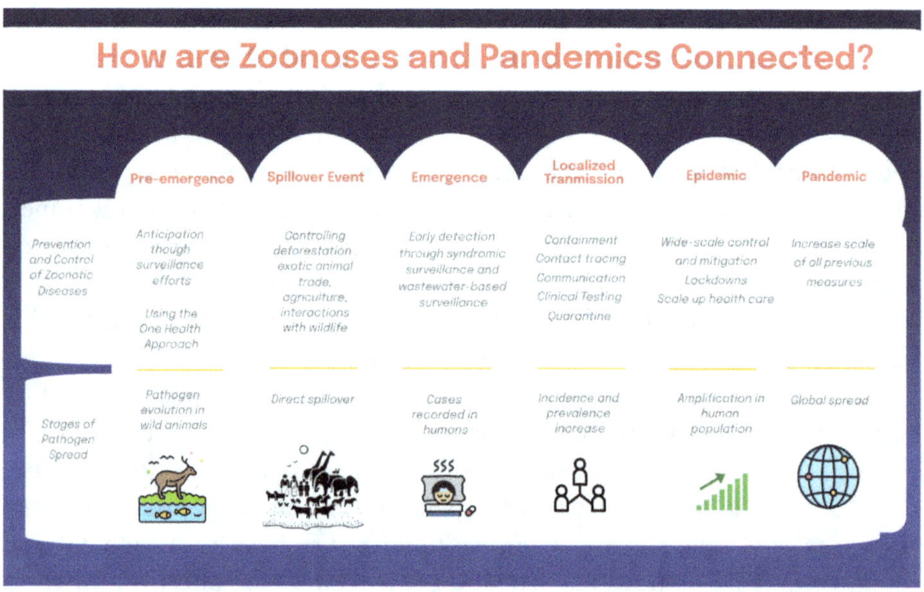


**FIGURE 1.3** The Connection Between Zoonoses, Epidemics, and Pandemics (Adapted from Bernstein et al., 2022)

in sequelae leading to an advanced debilitative condition. For example, AIDS patients may develop fatal cases of pneumonia as a result of exposure to viral, bacterial, or fungal zoonotic infections. Scientists, engineers, public health professionals, and medical professionals all play significant roles in the identification and risk management of EID agents.

# 2.0 DEFINITIONS YOU SHOULD KNOW

## 2.1 Public Health Definitions

As public health specialists, our goal is to create and enhance the conditions in which our community can be at its healthiest. Our understanding of infectious disease has undergone considerable transformation, but terminology continues to stay constant within the disciplines of epidemiology and public health. Whether to address foodborne illness or the involvement of zoonotic infections, key terms will appear. Here are some of the key terms and definitions (CDC, 2015):

- **Anthroponosis:** An infection or disease originating from a human and transmitted to human or nonhuman animals.
- **Communicable:** A disease that can be transmitted from one person to another.
- **Contagion:** A general term for any disease-causing infectious agent spread by direct or indirect contact.
- **Contamination:** The soiling or pollution of inanimate objects or living material with harmful, potentially infectious, or other unwanted material, for example organic matter or microorganisms.
- **Disease cluster:** Aggregation of relatively uncommon events or diseases in space and/or time in amounts that are believed or perceived to be greater than could be expected by chance.
- **DNA:** See Section 2.4: "Molecular Biology Definitions."
- **Emerging disease:** Appearance of a disease in a new host.
- **Endemic:** Consistently present within a localized area or population, making disease spread and rates more predictable.
- **Epidemic:** Attacking or affecting many persons simultaneously in a community or area; a widespread occurrence of disease.
- **Fomite:** An inanimate object that transmits infection (e.g., a towel or doorknob).
- **Incidence rate:** A measure of the frequency with which an event, such as a new case of illness, occurs in a population over a period of time.

The denominator is the population at risk; the numerator is the number of new cases occurring during a given time period.

- **Incubation time:** The period of time between exposure to an infectious agent and the appearance of symptoms of the infection or disease it causes.
- **Infectious disease:** A type of illness caused by a pathogenic agent, including viruses, bacteria, fungi, protozoa, parasites, or abnormal proteins known as prions.
- **Outbreak:** An unexpected increase in the incidence of a particular disease over a given time period and geographic range.
- **Pandemic:** An increase in the occurrence of a particular disease over a very large region, such as a continent or the entire globe, that is greater than what is expected over a given period of time.
- **Prevalence:** The number of cases or conditions in a given population.
- **Prevalence rate:** The number of persons in a population who have a particular disease or condition at a specified point in time or over a specified period of time.
- **Reservoir:** A place where potentially pathogenic microorganisms can survive and may be transferred onto patients.
- **Sequela** (plural **sequelae**): A condition that is the consequence of a previous disease or injury.
- **Soiled:** A term used to describe objects or items contaminated with debris, organic matter, which potentially can harbor pathogenic organisms.
- **Source:** The part of the reservoir that provides the organisms that have infected or colonized patients (i.e., where the organisms have come from).
- **Spillover** (or **host switching**): Process by which a pathogen adapted to one host species becomes adapted to another host species (Morens & Fauci, 2020).
- **Variants:** When errors occur in the replication of the viral genetic material, similar but not exact copies of the original virus are produced. These errors are called mutations, and viruses with these mutations are called variants. Variants could differ by a single or many mutations (Lauring & Hodcroft, 2021).
- **Vector:** An organism, usually an insect, that carries a disease-producing material from one host to another, either within or on the surface of its body.

- **Zoonosis:** A human infection caused by an animal pathogen that either may be a dead-end infection or may initiate person-to-person spread (Morens & Fauci, 2020).

## 2.2 Microbiology Definitions

Relevant microbiology definitions that are of interest to water professional include the following (Mara & Horan, 2003):

- **Aerobe:** An organism that requires free oxygen for respiration.
- **Anaerobe:** An organism that can thrive in the absence of free oxygen, nitrate, and nitrite.
- **Autotroph:** Organism that derives its cell carbon from carbon dioxide.
- **Bacterium:** Singular form of a large group of unicellular microorganisms that lack a cell nucleus. Some bacteria are pathogenic and harmful to humans, some have no effect at all on humans, and some are beneficial.
- **Bioaerosol:** A mist that could transport biological material.
- **Biological oxidation:** A process in which living organisms oxidize organic matter into a more stable or mineral form.
- **Colony-forming units:** The number of bacteria present in a sample, as determined in a laboratory plate-count test. In this test, the number of visible bacteria colony units present is counted.
- **Cryptosporidiosis:** Gastrointestinal disease caused by ingesting waterborne *Cryptosporidium* spp., often the result of drinking contaminated runoff from pastures or farmland.
- ***Cryptosporidium parvum* and *Cryptosporidium hominis*:** Two species of the *Cryptosporidium* genus known to be infectious to humans. A protozoan parasite that can live in the intestines of humans and animals.
- **Culture:** A microbial growth developed by furnishing enough nutrients in a suitable environment.
- **Cyst:** A resting stage formed by some bacteria and protozoa in which the whole cell is surrounded by a protective layer.
- **Enteric bacteria:** Bacteria that inhabit the gastrointestinal tract of warm-blooded animals.
- ***Escherichia coli*:** A fecal coliform bacteria species used as an indicator of wastewater pollution.
- **Facultative bacteria:** Microbes that can survive with or without oxygen.

- **Fecal coliform:** Coliforms present in the feces of warm-blooded animals. Aerobic and facultative, Gram-negative, non-spore-forming, rod-shaped bacteria capable of growth at 44.5 °C (112 °F) and associated with the fecal matter of warm-blooded animals.
- **Fecal indicators:** Fecal coliform, fecal *Streptococci*, and other bacterial groups originating in human or other warm-blooded animals, indicating contamination by fecal matter.
- **Fungi:** A taxonomic kingdom of spore-forming organisms distinct from plants, animals, and bacteria that includes microorganisms such as yeast and molds, as well as mushrooms.
- ***Giardia lamblia*:** A protozoan parasite; responsible for giardiasis.
- **Giardiasis:** A gastrointestinal disease caused by ingesting waterborne *Giardia lamblia*, often resulting from the activity of beavers, muskrats, or other warm-blooded animals in surface water used as a potable water source.
- ***Heliobacter pylori*:** A bacterium that causes stomach ulcers and has been identified as an emerging waterborne health threat.
- **Helminths:** Parasitic worms that live and feed on living hosts to receive nourishment and protection, which results in illness of the host.
- **Heterotrophic bacteria:** A type of bacteria that derives its cell carbon from organic carbon; most pathogenic bacteria are heterotrophic bacteria.
- **Indicator organism:** A microbe whose presence indicates the absence or presence of a specific pollutant.
- **Infectious agent:** Any organism that can be communicated in body tissues and can cause disease or other adverse health effects in humans.
- **Most probable number (MPN):** A statistical analysis technique based on the number of positive and negative results acquired when testing multiple portions of equal volume; typically used to count pathogens in solids samples.
- **Ovum** (plural **ova**): A mature parasite egg ready to be fertilized.
- **Pathogen:** Highly infectious, disease-producing microbes typically found in sanitary wastewater.
- **Prion:** A causative agent of infectious disease that is composed primarily of protein.
- **Protozoan** (plural **protozoa**): A taxonomic group of single-celled microorganisms that live in almost every kind of habitat and include some pathogenic parasites of humans and other animals.

- **Spore:** A reproductive cell or seed of a microbe, often dormant or environmentally resistant.
- ***Streptococcus*:** A genus of bacteria that includes some of the most common human pathogens.
- **Total coliforms:** A group of bacteria that are naturally found in soil, water, plants, and the intestinal tract of humans and warm-blooded animals. Because they are commonly found in the environment, they are often used to provide an indication of drinking water safety or changes in drinking water quality. Fecal coliforms and *E. coli* are total coliforms.
- **Virus:** The smallest biological structure capable of reproduction; it can only grow and reproduce inside a host organism; infects its host, producing disease.

## 2.3 Water and Wastewater Treatment Definitions

The main source for the definitions provided below are the Water Environment Federation (WEF) Fact Sheets and *Standards for Design of Water Resource Recovery Facilities* (WEF 8) (WEF, 2023).

- **Aerobic:** Condition characterized by the presence of free oxygen.
- **Aerobic digestion:** Solids stabilization process involving direct oxidation of biodegradable matter and oxidation of microbial cellular material.
- **Alkaline stabilization:** The process by which lime or other alkaline materials are added to solids to raise the pH above 12 for 2 hours to reduce pathogens.
- **Alkalinity:** The ability of a water to neutralize an acid via the presence of carbonate, bicarbonate, and hydroxide ions.
- **Anaerobic:** Condition characterized by the absence of free oxygen and other electron receptors such as nitrate and sulfate. As a process, *anaerobic* implies the active presence of strictly anaerobic organisms. As an activated sludge process component, *anaerobic* implies the active presence of anaerobic and facultative organisms.
- **Anaerobic digestion:** Solids stabilization process operated specifically without oxygen in which much of the organic waste feed is converted to methane and carbon dioxide.
- **Autothermal thermophilic aerobic digestion:** A biological digestion system that converts soluble organics to lower-energy forms

via anaerobic, fermentative, and aerobic processes at thermophilic temperatures.

- **Average daily flow:** The total flow past a physical point over a period of time divided by the number of days in that period, typically taken to mean a yearly (annual) average.
- **Average day, maximum month:** The average daily flow or mass of a constituent during the month of maximum measurements for that constituent.
- **Average flow:** The arithmetic average of flows measured at a given point, typically taken to mean a yearly (annual) average.
- **Biological process:** Any process in which microorganisms metabolically break down complex organic materials into simple, more stable substances.
- **Biosolids:** Solids that have been removed from wastewater and stabilized (e.g., digested or composted) to meet the criteria in the U.S. Environmental Protection Agency's (U.S. EPA's) 40 CFR 503 regulations and, therefore, can be beneficially used.
- **Chemical treatment:** Any treatment process (e.g., precipitation, coagulation, flocculation, sludge conditioning, disinfection, or odor control) involving the addition of chemicals to obtain a desired result.
- **Class A biosolids:** Biosolids that contain less than 1000 most probable number (MPN)/g of fecal coliforms and less than 3 MPN/4g of *Salmonella* bacteria and meet one of six stabilization alternatives given in 40 CFR 503. The material also must meet the pollutant limits and vector-attraction reduction requirements set forth in Part 503.
- **Class B biosolids:** Biosolids that contain less than 2 million colony-forming units (CFU; most probable number [MPN]) of fecal coliforms per gram of dry biosolids. The material also must meet the pollutant limits and vector-attraction reduction requirements set forth in 40 CFR 503.
- **Co-digestion:** A process in which two or more types of substrates (feedstocks) are digested together in the same reactor. Frequently refers to anaerobic digestion of solids together with food wastes, food processing wastes, fats, oils and grease (FOG), or other organic waste material.
- **Compliance standards:** The water-quality and biosolids-quality requirements specified in a treatment plant's National Pollutant Discharge Elimination System permit that must be met before the effluent can be discharged and the biosolids beneficially used (or disposed).

- **Composite variable:** A combination of state variables typically to form variables that can actually be measured in the facility (e.g., biochemical or biological oxygen demand [BOD], chemical oxygen demand [COD], total Kjeldahl nitrogen [TKN], total phosphorus [TP], total suspended solids [TSS], volatile suspended solids [VSS]).
- **Digested solids:** Solids in which the concentration of volatile solids has been significantly reduced via oxidation by microbes in an aerobic or anaerobic reactor. The digested material is now relatively nonputrescible and inoffensive.
- **Digester:** A tank or other vessel used to store and anaerobically or aerobically decompose the organic matter in solids. See also *aerobic digestion* and *anaerobic digestion*.
- **Digestion:** The process of biologically oxidizing the organic matter in solids, thereby reducing the concentrations of volatile solids and pathogens.
- **Disinfectant:** A substance used to disinfect water, wastewater, or solids.
- **Disinfection:** The selective destruction of disease-causing microbes via the application of chemicals or energy.
- **Extended aeration:** A modification of the activated sludge process that uses long aeration periods to promote aerobic digestion of the biological mass via endogenous respiration. The process also stabilizes organic matter under aerobic conditions and emits gaseous products. The effluent contains finely divided suspended matter and soluble matter.
- **Hydrophilic:** Having a strong affinity for water.
- **Hydrophobic:** Having an aversion to water.
- **Indirect potable reuse:** The use of highly treated reclaimed water to augment a surface water or groundwater intended to serve as a potable water supply.
- **Lime stabilization:** A process in which lime is added to solids to raise the pH to 12 for at least 2 hours to chemically inactivate pathogens.
- **Mesophilic:** An operating temperature range (typically 30 to 40 °C) for anaerobic digestion; it affects the microbial population in the digester and the reaction rates.
- **Mesophilic digestion:** A process in which solids are digested by microorganisms that thrive in the mesophilic temperature range (about 30 to 40 °C).

- **Microconstituents:** Natural and anthropogenic substances (e.g., elements and inorganic and organic chemicals) detected in water and the environment, for which a prudent course of action is suggested for the continued assessment of the potential effect on human health and the environment.
- **Onsite (wastewater treatment) system:** A system designed to collect and treat wastewater from one or more dwellings, buildings, or structures; the resulting effluent is dispersed on property owned by the individual or entity.
- **Oxidation:** (1) A chemical reaction in which an element or ion loses an electron or increases in oxidation state (i.e., the opposite of reduction). (2) The biological or chemical conversion of organic matter to simpler, more stable forms.
- **Ozonation:** A treatment process that uses ozone for oxidation, disinfection, or odor control.
- **Ozone:** A strong oxidizing agent with disinfection properties similar to chlorine. Also used in odor control and solids processing. Chemical formula is $O_3$.
- **Pasteurization:** A process in which heat is applied for a specific period of time to kill pathogens.
- **Physical treatment:** A water or wastewater treatment process that uses only physical methods (e.g., filtration or sedimentation).
- **Physical–chemical treatment:** A water or wastewater treatment process that uses both physical and chemical methods.
- **Preliminary treatment:** Treatment steps (e.g., comminution, screening, grit removal, pre-aeration, and/or flow equalization) that prepare wastewater influent for further treatment.
- **Pretreatment:** (1) The initial water or wastewater treatment process that precedes primary treatment processes. (2) The treatment of industrial wastes to reduce or alter the characteristics of pollutants before the wastes are discharged to a wastewater treatment facility.
- **Primary clarifier:** A sedimentation basin that precedes secondary wastewater treatment.
- **Primary distribution:** The high-voltage section (at least 2400 V) of the electrical distribution system.
- **Primary residuals:** Solids produced via sedimentation.
- **Primary sedimentation:** A gravity-based process for removing settleable suspended solids from water or wastewater; typically occurs in a

quiescent basin or clarifier. The principal form of primary wastewater treatment, which is used to reduce the solids loading on subsequent treatment processes.

- **Primary sludge:** Solids produced during primary wastewater treatment (e.g., sedimentation).
- **Primary treatment:** Treatment processes (e.g., sedimentation and/or fine screening) designed to produce an effluent suitable for biological treatment.
- **Residuals:** The nonliquid components of wastewater that are removed from the liquid during various treatment processes.
- **Retention time:** The length of time that water or wastewater will be retained in a unit treatment process or facility.
- **Return activated sludge:** Settled activated sludge that is returned to the beginning of the activated sludge process to mix with raw or primary settled wastewater.
- **Secondary clarifier:** A vessel in which suspended matter is removed from wastewater via gravity; this vessel is located after a secondary treatment process.
- **Secondary distribution:** The low-voltage (480 V) section of the electrical-distribution system.
- **Secondary effluent:** Wastewater that has received preliminary, primary, and secondary treatment.
- **Secondary sludge:** Solids generated during a secondary treatment process.
- **Secondary treatment:** Any process designed to degrade the biological content of wastewater; typically follows primary treatment.
- **Sedimentation:** A gravity-based process for removing settleable suspended solids from water or wastewater; typically occurs in a quiescent basin or clarifier.
- **Sedimentation basin:** A quiescent tank in which suspended solids are removed from water or wastewater via gravity; they typically are equipped with a motor-driven rake to collect settled solids and move them to a central discharge point. Also called *clarifiers* or *settling tanks*.
- **Sludge:** Any residual produced during primary, secondary, or advanced wastewater treatment that has not undergone any process to reduce pathogens or vector attraction. Also called *raw sludge*. The term *sludge* should be used with a specific process descriptor (e.g., primary sludge, waste activated sludge, or secondary sludge).

- **Solids thickening:** A process designed to increase the solids concentration in residuals by removing a portion of the liquid; such processes include a sedimentation tank, dissolved air flotation (DAF), gravity thickener, centrifuge, gravity belt thickener, and membrane thickener.
- **Storm sewer:** An underground pipe used to transport stormwater, not wastewater.
- **Stormwater:** Water produced during precipitation events (e.g., snowmelt and storm-water runoff).
- **Thermal conditioning:** A process in which heat and pressure are applied to solids simultaneously to enhance their dewaterability without the addition of conditioning chemicals.
- **Thermal oxidation:** A process in which organic solids are converted to oxidized products by heating them in the presence of oxygen or air.
- **Thermophiles:** Bacteria that grow best at temperatures between 45 and 60 °C.
- **Thermophilic:** An operating temperature range (typically 50 to 60 °C) for an aerobic digester. It affects the microbial population in the digesters, as well as the reaction rates.
- **Thermophilic aerobic digestion:** An aerobic digester that operates in the range of 40 to 80 °C.
- **Thermophilic digestion:** A process in which solids are digested by microorganisms that thrive in the thermophilic temperature range (about 50 to 60 °C).
- **Thickener:** A tank vessel, or apparatus where residuals or a slurry is thickened by reducing its water content.
- **Turbidity:** Suspended matter in water or wastewater that scatters or otherwise interferes with the passage of light through the water.
- **Ultraviolet light:** The portion of the electromagnetic spectrum extending from the violet end of visible light to the X-ray region. Its wavelengths are between about 400 and 10 nm, corresponding to frequencies of $7.5 \times 10^{14}$ to $3 \times 10^{16}$ Hz.
- **U.S. Environmental Protection Agency:** The governmental agency in the United States with primary responsibility for enforcing federal environmental laws.
- **Waste activated sludge:** Excess activated sludge that is discharged from an activated sludge treatment process.
- **Wetlands treatment:** A wastewater treatment system in which the aquatic root system of cattails, reeds, and similar plants are used to

treat wastewater applied either above or below the soil surface. The vegetation, soil, and microbial environment filter and remove many contaminants from wastewater through natural processes.

## 2.4 Molecular Biology Definitions (Kaushansky, 2001; Mara & Horan, 2003)

- **Complementary DNA (cDNA):** A DNA copy of a messenger RNA (mRNA) molecule produced by reverse transcription. The term is used to reflect the fact that its sequence is a complement of the original mRNA sequence. Scientists often use cDNA deployed in gene cloning or as gene probes or in the creation of a cDNA library. Some viruses also use cDNA to convert their viral RNA into mRNA, which allows viral proteins to take over the host cell.
- **Digital PCR:** Measures the fraction of negative micro-reactions to determine absolute copies.
- **DNA** (or **Deoxyribonucleic acid**): A complex chemical found in the nucleus and mitochondria of a cell. It provides the genetic instructions needed for an organism to develop, survive, and reproduce.
- **DNA probe:** A fragment of single-strand DNA that contains a specific nucleotide sequence that is used to detect a complementary sequence in a sample genome. The probe is labeled with a radioactive or chemical tag, which provides a signal when it binds to the complementary DNA.
- **DNase:** Deoxyribonuclease, a class of enzymes that digest DNA. The most common is DNase I, an endonuclease that digests both single and double-stranded DNA.
- **Expression:** To "express" a gene is to cause it to function. A gene that encodes a protein will, when expressed, be transcribed and translated to produce that protein. A gene that encodes an RNA rather than a protein (e.g., a rRNA gene) will produce that RNA when expressed.
- **Gene:** The determinant of an observable trait or characteristic of an organism (pigmentation) and the DNA sequence that determines the chemical structure of a specific polypeptide molecule or RNA molecule.
- **Genome:** A complete set of genes in an organism.
- **Genomics:** A branch of genetics established in the 1980s directed toward studying the structure, function, evolution, and mapping of genomes in organisms.

- **Molecular biology:** A branch of science that studies the composition, structure, function, and interactions of cellular molecules, such as proteins and nucleic acids.
- **mRNA (messenger RNA):** An RNA molecule in cells that conveys genetic information from DNA to the ribosome, the site of protein synthesis. Typically about 2% of the total cellular RNA is mRNA.
- **Nucleic acid:** A long molecule made up of smaller molecules called nucleotides that are chemically linked together in a chain. Nucleotides are instrumental in transferring genetic information from one generation to another. There are two types of nucleic acids: deoxyribonucleic acid (DNA) and ribonucleic acid (RNA).
- **Nucleotides:** Molecules present in all cells of the body. They serve many purposes, including acting as cellular messengers between the outside and the inside of a cell's nucleus, storing energy, and serving as physiological mediators. Nucleotides are also necessary to the construction of the nucleic acids DNA and RNA. DNA is made up of four base nucleotides: adenine (A), cytosine (C), guanine (G), and thymine (T). RNA is made up of A, G, and C, plus uracil.
- **Oligonucleotide:** Commonly made in a laboratory, an oligonucleotide is a short sequence of DNA (usually 2 to 50 bases). Oligonucleotides are important in artificial gene synthesis, polymerase chain reactions, DNA sequencing, and library construction and can be used as molecular probes.
- **Polymerase chain reaction** (or **PCR**): A technique for replicating a specific piece of DNA in-vitro, even in the presence of excess nonspecific DNA. Primers are added (which initiate the copying of each strand) along with nucleotides and Taq polymerase. By cycling the temperature, the target DNA is repetitively denatured and copied. A single copy of the target DNA, even if mixed in with other undesirable DNA, can be amplified to obtain billions of replicates. PCR can be used to amplify RNA sequences if they are first converted to DNA via reverse transcriptase. This two-phase procedure is known as "RT-PCR."
- **Primer:** A small oligonucleotide (anywhere from 6 to 50 nucleotides [nt] long) used to prime DNA synthesis. The DNA polymerases are only able to extend a preexisting strand along a template; they are not able to take a naked single strand and produce a complementary copy of it de-novo. A primer that sticks to the template is therefore used to initiate the replication. Primers are necessary for DNA sequencing and PCR.
- **Real-time PCR:** Measures amplification as it occurs.

- **Reverse transcription:** The process of copying an RNA molecule back into its DNA complement. This is done with the help of an enzyme called reverse transcriptase. In nature, retroviruses use reverse transcription to insert themselves into an organism's genome. In the laboratory, reverse transcription is used to translate unstable mRNA molecules into stable cDNA.
- **RNA** (or **Ribonucleic acid**): A type of molecule that plays a role in the coding, decoding, regulation, and expression of genes. RNA is made up of nucleotides—compound of sugars, phosphates, and derivatives of the four bases adenine (A), guanine (G), cytosine (C), and uracil (U). RNAs function in cells as messengers of information from DNA that are translated into protein or as molecules that have certain structural or catalytic functions in the synthesis of proteins. RNA is also the carrier of genetic information for certain viruses. RNAs may be single or double stranded.
- **RNase** (or **Ribonuclease**): An enzyme that degrades RNA. It is ubiquitous in living organisms and is exceptionally stable. The prevention of RNase activity is the primary problem in handling RNA.
- **Sequence:** As a noun, the sequence of a DNA is a buzz word for the structure of a DNA molecule in terms of the sequence of bases it contains. As a verb, *to sequence* is to determine the structure of a piece of DNA (i.e., the sequence of nucleotides it contains).
- **Taq polymerase:** A DNA polymerase isolated from the bacterium *Thermophilis aquaticus* that is very stable to high temperatures. It is used in PCR procedures and high temperature sequencing.
- **Variants:** See Section 2.1.
- **Whole genome sequencing (WGS):** Laboratory process to determine the complete DNA sequence of an organism's genome or a mixed environmental sample.

## 3.0 IMPACTS ON ECONOMIES, COMMUNITIES, AND WATER RESOURCE RECOVERY FACILITIES

### 3.1 Impacts on the Economy, Health Care, and Other Areas

Outbreaks, epidemics, and pandemics all pose major health, social, and economic risks. In 2017, the World Bank estimated that a moderately severe to severe pandemic could result in millions of deaths and destroy up to 1% of the global GDP (roughly US $570 billion, or 0.7% of global income; Maal-Bared et al., 2020). Similarly, the National Academy of Sciences calculated that novel infectious diseases will cost the global economy US $60 billion per year, amounting to US $6 trillion in the 21st century, highlighting the vulnerability of our global economy to emerging disease outbreaks and our lack of preparedness (Gostin, 2021).

Economic analyses of past outbreaks correspond with these predictions. Gostin (2021) reported that the 2003 SARS-CoV-1 outbreak, annual seasonal influenza, and the West African Ebola outbreak cost US $54 billion, US $30 billion, and at least US $2.2 billion in lost economic growth, respectively. These figures may be shocking, but it is important to understand that novel or emerging pathogens have unexpected clinical and often long-term outcomes and complications. Thus, cost predictions can be difficult, and cost estimates are frequently limited to short-term medical spending, health burden, or mortality (Smith et al., 2019).

A good example of this is Zika virus. Although typical Zika infections without sequelae are unlikely to result in significant burden, disease manifestation in infants can have extensive impacts on direct medical expenses during pregnancy and postnatal costs. Studies have estimated the lifetime direct medical cost associated with sequelae related to microcephaly to be US $179,760, and the cost per case of complications occurring in 1% of patients resulting in Guillain-Barre Syndrome to be US $56,863 in the United States (Alfaro-Murillo et al., 2016). It was also estimated the Zika epidemic thus far has cost Latin America and the Caribbean US $7 to 18 billion from 2015 to 2017 (United Nations Development Program, 2017). Zika cost estimates vary widely by country and often do not include indirect costs, such as specialized child-care support, parental productivity losses, psychological toll on families with children with microcephaly, or losses in productivity of the child once an adult and the support services required for that individual throughout life (Smith et al., 2019). The limitations seen with Zika burden-of-disease estimates apply to many other infectious disease estimates.

In addition, authors have cautioned that cost tends to be related to infectious agent transmissibility. A study from the Democratic Republic of Congo reported that the cost of COVID-19 far exceeded the cost of Ebola in the country due to disease transmissibility and resulting public health interventions (Stoop et al., 2021). Cost estimates are also impacted by infectious disease underreporting and whether the infectious agent is reportable (i.e., part of the public health monitoring and surveillance system).

Other infectious disease outbreak impacts on the economy include

- *Reduced food production and food security.* The Saudi Arabian and Yemeni Rift Valley fever outbreak in 2000 resulted in banning the import of live animals from at least nine African countries, causing the Somalian livestock market to completely collapse. Ninety percent of Somalia's total income had been from livestock export, and the ban resulted in a loss of more than 75% in exports and US $300 million. This caused social and financial instability, loss of livelihoods and food security, and ultimately instability of the Somalian government, with a 25% to 36% reduction in GDP (Smith et al., 2019). Both Ebola and COVID-19 have been reported to impact food security in the Democratic Republic of Congo (Stoop et al., 2021).

- *Reduced tourism.* The 2003 SARS outbreak resulted in 68% less travel to Hong Kong just 2 months after the WHO issued a warning. Simultaneously, Asia-Pacific and North American carriers lost US $6 billion and US $1 billion in revenues (Begley, 2018).
- *Environmental impacts.* Quarantine and travel restriction measures and enforcement during the West Africa Ebola outbreak led to illegal poaching, logging, and mining and negatively impacted water resources and wetlands. Similarly, the H5N1 influenza outbreak resulted in misguided national campaigns to cull wild bird populations, close protected wetlands, and destroy waterbird habitats to limit virus spread (Smith et al., 2019).

Being unprecedented in its scale and impact on global health care systems and economies (Armocida et al., 2020; Hutchins Coe et al., 2020), COVID-19 provides a good case study to explore the impacts of pandemics. The Organisation for Economic Co-operation and Development and the United Nation's Trade and Development Agency cautioned that the COVID-19 pandemic poses the biggest threat to the international economy after the recession in 2008 to 2009, reducing the global growth rate to below 2% in 2020, which may eventually pose a total cost of US $1 trillion to the global economy (Bagchi et al., 2020). The pandemic has especially impacted the Chinese economy since December 2019. China's GDP grew 5.94% in 2019 to $142.29.94 billion, but China's economy shrank 6.8% in Q1 2020 due to nationwide lockdowns during the COVID-19 outbreak. It was the first decline since Beijing began reporting quarterly GDP in 1992 (Habibi et al., 2022).

COVID-19 resulted in an unprecedented increase in demand on the health care system and specialized acute care, coupled with a major reduction in routine health services (Canadian Institute for Health Information, 2021; Ducarme, 2020). In France, consultations have fallen by 40% among general practitioners and by 50% among specialists since the beginning of the epidemic, even when accounting for the surge of teleconsultations (Ducarme, 2020). In Canada, almost 560 000 fewer surgeries were performed over the first 16 months of the pandemic compared with 2019 (Canadian Institute for Health Information, 2021). Blumenthal et al. (2020)

reported that COVID-19 disproportionately impacted minority communities that already struggled with health services and insurance access. Hutchins Coe et al. (2020) estimated that although the direct impact of COVID-19 has already been substantial, additional layers of delayed or indirect impacts could result in US $125 billion to US $200 billion in incremental annual U.S. health care system costs. In addition, impacts on health care staff were significant, resulting in high rates of burnout, psychological stress, increased occupational infections, excessive financial hardships, stress related to known and particularly unknown information, fear of uncertainty regarding continued impact, and even suicide (Gupta et al., 2021; Sridhar, 2022).

## 3.2 Impacts on Water Resource Recovery Facilities—The COVID-19 Case Study

### *3.2.1 Lockdowns and Changes in Consumption Patterns*

During the COVID-19 lockdown, residential water demand increased, while nonresidential demand (i.e., commercial, industrial, and irrigation) decreased, potentially resulting in a reduction in utility income (Balamurugan et al., 2021; Cooley et al., 2020). Large commercial buildings were empty for months, compromising water quality and requiring large flushing exercises before the buildings could be populated again (Rhoads et al., 2020). Lockdown also changed diurnal consumption patterns and flows to water resource recovery facilities (WRRFs). Traditionally, WRRFs receive daily peak flow at the beginning of the day (6:00–8:00 AM). With working from home and modified schedules, modeling studies reported that the peak was less pronounced and use was averaged out throughout the day (Dobson et al., 2021). Another study highlighted an increase of up to 20% in influent water flows into WRRF (Yazdian & Jamshidi, 2021). A study from London also suggested the possibility of overall increased ammonia and phosphate loading into waterbodies receiving final treated wastewater effluent during lockdown due to dilution of the wastewater matrix and reduced treatment efficiency (Dobson et al., 2021). However, some studies have reported reduced heavy metal inputs into aquatic environments and WRRFs due to the shutdown of industries (Karunanidhi et al., 2021).

### *3.2.2 Increased Use of Wipes*

There is mounting evidence that the COVID-19 pandemic resulted in an unusual surge in the use of disinfecting wipes (Shruti et al., 2021). Clorox alone has increased its daily production of wipes from 1 to 1.5 million canisters of wipes per day between 2020 and the first quarter of 2021.

Wipes, like masks and other personal protective equipment (PPE), can be major contributors of microplastics in the aquatic environment and have been known to clog the sewage system and pretreatment infrastructure at WRRFs for many decades (Pantoja Munoz et al., 2018). In 2019, Northumbrian Water in Durham (England) reported that 64% of the 15 600 collection-system blockages uncovered by its field staff in 2019 resulted directly from customers flushing products improperly labeled as "flushable" (Northumbrian Water, 2020). This situation could only have been made worse by COVID-19.

### 3.2.3 Increased Consumption of Antibiotics, Disinfectants, and Cleaning Products

The medical community has reported that the lack of specific antiviral treatments for COVID-19 has resulted in increased misuse of existing antimicrobials for the treatment of COVID-19 patients, potentially contributing to the worldwide antimicrobial resistance crisis (Nieuwlaat et al., 2021). Several studies have reported increased prevalence of microbial resistance in aquatic environments since the COVID-19 pandemic (Kumar et al., 2021; Seethalakshmi et al., 2021). Given that WRRFs are considered hotspots for microbial resistance, this is an area where impacts may be seen in the future. Some studies have reported increased concentrations of hydroxychloroquine, disinfectants, drugs, antidepressants, UV filters (used in sunscreens), and benzotriazole (traffic marker) (Adhikari et al., 2022; Marteinson et al., 2022; Nason et al., 2022). Several studies looked at increases in quaternary ammonium compounds (commonly known as quats or QACs) in sludge and wastewater. QACs are cationic surfactants that combine bactericidal and virucidal activity with cleaning ability that have been reported to impact nitrification at WRRFs (Hora et al., 2020; Marteinson et al., 2022). Some QACs precipitate or denature proteins and destroy enveloped viruses, which is why QACs were listed on the EPA-approved disinfectant list for COVID-19 infection control at the beginning of 2020.

These are just some examples of the possible direct and indirect impacts of pandemics on communities and WRRFs. Utilities also were impacted through standard operating procedure changes, staff schedule and work arrangement modifications to ensure business continuity, and supply chain shortages that delayed maintenance and infrastructure projects, some of which will be discussed in Chapters 5 and 8 in this book. A general overview of impacts of infectious disease outbreaks on communities is provided in Figure 1.4. Overall, given the scale, severity, and potential long-term consequences of pandemics, it is critical that the water sector move from panic to preparedness.

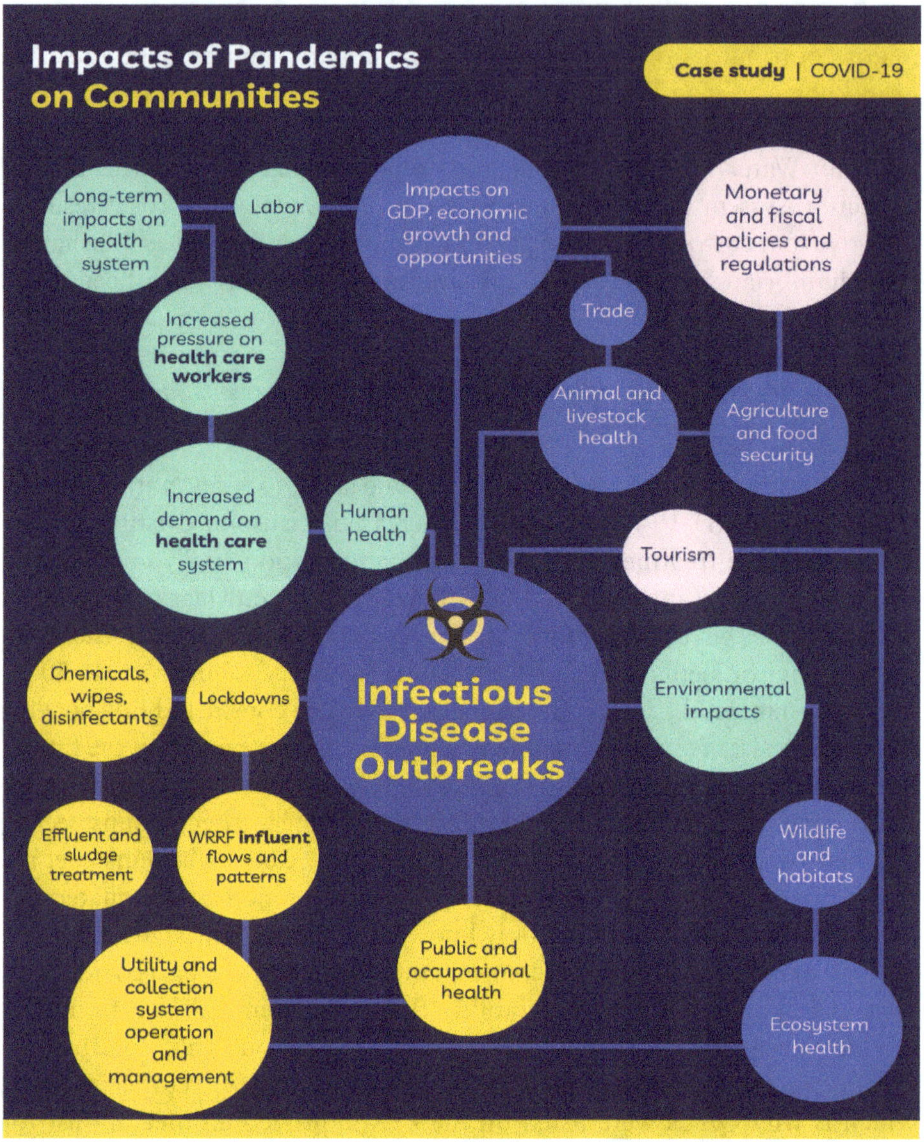


**FIGURE 1.4** Potential Impacts of Infectious Disease Outbreaks on Communities

# 4.0 RESOURCES

## 4.1 Publicly Available

Centers for Disease Control and Prevention. (2015). *Epidemiology glossary*. Division of Reproductive Health, National Center for Chronic Disease Prevention and Health Promotion. https://www.cdc.gov/reproductivehealth/data_stats/glossary.html

Johns Hopkins University. (2023). Coronavirus Resource Center. Retrieved November 2022 from https://coronavirus.jhu.edu/map.html

National Academy of Sciences. (n.d.). *What you need to know about infectious disease.* Retrieved February 2, 2023, from http://needtoknow.nas.edu/id/glossary

Organisation for Economic Co-operation and Development. (n.d.). *The impact of COVID-19 on health and health systems.* Retrieved February 2, 2023, from https://www.oecd.org/health/covid-19.htm

Stanford University. (2019). *Stanford encyclopedia of philosophy—molecular biology.* https://plato.stanford.edu/entries/molecular-biology

World Bank. (2021). *Pandemic preparedness and health system strengthening.* Last updated May 13, 2021. https://www.worldbank.org/en/topic/pandemics

World Health Organization. (2020a). Novel Coronavirus - China. Retrieved Feb 28, 2023, from https://www.who.int/emergencies/disease-outbreak-news/item/2020-DON233

World Health Organization. (2020b). Statement on the second meeting of the International Health Regulations (2005) Emergency Committee regarding the outbreak of novel coronavirus (2019- nCoV). Retrieved Feb 28, 2023, from https://www.who.int/news-room/detail/30-01-2020-statement-on-the-second-meeting-of-the-international-health-regulations-%282005%29-emergency-committee-regarding-the-outbreak-of-novel-coronavirus-%282019-ncov%29

## 4.2 Require Access

Adhikari, S., Kumar, R., Driver, E. M., Perleberg, T. D., Yanez, A., Johnston, B., & Halden, R. U. (2022). Mass trends of parabens, triclocarban and triclosan in Arizona wastewater collected after the 2017 FDA ban on antimicrobials and during the COVID-19 pandemic. *Water Research*, 118894.

Alfaro-Murillo, J. A., Parpia, A. S., Fitzpatrick, M. C., Tamagnan, J. A., Medlock, J., Ndeffo-Mbah, M. L., Fish, D., Ávila-Agüero, M. L., Marín, R., & Ko, A. I. (2016). A cost-effectiveness tool for informing policies on Zika virus control. *PloS Neglected Tropical Diseases, 10*, e0004743.

Armocida, B., Formenti, B., Ussai, S., Palestra, F., & Missoni, E. (2020). *The Italian health system and the COVID-19 challenge.* The Lancet Public Health.

Bagchi, B., Chatterjee, S., Ghosh, R., & Dandapat D. (2020). Impact of COVID-19 on global economy. In *Coronavirus Outbreak and the Great Lockdown* (pp. 15–26). Springer.

Balamurugan, M., Kasiviswanathan, K. S., Ilampooranan, I., & Soundharajan B-S. (2021). COVID-19 lockdown disruptions on water resources, wastewater, and agriculture in India. *Frontiers in Water*, 24.

Begley S. (2018, January 21). Flu-conomics: The next pandemic could trigger global recession. *Reuters*. https://www.reuters.com/article/us-reuters magazine-davos-flu-economy/flu-conomics-the-next-pandemic-could-trigger-global-recession-idUSBRE90K0F820130121

Bernstein, A. S., Ando, A. W., Loch-Temzelides, T., Vale, M. M., Li, B. V., Li, H., Busch, J., Chapman, C. A., Kinnaird, M., & Nowak, K. (2022). The costs and benefits of primary prevention of zoonotic pandemics. *Science Advances, 8*, eabl4183.

Blumenthal, D., Fowler, E. J., Abrams, M., & Collins, S. R. (2020). Covid-19—Implications for the health care system. *New England Journal of Medicine, 383*, 1483–1488.

Canadian Institute for Health Information. (2021). COVID-19's impact on health care systems. Retrieved on November 19, 2022, from https://www.cihi.ca/en/covid-19-resources/impact-of-covid-19-on-canadas-health-care-systems/the-big-picture

Collier, S. A., Deng, L., Adam, E. A., Benedict, K. M., Beshearse, E. M., Blackstock, A. J., Bruce, B. B., Derado, G., Edens, C., & Fullerton, K. E. (2021). Estimate of burden and direct healthcare cost of infectious waterborne disease in the United States. *Emerging Infectious Diseases, 27*(1), 140.

Cooley, H., Gleick, P., Abraham, S., & Cal, W. (2020). How the Coronavirus pandemic is affecting water demand changes in water demand. https://pacinst.org/publication/coronavirus-impacts-on-municipalwater-demand

Corso, P. S., Kramer, M. H., Blair, K. A., Addiss, D. G., Davis, J. P., & Haddix, A. C. (2003). Costs of illness in the 1993 waterborne Cryptosporidium outbreak, Milwaukee, Wisconsin. *Emerging Infectious Diseases, 9*, 426.

Dobson, B., Jovanovic, T., Chen, Y., Paschalis, A., Butler, A., & Mijic, A. (2021). Integrated modelling to support analysis of COVID-19 impacts on London's water system and in-river water quality. *Frontiers in Water, 3*, 641462.

Ducarme T. (2020). What will be the impact of the Covid-19 pandemic on healthcare systems? Deloitte, France. Retrieved on November 28, 2022, from https://www2.deloitte.com/content/dam/Deloitte/fr/Documents/covid-insights/deloitte_impact-covid19-on-healthcare-systems.pdf

Gostin, L. O. (2021). *Global health security: A blueprint for the future.* Harvard University Press.

Gupta, N., Dhamija, S., Patil, J., & Chaudhari, B. (2021, January). Impact of COVID-19 pandemic on healthcare workers. *Industrial Psychiatry Journal, 30*, S282.

Habibi, Z., Habibi, H., & Mohammadi, M. A. (2022). The potential impact of COVID-19 on the Chinese GDP, trade, and economy. *Economies, 10*, 73.

Hora, P. I., Pati, S. G., McNamara, P. J., & Arnold, W. A. (2020). Increased use of quaternary ammonium compounds during the SARS-CoV-2 pandemic and beyond: Consideration of environmental implications. *Environmental Science & Technology Letters, 7*, 622–631.

Hutchins Coe, E., Enomoto, K., Finn, P., & Weber, K. (2020). *Understanding the hidden costs of COVID-19's potential impact on US healthcare.* McKinsey & Company. https://www.mckinsey.com/industries/healthcare/our-insights/understanding-the-hidden-costs-of-covid-19s-potential-impact-on-us-healthcare

Karunanidhi, D., Aravinthasamy, P., Subramani, T., & Setia, R. (2021). Effects of COVID-19 pandemic lockdown on microbial and metals contaminations in a part of Thirumanimuthar River, South India: A comparative health hazard perspective. *Journal of Hazardous Materials, 416*, 125909.

Kaushansky, K. (2001). Glossary of molecular biology terminology. *ASH Education Program Book*, 522–540.

Kumar, M., Dhangar, K., Thakur, A. K., Ram, B., Chaminda, T., Sharma, P., Kumar, A., Raval, N., Srivastava, V., & Rinklebe, J. (2021). Antidrug resistance in the Indian ambient waters of Ahmedabad during the COVID-19 pandemic. *Journal of Hazardous Materials, 416*, 126125.

Lauring, A. S., & Hodcroft, E. B. (2021). Genetic variants of SARS-CoV-2—What do they mean? *Journal of the American Medical Association, 325*(6), 529–531.

Maal-Bared, R., Bibby, K., Brisolara, K., Gary, L., Reimers, R. S., Schaefer, S., et al. (2020). Lessons from the Ebola outbreak: How the water sector can transition from panic to preparedness. *Water Environment Technology, 32*(1), 38–45.

MacIntyre, C. R., Dyda, A., Bui, C. M., & Chughtai, A. A. (2018). Rolling epidemic of Legionnaires' disease outbreaks in small geographic areas. *Emerging Microbes & Infections, 7*, 1–10.

Mara, D., & Horan, N. (2003). *Handbook of water and wastewater microbiology.* Academic Press.

Marteinson, S. C., Lawrence, M. J., Taranu, Z. E., Kosziwka, K., Taylor, J. J., Green, A., Winegardner, A. K., Rytwinski, T., Reid, J. L., Dubetz, C., Leblanc, J., Galus, M. D., & Cooke, S. J. (2022). Increased use of sanitizers and disinfectants during the COVID-19 pandemic: Identification of antimicrobial chemicals and considerations for aquatic environmental contamination. *Environmental Reviews*. https://doi.org/10.1139/er-2022-0035

Morens, D. M., & Fauci, A. S. (2013). Emerging infectious diseases: Threats to human health and global stability. *PLoS pathogens, 9*(7), e1003467.

Morens, D. M., & Fauci, A. S. (2020). Emerging pandemic diseases: How we got to COVID-19. *Cell, 182*, 1077–1092.

Nabarro, D., & Wannous, C. (2014). The potential contribution of livestock to food and nutrition security: The application of the One Health approach in livestock policy and practice. *OIE Revue Scientifique et Technique, 33*, 475–485.

Nason, S. L., Lin, E., Eitzer, B., Koelmel, J., & Peccia, J. (2022). Changes in sewage sludge chemical signatures during a COVID-19 community lockdown, part 1: Traffic, drugs, mental health, and disinfectants. *Environmental Toxicology and Chemistry, 41*, 1179–1192.

Nieuwlaat, R., Mbuagbaw, L., Mertz, D., Burrows, L. L., Bowdish, D. M., Moja, L., Wright, G. D., & Schünemann, H. J. (2021). Coronavirus disease 2019 and antimicrobial resistance: Parallel and interacting health emergencies. *Clinical Infectious Diseases, 72*(9), 1657–1659.

Northumbrian Water. (2020). North East Water Company calling for a total wipe out. North East Water Group 2020. Retrieved November 28, 2022, from https://www.nwg.co.uk/news-and-media/news-releases/north-east-water-company-calling-for-a-total-wipe-out/

O'Brien, E., & Xagoraraki, I. (2019). A water-focused one-health approach for early detection and prevention of viral outbreaks. *One Health, 7*, 100094.

Ogden, N., AbdelMalik, P., & Pulliam, J. (2017). Emerging infections: Emerging infectious diseases: Prediction and detection. *Canada Communicable Disease Report, 43*, 206.

Pantoja Munoz, L., Gonzalez Baez, A., McKinney, D., & Garelick, H. (2018). Characterisation of "flushable" and "non-flushable" commercial wet wipes using microRaman, FTIR spectroscopy and fluorescence microscopy: To flush or not to flush. *Environmental Science and Pollution Research, 25*, 20268–20279.

Rhoads, W., Prévost, M., Pieper, K., Keane, T., Whelton, A., Rölli, F., Proctor, C. R., & Grimard-Conea, M. (2020). *Responding to water stagnation in buildings with reduced or no water use*. American Water Works Association.

Rose, J. B. (1997). Environmental ecology of Cryptosporidium and public health implications. *Annual Review of Public Health, 18*, 135–161.

Schwake, D. O., Garner, E., Strom, O. R., Pruden, A., & Edwards, M. A. (2016). Legionella DNA markers in tap water coincident with a spike in Legionnaires' disease in Flint, MI. *Environmental Science & Technology Letters, 3*, 311–315.

Seethalakshmi, P., Charity, O. J., Giakoumis, T., Kiran, G. S., Sriskandan, S., Voulvoulis, N., & Selvin, J. (2021). Delineating the impact of COVID-19 on antimicrobial resistance: An Indian perspective. *Science of the Total Environment, 818*, 151702.

Shruti, V., Pérez-Guevara, F., & Kutralam-Muniasamy, G. 2021. Wet wipes contribution to microfiber contamination under COVID-19 era: An important but overlooked problem. *Environmental Challenges, 5*, 100267.

Smith, K. M., Machalaba, C. C., Seifman, R., Feferholtz, Y., & Karesh, W. B. (2019). Infectious disease and economics: The case for considering multisectoral impacts. *One Health, 7*, 100080.

Spyrou, M. A., Bos, K. I., Herbig, A., & Krause, J. (2019). Ancient pathogen genomics as an emerging tool for infectious disease research. *Nature Reviews Genetics, 20*, 323–340.

Sridhar, D. (2022, August 12). A scientist in the public eye has taken her own life. This has to be a wake-up call. *The Guardian*. https://www.theguardian.com/commentisfree/2022/aug/12/scientist-taken-own-life-health-workers-covid

Stoop, N., Desbureaux, S., Kaota, A., Lunanga, E., & Verpoorten, M. (2021). Covid-19 vs. Ebola: Impact on households and small businesses in North Kivu, Democratic Republic of Congo. *World Development, 140*, 105352.

United Nations Development Program. (2017). *A socio-economic impact assessment of the Zika virus in Latin America and the Caribbean: With a focus on Brazil, Colombia and Suriname.*

Water Environment Federation. (2023). *Standards for design of water resource recovery facilities*. WEF 8.

Yazdian, H., & Jamshidi, S. (2021). Performance evaluation of wastewater treatment plants under the sewage variations imposed by COVID-19 spread prevention actions. *Journal of Environmental Health Science and Engineering, 19*, 1613–1621.

# Chapter 2

# Pathogens in Wastewater

*Chuck Gerba; Tiong Gim Aw, PhD; Bo Li; Prakit Saingam; Samendra Sherchan; & Mark Sobsey*

# 1.0 CLASSIFICATIONS AND CHARACTERISTICS RELEVANT TO REMOVAL, INACTIVATION, OR CONTROL MICROORGANISMS

## 1.1 Classes and Their Properties

Microorganisms may be classified into the following large biological groups: algae, protozoa, slime molds, fungi, bacteria, archaea, and viruses (Pitt & Barer, 2012). Parasitic worms (or helminths) can also be of relevance in wastewater systems but are not considered microorganisms due to their size. Algae (except blue-green algae or cyanobacteria), protozoa, slime molds, and fungi include the larger eukaryotic microorganisms. These have cells with the same general structure and organization as those found in plants and animals (i.e., a nucleus and organelle). Bacteria (e.g., *E. coli, Salmonella, Campylobacter*), blue green-algae, and archaea are prokaryotes. Prokaryotes are organisms whose cells lack a nucleus and other organelles. They are smaller in size and less complex than eukaryotes. Some bacteria can sporulate, which confers some levels of environmental persistence. Archaea are a distinct phylogenetic group of prokaryotes that bear only a remote ancestral relationship to other organisms; many archaea used to be referred to as extremophiles (Pitt & Barer, 2012). Viruses are the smallest of the infectious agents; they have a relatively simple structure that is not comparable with that of a cell, and their mode of reproduction is fundamentally different from that of cellular organisms. Figure 2.1 shows the different categories of pathogens, their size ranges, and examples of diseases they cause, along with sample images.

Viruses usually consist of little more than a strand of DNA or RNA enclosed in a simple protein shell known as a capsid. In enveloped viruses (e.g., SARS-CoV-2, Ebola virus), the nucleocapsid is enclosed in a lipoprotein envelope derived largely from the host cell. This is not the case for non-enveloped viruses (e.g., norovirus, rotavirus). Viruses can grow only within the living cells of an appropriate animal, plant, or bacterial host; none can grow in an inanimate nutrient medium. The viruses that infect and parasitize bacteria are termed bacteriophages or phages. Despite the name, enveloped viruses are more susceptible to disinfection and environmental stressors compared to non-enveloped viruses (see Chapter 3, Figure 3.1).

| Category | Description | Size range | Examples | Resulting diseases | Images |
|---|---|---|---|---|---|
| **Prion** | Non-cellular | <20 nm | Prion protein (image, IM) | Creutzfeldt -Jacob | 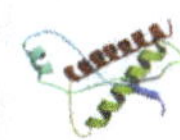 |
| **Viruses** | Non-cellular | 20 - 400 nm | Poliovirus (IM)<br>SARS-CoV-2 | Poliomyelitis<br>COVID-19 |  |
| **Bacteria** | Prokaryotic cells | 0.2 - 15 µm | *Campylobacter jejuni*<br>*Pathogenic E. coli* | Campylobacteriosis<br>Bacteremia (IM) | 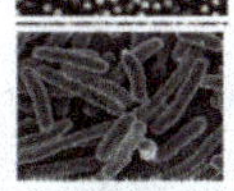 |
| **Fungi** | Eukaryotic cells | 2 - 200 µm | *Microsporidia*<br>*Aspergillus* (IM) | Microsporidiosis<br>Aspergillosis | 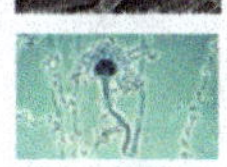 |
| **Protozoa** | Eukaryotic cells | 1 - 50 µm | *Acanthamoeba keratitis*<br>*Cryptosporidium* | Keratitis (IM)<br>Cryptosporidiosis |  |
| **Helminths** | Eukaryotic cells | 3 mm - 10 m | *Schistosoma*<br>Hookworms (IM) | Schistosomiasis<br>Hookworm disease | 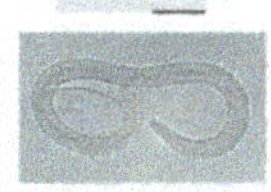 |

**FIGURE 2.1** Categories of Different Pathogens, Along With Their Size Ranges and Examples

Blue-green algae (i.e., cyanobacteria) do not cause infection, but certain species produce potent neurotoxins (e.g., microcystin, anatoxin). As the algae, slime molds, and archaea are not currently thought to contain species of medical or veterinary importance, they will not be considered in this chapter. In addition, we will not discuss prions, which are a class of infectious particles that cause fatal neurodegenerative disorders in animals and humans (e.g., Jacob-Creutzfeldt disease).

## 1.2 The Chain of Infection

Despite the variety of microorganisms that can cause infections in humans generally and wastewater workers specifically, microorganisms spread from person to person through a common series of events. These events are called the chain of infection (shown in Figure 2.2). Regardless of the microorganisms, there are six points at which the chain can be broken to protect public or occupational health. The six links include: the infectious agent, reservoir, portal of exit, mode of transmission, portal of entry, and susceptible host. In

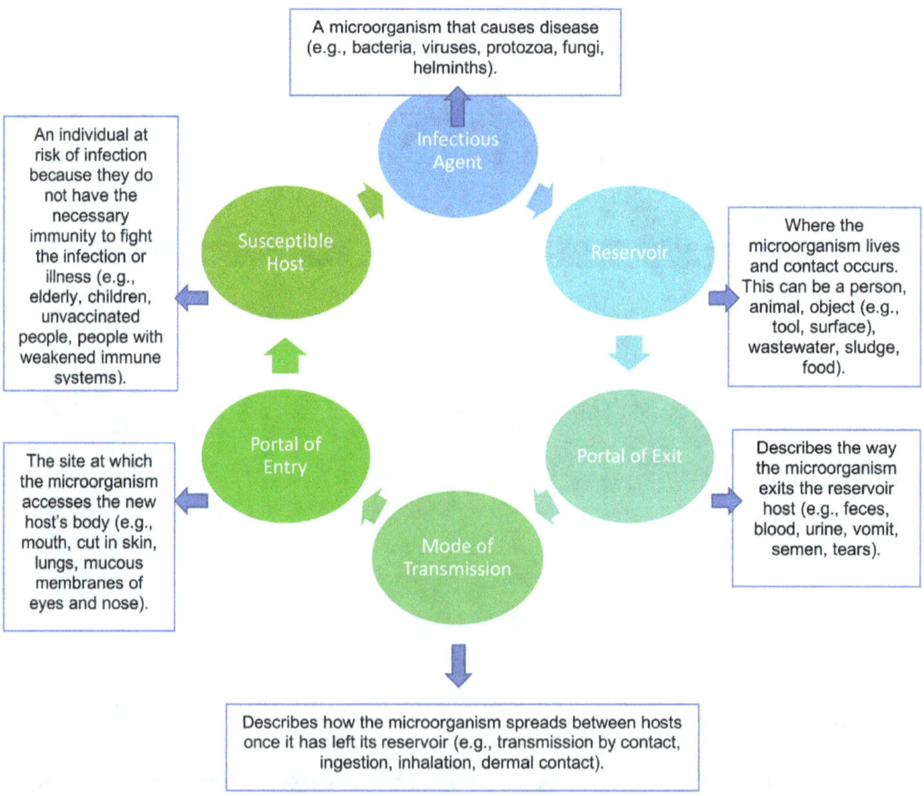


**FIGURE 2.2** Chain of Infection Involved in Infectious Agent Transmission From One Host to the Next

this chapter we will discuss the infectious agents that may be of relevance in wastewater systems.

Although discussions about health effects are mainly focused on exposure to pathogens, it is highly important to understand that disease transmission is the result from coexistence of multiple components. The transmission occurs through an event series: spreading of pathogen from the reservoir, pathogen exposure, pathogen entry to infection site, and successful infection in a susceptible host (Krämer et al., 2009). *Infection* describes a scenario that pathogens are able to establish in a host system and grow/multiply. This transmission event series is also called the "chain of infection." Major components of the chain of infection are reservoir of pathogen, portal of exit, mode of transmission, portal of entry, and a susceptible host.

The reservoir is defined as the habitat, in which the infectious disease-causing pathogens live, grow, and multiplies. The reservoir could be human, animal, or environmental, and transmission of a pathogen can occur within only one reservoir or among all three reservoirs. Within the human reservoir category, the complex interaction between infection and immunity results in variations in health effect and transmission capability. Although being successfully infected, some individuals might never develop symptoms or illnesses, and they become asymptomatic or passive carrier of pathogens to be transmitted to other hosts. There are other types of carriers, including incubatory carriers, who transmit during disease incubation (no observed symptoms); convalescent carriers, who transmit after recovery from symptoms; and chronic carriers, who show a long period of transmission. Environmental reservoirs include soil, water, plants, and air.

To transmit from a host/reservoir to the another, pathogens leave from the portal of exit, which is usually the site of infection. For instance, the respiratory tract is the portal of exit for influenza viruses.

Pathogens are transmitted between hosts/reservoirs in a variety of routes (also known as mode of transmission). There are direct and indirect transmissions. Direct transmission, in which pathogens are passed directly from one host/reservoir to the other, includes direct contact and droplet spread. Examples of direct contact are skin-to-skin contact between human hosts or human hosts contacting soil or vegetation, which confers direct transmission from environment to human. In contrast, indirect transmission describes the route with components in addition to the two hosts/reservoir under a circumstance of interest. With the indirect approach, infectious disease pathogens transfer between two hosts/reservoirs through inanimate objects (vehicles) and organisms (vectors). Scenarios portraying intermediate inanimate objects include pathogens in aerosols (small airborne particles), contamination of pathogens in food ingredients, consumption of water contaminated with pathogens, and touching objects or materials having the presence of pathogens (fomites). Examples of vector organisms are mosquitoes, flies, fleas, and ticks. Most vectors mechanically carry and transport pathogens to hosts, whereas some might also serve as the intermediate host, where pathogens go through changes in physiology and growth development while being on or in them (e.g., mosquitoes).

Portal of entry provides pathogens access to the site of infection, where they multiply and carry out pathogenicity. In many cases, the portal of entry is the infection site, as observed among respiratory virus transmissions through droplets. In contrast, transmission of enteric diseases usually occurs through consumption of food or water contaminated with pathogens, which results in infection in the gastrointestinal tract. In this scenario, the portal of entry is the mouth, which is different from the infection site. Other portals of entry include the skin, eye, mucous membrane, and blood.

Susceptibility to pathogens varies among host individuals due to different genetics, health conditions, living environments, and behaviors, which would affect defense mechanisms against pathogen infection. Examples of defense mechanisms include pathogen-specific antibodies, nonspecific immune response, gastric acid, skin, mucous membranes, and cilia in the respiratory tract. The hosts with infections might or might not develop symptoms and illnesses and become reservoirs, continuing the chain of infection to transmit diseases.

Occupation-related health effects among wastewater workers are associated with the indirect mode of infectious disease transmission from human reservoirs to human hosts (i.e., wastewater workers) mainly through inanimate vehicles including wastewaters, bioaerosols, sewage sludge, biosolids, and contaminated surfaces. Environmental conditions of the vehicles have also shown effects on persistence and multiplication of pathogens (further details in Chapter 4). The portals of exit are urine, feces, and respiratory exudates (for respiratory pathogens) from infected human hosts, and the major portals of entry are mouth, cuts or wounds on skin, and the respiratory tract. As described in a previous review document by the Water Environment Federation (WEF, 2020), the transmissions would be likely through inhalation, ingestion, and dermal exposure.

With inhalation mechanisms, the wastewater workers breathe in droplets or aerosols carrying pathogens and provide the pathogens access to the respiratory tract, and some bioaerosols and droplets from respiratory tract might also get swallowed into the digestive and gastrointestinal tract.

The ingestion mechanism portrays the process in which the wastewater workers accidentally intake wastewater during incidences such as splashes. In addition, the ingestion might occur through contaminated hands. When the workers do not sanitize their hands after work, there might be pathogens on their hands from touching fomite surfaces or handling wastewater, sewage sludge, and biosolids. They could contaminate food items, beverages, and cigarettes, and the pathogens can enter the gastrointestinal tract through eating, drinking, and smoking. This mechanism is also called "fecal-oral" transmission, in which pathogens are shed in feces, are transported in wastewater, and eventually enter a host through the mouth.

With dermal exposure, the wastewater workers might accidentally expose skin to wastewaters, sewage sludge, and biosolids during work activities. The wastewater, sewage sludge, and biosolids are vehicles in the chain of infection, and in many cases, these provide environments influencing persistence and growth of pathogens (detail in the Chapter 4). Some pathogens can enter their host through the skin and eyes. In addition, pathogens can also enter the bloodstream through cuts and other wounds on the skin.

LeChevallier and colleagues (2020) have provided a list of activities and their risks for workers in wastewater collection systems, treatment facilities, maintenance, and biosolid handling. Personal protective equipment (PPE) is recommended for all operations, and activities in several areas need more protection for specific parts of the body. For instance, operations in collection sewers are indicated with risks of splashes into the face with transmission through ingestion, inhalation, and blood transmissions through abrasions, cuts, or punctures while working.

Weldon et al. (2000) also described the major routes of disease transmission among wastewater workers. The study (performed in Texas) identified significant association between specific behavior and routines of 359 wastewater workers in relation to their antibody levels against hepatitis A. Those included eating outside of the lunchroom, never wearing face protection, and skin contact with wastewater at least once per day. This documented the transmission route of infectious disease from wastewaters through the three main mechanisms of ingestion, inhalation, and dermal contact.

# 2.0 EPIDEMIOLOGY OF INFECTIOUS DISEASES IN WATER RESOURCE RECOVERY FACILITY WORKERS

## 2.1 Major Routes of Exposure in Wastewater Systems

Collection systems and water resource recovery facility (WRRF) workers provide an essential public service by cleaning, maintaining, operating, and/or emptying sanitation technologies in various steps of the sanitation chain (World Bank et al., 2019). They face operational health and safety risks due to exposure to hazardous gases and biological and chemical agents in sewers, pumping stations, and WRRFs, if these risks are not mitigated. Wastewater workers are a critical and essential part of the public health system. Wastewater, biosolids and fecal sludge contain a broad spectrum of microbial pathogens, with the type and concentrations being highly dependent on the socioeconomic conditions of communities. The Ebola epidemic in 2014 highlighted a critical need to evaluate infectious disease risks for sanitation workers who could come into contact with untreated sewage (Haas et al., 2017; Zabinski et al., 2018).

### 2.1.1 Untreated Wastewater

There are two major sources of wastewater: municipal and industry. Municipal wastewater, also termed *domestic wastewater*, is contributed to by houses, public restrooms, businesses, sport centers, hotels, health care facilities, and other facilities with sanitation facilities. Industrial wastewater is from manufacturing and commercial units and contains various chemicals that might require different treatment processes than municipal wastewater systems. Municipal wastewater receives toilet flushes and discharges from bathroom and kitchen sinks, bathing, and laundry. In contrast to industrial wastewater, municipal wastewater would have more infectious disease pathogens present. These inlets introduce bodily fluids and excreta, including blood, serum, wound exudates, semen, vaginal fluid, saliva, urine, feces, sweat, tears, and breast milk into municipal wastewater. Body fluid and excreta, which can contain different types and levels of infectious disease pathogens, are considered infectious materials and need to be handled with PPE in health care facilities and other high-risk settings (Prieto et al., 2014). High levels of pathogens have been detected in feces, such as an estimated level of $10^3$–$10^7$ RNA copies of SARS-CoV-2 per gram of feces of infected individuals (Foladori et al., 2020), making feces a concerning entity in wastewater disease transmission.

The inlets to municipal wastewaters can also be separated into black water (the presence of feces and urine), grey water (absence of feces and urine but presence of wastes from sinks, showers, baths, washing machines, or dishwashers), and yellow water (presence of only urine). However, municipal wastewater in its entirety contains feces that is mixed with other wastewater inputs such as grey water, groundwater infiltrations due to sewer cracks, and stormwater runoffs. Wastewater is collected from two types of sewer systems: combined or separate. A separate sewer system (SSS) collects wastewater in sewer pipes separated from other pipes for stormwater. A combined sewer system (CSS) receives both community wastewater and stormwater runoff in the same sewer pipes. In the United States, CSSs are prevalent in 772 communities in the Midwest, Southeast, and Pacific Northwest, serving 40 million people (U.S. Environmental Protection Agency [U.S. EPA], 2004). Stormwater runoffs contain water and other materials that are scoured and collected from terrestrial surface areas, and often carry environmental debris (e.g., leaves and litter), animal feces, sometimes human feces (e.g., from homeless people) transport infectious contaminants into wastewater collection systems.

Sidhu et al. (2012) showed detections of gastrointestinal disease pathogens (adenovirus, *Campylobacter jejuni*, *Campylobacter coli*, *Salmonella enterica*) in river and creek samples receiving stormwater runoffs. The wastewater flow to the treatment facilities depends on served population sizes. Weirich et al. (2011) found wastewater influent flowrates varying from 1 $m^3$/d to 335 000 $m^3$/d across 209 facilities in the United States. Previous analyses of wastewater influent samples have identified genetic materials of gastrointestinal diseases pathogens (Cai & Zhang, 2013), respiratory disease pathogens (Heijnen & Medema, 2011), and waterborne outbreak disease pathogens such as SARS-CoV-2 (Markt et al., 2022).

### 2.1.2 Bioaerosols

Bioaerosols are aerosols that contain microbiological organisms. Aerosols are small solid or liquid particles that are suspended in the air. The word *droplet* has also been used to define airborne particles. However, the distinction between *aerosol* and *droplet* has not been clearly demonstrated and particle size has been a well-known aspect of their differentiation with debatable agreement on threshold values (Randall et al., 2021). However, the World Health Organization (WHO) now considers airborne particles to be a continuum of sizes, with the larger ones settling rapidly and the progressively smaller ones remaining airborne for an extended period of time. Bubble-jet-droplet mechanism is one approach of bioaerosol generation, in which air bubbles moving through water and bursting of bioaerosol particles from the air-water surface to the air (Keene et al., 2007; Mirskaya & Agranovski, 2018; Warneck, 1999; Woodcock, 1955). Small particles are generated initially, followed by larger droplets following the retreat of water surface. Particle size of small bioaerosols (also known as "film droplets") is in the range of $< 1$ µm to 20 µm (Garner et al., 1954).

Bioaerosols in WRRFs could be generated in areas where matrices undergo rapid movement, mechanical agitation, and forced aeration (Fracchia et al., 2006; Pascual et al., 2003; Sánchez-Monedero et al., 2008). The aerosols captured within one activated sludge pilot facility showed 99.99% prevalence of $< 6.12$ µm and 99.9% prevalence of $< 2$ µm in size (Bauer et al., 2002). The aerosols at different areas of WRRFs are varied in size and level of contained microorganisms. The aerosols of $> 2$ µm particle size were dominant in the sludge-thickening basin, whereas those $< 2$ µm were most prevalent at the biological reaction basin, followed by screen room, effluent outlet, and downwind facility outdoor boundary (Li et al., 2016). Within trickling filter wastewater treatment, the levels of culturable airborne mesophilic bacteria were higher at sewage inflow, equalization tank, trickling filters (average range of 2000 to > 40 000 CFU/$m^3$) than the sludge-thickening area (average range of 539 to > 6190 CFU/$m^3$) during both summer and winter seasons (Fracchia et al., 2006). In the same study, the mesophilic bacteria were higher in the aeration tank of the secondary treatment and grit collection tank (average range of 4730 to 5320/$m^3$) than in sewage inflow and the biomass settler (average range of 1790 to 2060/$m^3$) in summer, whereas bacterial numbers in the winter were dominant in sewage inflow and the grit collection tank. The areas with high bacteria counts presented bioaerosol generation mechanisms of splashing, bubbling, and high spraying. Maintenance activities, which were performed using high-pressure water for cleaning equipment, could also contribute to bioaerosol generation. The particles (also called particulate matter) with < 10 µm (PM10) and < 2.5 µm (PM2.5) in size are respirable and could enter the respiratory tract (Górny, 2020). According to previous studies, the bioaerosols generated in WRRFs are prevalent with respirable particles, and this raises health effects concerns among wastewater workers. An analysis of PM2.5 and PM10 bioaerosol particles collected from two biological WRRFs found human opportunistic pathogenic bacteria (*Pseudomonas*, *Bacillus*, *Clostridium*, *Mycobacterium*, *Legionella*, and *Staphylococcus* spp.) (Liu et al., 2020).

### 2.1.3 Sludge and Biosolids

Water resource recovery facilities also produce sewage sludge, which is solid, semisolid, and other waste residue that are further processed for proper disposal or applications (National Research Council [NRC], 2002). The sludge from primary treatment is called primary sludge, and the sludge from secondary treatment is called secondary sludge. Primary sludge is mainly composed of solids, and secondary sludge contains large number of microbial cells resulting from biological treatment such as anaerobic or aerobic digestion (Xu & Li, 2017). The sewage sludge could be processed for landfilling disposal, incineration, composting or other biosolids production processes (Epstein, 2002). Among those options, biosolids present not only an approach of sewage sludge disposal but also a beneficial use of organic matter and nutrients in sewage sludge for land-use applications such as agricultural soil amendment. The biosolid is the sewage sludge that has been treated and meets standards for downstream applications, which is the key distinction between biosolids and sewage sludges (NRC, 2002). Untreated sludge samples contain respiratory and gastrointestinal disease pathogens as well as genetic materials of emerging viruses (e.g., Coronaviruses; Bibby & Peccia, 2013; Brisolara et al., 2021; Dudley et al., 1980).

However, the treatment processes of biosolids production achieve reductions in microbial and pathogen loads (Dumontet et al., 1999). For instance, 50% to 90% inactivation of viruses was observed after thermal treatment of sewage sludge (Berg et al., 1988). The biosolids production processes combines various treatments including thickening, anaerobic digestion, alkaline stabilization, conditioning, dewatering, composting, and heat drying (Turovskiy & Mathai, 2006). The treatment processes reduce water content, organic matter, and microbial pathogens.

In the United States, there are class A biosolids and class B biosolids, which are regulated with different treatment conditions and microbial assessment levels and used in different applications (NRC, 2002; U.S. EPA, 1994). In terms of microbial risk levels, class A is assigned to biosolids with undetectable levels of pathogens (i.e., *Salmonella* sp.), whereas class B showed detectable levels that have been reduced to pose reduced risks to public health. The use of class B biosolids is more restricted and needs a barrier to prevent reduce public exposure, such as long holding times before crop harvest (U.S. EPA, 1994).

Wastewater workers who operate and handle sludge and biosolids throughout the production process could be exposed to health risks of infectious pathogens and their diseases. Employees at the National Institute for Occupational Safety and Health developed symptoms of gastrointestinal illness after handling biosolids, and enteric bacteria (*Mycobacterium*, *Pseudomonas*, and *Staphylococcus*) were detected from a bulk sample of biosolids and an air sample at the land application and storage site (Burton & Trout, 1999). Multiple studies have found evidence of health effects observed among residents living close to land areas treated with biosolids (i.e., land-use farming; Harrison & Oakes, 2003; Khuder et al., 2007; Lewis et al., 2002; Virginia Department of Health, 2019). Studies of health effects from sewage sludge and biosolids among wastewater workers have been scarce (Jenkins et al., 2007), and infectious disease transmission risks are mainly from risk assessment modeling study (Snyder, 2003) and detection of pathogens in sludge and biosolid samples (Sidhu & Toze, 2009). Jenkins et al. (2007) stated, "The absence of evidence of a health effect is NOT the same as evidence for the absence of a health effect." Microbiological analysis of class A and class B treated biosolids have detected genetic material for a wide variety of pathogens (Viau & Peccia, 2009), and in some cases infectivity of recovered pathogens was also reported (Wong et al., 2010).

## 2.2 Epidemiology of Infectious Diseases Among Wastewater Workers

Epidemiological studies exploring whether wastewater workers had a higher burden of infectious diseases compared to the general populations have often reported contradictory results. Industry reports suggest that the majority of infections in workers occur in younger, new operators and collection systems workers, possibly due to lack of immunity to certain infectious agents or due to lack of experience resulting in higher exposure.

Cohort-type analytical studies, which study health effects among a group of individuals exposed to the entity of interest and a nonexposed group or control, have been prevalent in the epidemiology of health effects among wastewater workers (Thorn & Kerekes, 2001). In those studies, recruited participants were wastewater workers who maintain collection sewers or operate treatment systems with exposure to wastewater, bioaerosols, and biosolids, and the non-exposed control group. The control groups are individuals with similar job duties but without exposure to wastewater such as industrial workers, drinking water facility workers, gardeners, municipal office workers, and workers in the WRRFs but not exposed to wastewaters. The study outcome is based on a comparison of health effects between these two groups of participants. Studied health effects of wastewater workers included reported symptoms of discomfort, signs of microbial infections, and mortality and cancer incidences. Data-collection methods generally range from self-reported questionnaires to general medical examination and laboratory testing of blood samples. In several studies, the comparison was also based on matching of gender, age, and health-influencing habits such as smoking. The differences of health effects were determined using statistical analyses with a threshold of significance level.

As reviewed by Thorn and Kerekes (2001), gastrointestinal symptoms, fatigue, headache, dizziness, fever, and symptoms from the airways, skin, and the central nervous system were observed with significantly increased risks among wastewater workers in comparison to the controls (Thorn & Kerekes, 2001). Some observed symptoms might not be associated with pathogen infection causes but rather with chemicals or dusts. For example, it is widely known that exposure to organic solvents could affect the central nervous system (Kraut et al., 1988). Endotoxin, a component of bacteria cell walls and a product during cell death shedding, could also induce these symptoms. Furthermore, the blood sample testing in several studies revealed significantly increased levels of pathogen markers or antibodies, which show signs of microbial infections, among wastewater workers. The studies highlighted increased risk of viral infections (enterovirus, reovirus, echovirus, Norwalk agent, hepatitis A virus, adenovirus, parainfluenza type 1 virus, and influenza type A virus), bacterial infections (*Legionella*, *Leptospira*, *Salmonella*), and parasitic diseases (*Entamoeba histolytica*, *Giardia lambia*). The infections were consistent with the development of gastrointestinal and respiratory diseases (Thorn & Kerekes, 2001). However, this review showed that the outcomes of health effects on mortality were not consistent among studies in this review. A separate review focused on respiratory health effects in wastewater workers (Muzaini et al., 2021). The potential respiratory health effects reported were microbiological (bioaerosols), chemical (hydrogen sulfide, volatile organic compounds [VOCs]), and particulate matter (PM 2.5). The exposure risks contributed to adverse outcomes on the respiratory health of sewer workers, with some inconsistent findings on the relationship between respiratory hazard exposure and respiratory health effects.

Level of exposure to wastewaters, bioaerosols, and biosolids vary across different stages of water and wastewater treatment processes, which might reflect on health effects. The typical wastewater treatment process starts with primary treatment for the removal of solids from untreated influent through a sedimentation tank. In secondary treatment, removal of organics and other impurities in wastewaters is achieved by aerobic bacteria prior to the tertiary treatment of UV or chemical (e.g., chlorine) disinfection. The generated solids can be treated by separate processes of anaerobic or aerobic digestion, composting, or landfilling. One epidemiological study found that workers at incoming water tunnels, sludge treatment areas, and aerated basins showed significantly higher odds of reported gastrointestinal symptoms than the workers in sites of no water agitation such as sedimentation tanks (Lundholm & Rylander, 1983). In another study (Khuder et al., 1998), participating workers were categorized following exposure levels, which were based on the job title and main work area. The workers in high-exposure areas showed higher odds of reporting at least one respiratory, gastrointestinal, or systemic symptom than the workers in low-exposure areas. However, the differences observed in the study were not significant. Throughout working shifts, the workers might have spent varied amounts of time in different areas of the treatment process. The authors pointed out that exposure level, which is based on job title and reported work area, could contribute to the inconsistent outcome of studies on varied health effects following different exposure levels.

Knowledge of health effects from wastewater exposure was also implied from the outcome of epidemiological studies among individuals exposed to land application of wastewater, which includes irrigation of reclaimed waters and farms with treated sludge from wastewater treatment processes, as reviewed by Clark (1987). Among several studies, no significant increase in microbial infections were observed among the experimental group of individuals exposed to the irrigation water and sludge landfilling (Brown, 1985; Clark, 1981; Dorn et al., 1985; Linnemann et al., 1984). However, individuals who cleaned the sprinkle nozzles of irrigation waters and were the most exposed to wastewater consistently showed high levels of antibodies to enterovirus infection (Linnemann et al., 1984).

Epidemiological studies on health effects during infectious disease pandemics have been lacking. Especially during pandemics, causative agents are shed into feces and municipal wastewaters, and concerns over health effects of disease transmission from wastewaters should be addressed together with discussions on transmission risk levels and preventative action plans to protect wastewater workers. Epidemiological study outcomes could be beneficial in the planning and response process. However, among the literature, only two studies mentioned outcomes regarding disease outbreak (Clark et al., 1986). The studies showed increased levels of antibodies to an echovirus among irrigation and fishpond workers, who handled and were exposed to wastewater effluent, and their immediate families, and the spikes were only observed during the period following a major national echovirus epidemic in Israel in 1980 and 1981 (Linnemann et al., 1984; Shuval et al., 1983).

Table 2.1 provides a list of studies that have explored occupational infections among wastewater workers. Higher levels of illness in wastewater workers may also be influenced by the concentration of the pathogen in the wastewater. This was suggested by several studies in India and one in Egypt, where there is a greater incidence of hepatitis E infection among workers compared to no association found in the control (Tschopp et al., 2009).

Given the limitations of epidemiological studies (e.g., short-term investigations, use of survey, confounding variables, ethical issues), many studies now rely on quantitative microbial risk assessments to evaluate and mitigate possible health effects of wastewater exposure. This will be discussed in detail in Chapter 4.

## 3.0 RELEVANT PATHOGENS

In this section, we divide pathogens into two groups, those related to water, sanitation, and hygiene (WASH), and other pathogens not related to WASH conditions. Pathogens in each group are then discussed individually. Figure 2.3 shows the microorganisms discussed in this chapter.

### 3.1 Water, Sanitation, and Hygiene-Related Pathogens

Much of the data for pathogen source, mode of transmission, and illness are summarized from the following resources:

- The WHO Pathogen Fact Sheets: https://www.who.int/news-room/fact-sheets

**TABLE 2.1** List of Studies Investigating the Burden of Infectious Diseases Among Wastewater Workers

| Infectious Disease Agent(s) | Country | Study Period | Occupational Exposure Risk | Reference |
|---|---|---|---|---|
| Hepatitis A virus (HAV) | Singapore | November 1992–January 1993 | • 436 out of 600 sewage workers (72.7%) tested positive for anti-HAV by enzyme immunoassay.<br>• The adjusted seroprevalence of sewage workers was 2.2 times higher than that of another non-occupationally exposed population group.<br>• Sewage workers have an increased occupational risk of acquiring HAV infection and should be protected by active immunization. | Heng et al., 1994 |
| HAV | Georgia, U.S. | January–March 1999 | • Anti-HAV prevalence was 38% (365 sewage workers), similar to that of the 166 roads and drainage workers.<br>• Low risk of HAV exposure among sewage workers. | Venczel et al., 2003 |
| HAV, hepatitis B virus (HBV) | Greece | January–March 2002 | • Anti-HAV prevalence was 65.7% of 108 sewage workers and 32.6% of control group ($n$ = 86).<br>• The adjusted seroprevalence of sewage workers was 3.5 times higher than the control group.<br>• Evidence of HBV infection was found in 32.4% of sewage workers and 5.8% of the control group.<br>• Support vaccination policy of young sewage workers and those sensitive to HAV and HBV. | Arvanitidou et al., 2004 |
| HAV | Parisian area, France | November 8–December 25, 1993 | • Anti-HAV prevalence was 12.9% of 155 sewage workers; no significant increase.<br>• The prevalence was significantly associated with duration of occupational exposure to sewage. | Cadilhac & Roudot-Thoraval, 1996 |

(*continued*)

**TABLE 2.1** List of Studies Investigating the Burden of Infectious Diseases Among Wastewater Workers (*Continued*)

| Infectious Disease Agent(s) | Country | Study Period | Occupational Exposure Risk | Reference |
|---|---|---|---|---|
| HAV | Quebec City, Canada | June–September 1995 | • Confirmed the detection of HAV in three sewage workers during a small community outbreak involving 16 cases.<br>• Had no risk factor for hepatitis A apart from the occupation. | De Serres & Laliberte, 1997 |
| HAV | Tuscany, Italy | 4-month | • Anti-HAV was detected in 51% of sewage workers ($n$ = 65) and 44% of other employees ($n$ = 160).<br>• Difference was not statistically significant. | Bonanni et al., 2000 |
| HAV | Israel | November 1996–April 1997 | • Compared hepatitis A seropositivity in 100 sewage workers.<br>• Hepatitis A seropositivity was highly prevalent (82% sewage workers, 91% controls), but nonsignificant. | Levin et al., 2000 |
| *Legionella pneumophila* | Finland | 2006 | • Two Legionnaires' disease cases associated with industrial wastewater systems.<br>• High concentrations of *Legionella pneumophila* were detected in activated sludge basin.<br>• The most likely source of infection was the exposure to aerosols generated from wastewater of the post-clarification basin. | Kusnetsov et al., 2010 |
| *Leptospira* | Pune, India | Not stated | • Out of the 78 sewage workers, 13 (16.6 %) demonstrated antibodies to leptospiral antigens at or above the significant titer. | Ambekar et al., 2004 |
| *Leptospira interrogans*, HAV | Quebec City Canada | Not stated | • A matched case-control study that showed that sewage workers exhibited significantly higher prevalence of Leptospirosis antibodies 12% ($p$ = 0.003) when compared with controls matched for sex and age (2%). | De Serres et al., 1995 |

*(continued)*

**TABLE 2.1** List of Studies Investigating the Burden of Infectious Diseases Among Wastewater Workers (*Continued*)

| Infectious Disease Agent(s) | Country | Study Period | Occupational Exposure Risk | Reference |
|---|---|---|---|---|
| | | | • Antibodies to HAV were not significantly associated with occupation ($p = 0.51$).<br>• Prevalence of HAV antibodies increased with age among sewage workers ($p < 0.001$). | |
| *Helicobacter pylori* | Sweden | Not stated | • Cross-sectional study involved examining serum samples for immunoglobulin G (IgG) antibodies against *H. pylori.*<br>• Prevalence of IgG antibodies against *H. pylori* among sewage workers ($n = 151$) did not differ from that of the controls ($n = 138$). | Friis et al., 1996 |
| *Helicobacter pylori*, hepatitis E virus (HEV) | Switzerland | June 2000–July 2002 | • Prospective cohort study, 349 sewage exposed workers, 429 municipal manual workers.<br>• Peptic ulcer disease and clinical hepatitis E did not occur more often among sewage workers. | Jeggli et al., 2004 |
| *Helicobacter pylori* | Belgium | September 2004–June 2005 | • Prevalence of *H. pylori* IgG antibodies in sewage workers ($n = 317$) was 16.7% compared to 13.6% in pharmaceutical employees ($n = 250$).<br>• No significant associations were found between *H. pylori* status and gastrointestinal symptoms, different tasks for sewage workers, or individual hygiene behaviors. | Van Hooste et al., 2010 |
| *Tropheryma whipplei* | Vienna, Austria | Not stated | • 16 of the 64 sewage workers (25%) tested positive for *T. whipplei* in stool; only 12 of 178 (7%) disease controls tested positive. | Schöniger-Hekele et al., 2007 |

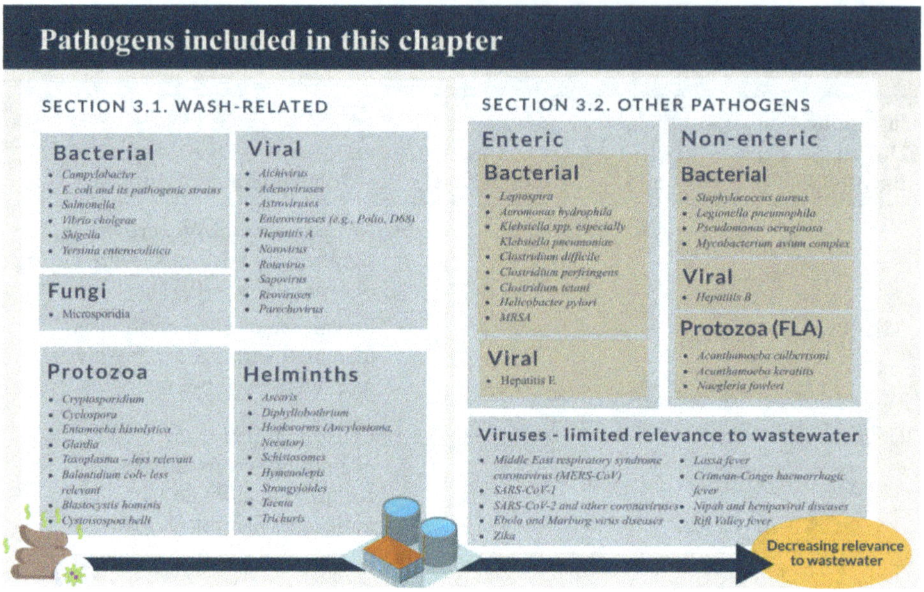


**FIGURE 2.3** List of Pathogens Explored in This Chapter Categorized by Relevance to Water, Sanitation, and Hygiene-Related Conditions

- The Health Canada Pathogen Safety Data Sheets: https://www.canada.ca/en/public-health/services/laboratory-biosafety-biosecurity/pathogen-safety-data-sheets-risk-assessment.html
- The CDC's National Center for Immunization and Respiratory Diseases, Division of Viral Diseases: https://www.cdc.gov/ncird/DVD.html
- The European Centre for Disease Prevention and Control (ECDC) Infectious Disease Fact Sheets: https://www.ecdc.europa.eu/en/all-topics

### 3.1.1 Bacterial

#### 3.1.1.1 Campylobacter

*Campylobacter jejuni* is the second leading cause of bacterial diarrhea in the United States and is usually associated with poultry and raw milk consumption. Although rare, some persons will develop reactive arthritis (joint pain) and rarely Guillain-Barre syndrome (weakness in the legs that may last several weeks and other systemic illness symptoms). It is thought that all surface waters, at one time or another, will contain *Campylobacter*, as it is a common infection of both wild and domestic birds, although it dies off rapidly in water compared to other enteric bacterial pathogens. The infection usually resolves in a few days and is easily treated with antibiotics. It is easily killed by chlorine and other disinfectants used to treat wastewater.

### 3.1.1.2 *Escherichia coli* and Its Pathogenic Strains

*E. coli* is a common inhabitant of the intestines of all warm-blooded animals, including birds. Only certain strains are capable of causing diarrheal illness in humans and animals. These *E. coli* carry certain genes that produce a toxin (enterotoxigenic), invasive (enteroinvasive), adherent (enteroadherent), or aggregative (Enteroaggregative) factors that can cause diarrhea and damage the kidneys or cause other intestinal pathology. The illness usually includes severe cramping and diarrhea. The strain that causes the most serious infection is enterohemorrhagic *E. coli* or EHEC, which, especially in the young and elderly, can lead to hemolytic-uremic syndrome, causing kidney failure. The disease results in permanent loss of kidney function. Most outbreaks of this illness are associated with undercooked red meat and raw vegetables or untreated water consumption.

### 3.1.1.3 *Salmonella*

In the early 20th century, before the disinfection of drinking water and improvements in individual hygiene, *Salmonella typhi* (typhoid) was a major killer in the United States. Typhoid is now under control in the United States, but different *Salmonella* and strains cause diarrhea, and are the leading cause of bacterial diarrhea in the United States today. The five most pathogenic strains are Salmonella Enteritidis, Typhimurium, Newport, Javiana, and Heidelberg. Food is the major route of transmission in the United States, with poultry and beef products being the most common sources of exposure. Wild animals are also another source that contaminate surface waters. Acute symptoms (diarrhea) usually only last 1 or 2 days with diarrhea, vomiting, and stomach cramps being common. There are now more than 2500 different known serotypes of *Salmonella*. Typhoid fever presents a very different clinical feature with fever, headache, and coughing and can lead to a chronic carrier state. Paratyphoid is similar but with milder symptoms. *Salmonella* species usually persist longer in the environment than *Campylobacter* and *E. coli* and may grow in untreated sewage sludge or biosolids that have become recontaminated.

### 3.1.1.4 *Vibrio cholerae*

This was the first bacterium to be recognized as being transmitted by drinking water. In 1854 it was shown to cause an outbreak associated with a fecally contaminated drinking water well in London. It caused large-scale outbreaks in large cities in Europe and North America in the 19th century. Today it is largely associated with epidemics and endemic regions in Asia, South America, and Africa. It causes a severe watery diarrhea and a significant mortality if untreated. It can easily be treated by rehydration of lost liquid and salts orally or by interveinous injection. The marine environment is its native habitant, and it can be transmitted via seafoods such as oysters and shrimp.

#### 3.1.1.5 *Shigella*

*Shigella* is closely related to *E. coli* but only infects humans and monkeys. There are four species, with *Shigella dysenteriae* causing the most severe diarrhea and *S. sonnei* the mildest. Diarrhea with abdominal pain is the common symptom. Transmission is mostly associated with foods and person-to-person contact, although recreational and drinking water outbreaks occur if the water is untreated. It does not survive long in water.

#### 3.1.1.6 *Yersiniae enterocolitica*

This bacterium is common in the environment, being isolated from a wide variety of animals and surface waters. Asymptomatic carriers exist who show no sign of infection. It, however, causes diarrhea with fever and abdominal pain being common, often with a bloody stool. It can also cause wounds, joint pain, and urinary tract infections. It is more common in Europe and Japan than the United States. Pigs are a major reservoir of this bacterium. Most cases have been associated with foodborne outbreaks, but waterborne outbreaks have occurred.

### 3.1.2 Viral

#### 3.1.2.1 Aichi Viruses

Aichi viruses (AiV) are small non-enveloped viruses. AiV belongs to the genus Kobuvirus of the family Picornaviridae. They have an icosahedral morphology that presents a single-stranded RNA genome with positive sense and a poly (A) chain. They have emerged as the agents responsible for gastroenteritis outbreaks associated with different foods and have been detected in wastewater. Clinical signs and symptoms of infection typically include diarrhea, abdominal pain, nausea, vomiting, and fever. AiV was first detected from clinical samples and in recent years has been involved in acute gastroenteritis outbreaks from different world regions. Furthermore, several studies conducted in Japan, Germany, France, Tunisia, and Spain showed a high prevalence of AiV antibodies in adults (between 80% and 99%), which is indicative of a large exposure to this virus (Rivadulla & Romalde, 2020).

#### 3.1.2.2 Adenoviruses

Human adenoviruses are members of the family Adenoviridae and genus Mastadenovirus. Within the almost 100 different serotypes of human adenovirus, 51 are known to be pathogenic in humans. The virus is non-enveloped with an icosahedral capsid at 70 to 90 nm in diameter, and each contains a single linear, double-stranded DNA genome. They are pathogens of respiratory and gastrointestinal mucous and eye membranes and can infect cells by the respiratory and fecal-oral routes. The majority of adenoviruses cause upper respiratory infections, but a group of fastidious strains known as serotypes 40 and 41 are responsible for gastroenteritis in young children. Infections can also spread through contaminated fomites, fingers, ophthalmic solutions, and airborne particulates. Infections tend to be self-limiting and asymptomatic. Adenovirus is of worldwide prevalence and is ubiquitous throughout the year, especially during late winter and early spring (Government of Canada, 2022).

#### *3.1.2.3 Astroviruses*

Human astroviruses (HAtVs) are positive-sense single-stranded RNA viruses that were discovered in 1975. Astroviruses infecting other species, particularly mammalian and avian, were identified and classified into the genera Mamastrovirus and Avastrovirus. Through next-generation sequencing, many new astroviruses infecting different species, including humans, have been described, and the Astroviridae family shows a high diversity and zoonotic potential (Bosch et al., 2014). Although astrovirus infections are often mild or asymptomatic, they are among the most common causes of pediatric viral gastroenteritis. Symptoms can include diarrhea, headache, malaise, nausea, occasionally vomiting, and a mild fever. Person-to-person spread by the fecal-oral route is thought to be the most common route of transmission. Food, water, and fomites have been implicated in astrovirus disease, which suggests that astroviruses persist in the environment; however, there are no confirmed cases of waterborne transmission. They have been isolated from water, wastewater, and groundwater.

#### *3.1.2.4 Enteroviruses*

Enterovirus is a small (~30 nm in diameter), non-enveloped virus group with a single-stranded positive-sense RNA genome (~7.4 kb) enclosed within an icosahedral capsid. It belongs to the species human enterovirus D in the genus Enterovirus and family Picornaviridae. Enteroviruses include numerous strains of coxsackievirus, echovirus, enterovirus, and poliovirus. Infections are highly contagious, typically affect many people in a community, and mostly affect children. These viruses are responsible for illness in 10 to 30 million people each year in the United States, primarily in the summer and fall. Transmission occurs by fecal-oral contamination, fomites, and inhalation. Symptoms of enterovirus infections include fever, headache, respiratory illness, sore throat, and sometimes mouth sores or a rash. Infections can be mild (e.g., resembling the common cold) or asymptomatic but can also result in more extreme symptoms (e.g., pneumonia, vomiting, and diarrhea). Some strains of enterovirus also cause a generalized, non-itchy rash on the skin or sores inside the mouth.

Hand-foot-and-mouth disease is caused by many different enteroviruses, such as coxsackievirus. Its symptoms include fever, sores in the mouth area, and a rash on the hands, feet, and mouth, most often in young children. It is most common in the spring, summer, and fall but can occur in winter. Transmission occurs through fecal-oral, droplet, and fomites transmission and person-to-person contact.

Poliomyelitis (or polio) is a highly contagious, sometimes fatal enterovirus infection that affects the nervous system and can cause permanent muscle weakness, paralysis, and other symptoms. Polio is spread by consuming contaminated food or water or touching a contaminated surface and then touching the mouth. Infections are only symptomatic 25% to 30% of the time and can result in one of three outcomes:

1. The mildest form called abortive poliomyelitis, which is flu-like and manifests as a fever, mild headache, sore throat, vomiting, and a general feeling of illness (malaise). These symptoms develop 3 to 5 days after exposure to the virus.
2. About 4% of people develop nonparalytic poliomyelitis, which manifests as a stiff neck and/or back and headache (aseptic meningitis) several days after the flu-like symptoms of abortive poliomyelitis. The symptoms last 2 to 10 days, but people do not develop paralysis.
3. Paralytic poliomyelitis, which fewer than 1% of infected people develop. People who have this form also develop paralysis. Symptoms usually appear 7 to 21 days after infection and include fever, severe headache, a stiff neck and back, and deep muscle pain. Some individuals report odd skin sensations (e.g., feeling pins and needles, unusual sensitivity to pain). Depending on which parts of the brain and spinal cord are affected, the disease may or may not progress. In some cases, weakness or paralysis develops in certain muscles (e.g., arms and legs making them limp and unable to contract, which is called flaccid paralysis).

Routine vaccination is available and can prevent the infection. However, where vaccination coverage is low, vaccine strains circulating in the population can back-mutate to paralytic forms that again cause poliomyelitis.

### 3.1.2.5 Hepatitis A

Hepatitis A (HAV) is a member of the Picornaviridae family and Hepatovirus genus. It is an icosahedral, non-enveloped, positive-sense RNA virus and is 27 to 32 nm in diameter. The World Health Organization estimates an annual total of 1.5 million cases of hepatitis A worldwide, but seroprevalence data suggest that tens of millions of HAV infections occur each year. Hepatitis A is very contagious, and people can spread the virus before they get symptoms, which include nausea, stomach pain, and yellow skin or eyes. Many infections are asymptomatic. People who get hepatitis A may feel sick for a few weeks to several months. Whereas most people recover and do not have lasting liver damage, some people need to be hospitalized. Hepatitis A can cause death, especially for people who are older or have other health problems. Transmission can occur through the fecal-oral route, fomites, or person-to-person contact. Vaccines are highly effective against HAV, and the Centers for Disease Control and Prevention (CDC) recommend vaccines for wastewater workers.

### 3.1.2.6 Norovirus

Noroviruses belong to the family Caliciviridae. They are a group of non-enveloped, single-stranded RNA viruses that primarily cause acute gastroenteritis (inflammation of the stomach and/or intestines).

Noroviruses were previously called Norwalk or Norwalk-like viruses. It is the leading cause of foodborne illness in the United States, causing about 20 million illnesses, 109,000 hospitalizations, and 900 deaths each year. Noroviruses are common in North America and are very contagious, affecting all age groups. Norovirus illness can happen year-round, but outbreaks are more common in fall and winter months. Symptoms of norovirus illness begin suddenly and usually include vomiting, diarrhea, nausea, and stomach pain. Infected people are contagious when they have symptoms but can also spread it a few days before and after they have symptoms. Most of the transmission is food- and waterborne, person to person, and through fomites.

### 3.1.2.7 Rotavirus

Rotavirus (HRV) is classified as a member of genus Rotavirus within the family Reoviridae. It is non-enveloped (70 nm diameter) and has a wheel-like appearance with an icosahedral shape and a three-layer capsid shell. HRV is the major cause of severe diarrhea (gastroenteritis) in children throughout the world, more so in developing countries, with about 95% of children contracting the infection by 5 years of age. Prevalence of disease ranges from ~30% during off season to ~70% during seasonal peaks. Disease outbreaks demonstrate a seasonal pattern in temperate climates, where the disease is more pronounced during drier and cooler months, while no specific seasonal trend is observed in tropical climates. Transmission is food- and waterborne, by fomites, and through the fecal-oral route. HRV infection is generally more severe and clinically significant in children aged 3 months to 35 months, with the first infection being the most severe. Adults tend to be asymptomatic and/or may demonstrate subclinical infection. Symptoms usually start about 2 days after a person is exposed to rotavirus. Vomiting and watery diarrhea can last 3 to 8 days. Additional symptoms may include loss of appetite and dehydration (loss of body fluids), which can be especially dangerous for infants and young children. A vaccine exists for children to protect them from severe infection.

#### 3.1.2.8 Sapovirus

Sapovirus is a genetically diverse genus of single-stranded positive-sense RNA, non-enveloped viruses within the family Caliciviridae. Sapovirus infections are responsible for both sporadic cases and occasional outbreaks of acute gastroenteritis both in low- and high- income settings. Although all age groups are affected, children younger than 5 years of age have the highest burden of disease. Sharing many similarities with the closely related noroviruses, common symptoms of sapovirus gastroenteritis include vomiting and diarrhea, which typically resolve within 1 week. Sapovirus has also been detected in asymptomatic individuals (Becker-Dreps et al., 2019). Transmission is food- and waterborne, by fomites, and through the fecal-oral route.

#### 3.1.2.9 Reoviruses

Reoviridae is a family of highly stable, non-enveloped, double-stranded RNA viruses. It was initially called "respiratory enteric orphan virus" before the name was changed. Reoviruses have a multilayered capsid and an approximate diameter of ~60 to 100 nm. Human infection primarily results in either asymptomatic infection or mild, self-limited symptoms such as upper respiratory tract illness and gastroenteritis. In rare circumstances, reoviruses have been identified as the causative agent in human cases of meningitis, encephalitis, pneumonia, and myocarditis. Transmission occurs by the fecal-oral and airborne routes. The virus can be readily detected in feces and may also be recovered from pharyngeal or nasal secretions, urine, cerebrospinal fluid, and blood. Reovirus serotypes 1, 2, and 3 are also found ubiquitously in the environment, and their sources include stagnant and river water and untreated sewage. Despite the ease of finding reoviruses in clinical specimens, their role in human disease and the effectiveness of water and wastewater treatment is still uncertain.

#### 3.1.2.10 Parechovirus

Human parechoviruses (PeVs), members of the Picornaviridae family, are common childhood pathogens associated with various clinical manifestations, ranging from asymptomatic or mild symptoms to severe illness. These were formerly known as echoviruses 22 and 23 but are now taxonomically distinct from enteroviruses. Both symptomatic and asymptomatic infected individuals can transmit PeV via the fecal-oral and respiratory routes. Shedding from the upper respiratory tract can occur for 1 to 3 weeks and from the gastrointestinal tract for as long as 6 months after infection. Parechoviruses are widespread and circulate worldwide. Some types show a clear seasonality of later summer and fall, similar to enteroviruses.

### 3.1.3 Protozoa

#### 3.1.3.1 Cryptosporidium

*Cryptosporidium* is an intracellular protozoan parasite of the family Cryptosporididae and phylum Apicomplexa. It has a complex lifecycle with an environmental, thick-walled oocyst stage that is highly resistant to chlorine. Oocysts are 4 to 6 µm, nearly spherical, which when ingested by the host, excyst within the lumen of the small intestine to release infective sporozoites stages that invade surrounding cells. Eventually, infective stages encyst and are shed in feces again. Oocysts can be excreted in feces up to 50 days after the diarrhea stops. It occurs worldwide and is ubiquitous in the environment. Cryptosporidiosis is in the top five most common causes of infectious diarrhea around the globe. Prevalence varies based on climate and level of development, accounting for 0.1% to 2% of diarrheal illness in cooler and developed areas and 0.5% to 10% in warmer and developing countries. It is transmitted through the fecal-oral route, direct contact with infected humans or animals, contaminated food or water, and aerosols. Outbreaks have been associated with contaminated food, drinking, and recreational water (e.g., the contaminated drinking water outbreak in Milwaukee, Wisconsin).

#### 3.1.3.2 Cyclospora

Cyclospora are coccidian in a genus isopora, family Eimeria, phylum Apicomplexa. Cyclospora have spherical, thick-walled, chlorine-resistant oocysts with a diameter of 8 to 10 µm, which contain infective sporozoites. Cyclospora oocysts, when freshly passed in the stool, are not sporulated and therefore are not infective. Oocysts require a few days to weeks to develop and mature in the environment into the infective sporulated oocyst. Upon ingestion, oocysts excyst in the gut and release the sporozoites, which invade the epithelial cells of the small intestine. Cyclospora spp. are present worldwide and endemic to Peru, Haiti, and Nepal. Direct person-to-person transmission of Cyclospora is unlikely to occur because the organism is not immediately infectious when shed, but consumption of untreated water or contaminated food or direct or indirect contact of contaminated soil can cause infection. Cyclospora causes opportunistic infection and causes diarrhea, abdominal cramps or bloating, nausea, low-grade fever, and weight loss.

#### 3.1.3.3 Entamoebah histolytica

*Entamoeba histolytica* is a pseudopod-forming, nonflagellate protozoan parasite and a member of the phylum Sarcomastigophora and subphylum Sarcodina and family Entamoebidae. The lifecycle consists of two stages: an ameboid trophozoite and infectious cyst forms. Cysts are then expelled with the feces and may remain viable in a moist environment for weeks to months. The invasive trophozoites are 10 to 60 µm in diameter and are facultative anaerobes. Infections with *E. histolytica* can be asymptomatic (90%), symptomatic without tissue invasion, or symptomatic with tissue invasion. Symptomatic patients present mainly with intestinal diseases. Amebic dysentery is characterized by diarrhea with severe cramping, lower abdominal pain, low-grade fever, and presence of blood or mucus in the stool for long periods of time. Complications occur in rare cases. *E. histolytica* is a worldwide distributed pathogen and is the second leading cause of parasitic death. Infection rates are also high in tropical and subtropical countries, and in temperate areas with poor sanitation. Most cases of amebiases in the Unites States occur in immigrants from endemic areas, HIV-infected patients, and people living in states that border Mexico. Transmission can occur through the fecal-oral route (ingestion of food and water, contaminated with feces containing *E. histolytica* cysts), but sexual transmission can also occur.

#### 3.1.3.4 Giardia

*Giardia* is a flagellated enteric protozoan parasite. There are two stages in the lifecycle: a motile vegetative form (trophozoite) that resides in the small intestine and is responsible for disease manifestations, and an infective resistant form (cyst) responsible for transmission. Giardia infections occur worldwide and are the most identified parasitic infection in the United States and Canada. The prevalence rate in temperate climates is 2% to 10% in adults and 25% in children, whereas in tropical countries, 50% to 80%

of people are carriers. The majority of infections are asymptomatic, but in symptomatic individuals, patients may experience nausea, chills, low-grade fever, epigastric pain, and sudden onset of watery diarrhea. Diarrhea is often explosive and presents a foul smell without the presence of blood, gas, bloating, mucus, or cellular exudate. Most infections resolve spontaneously within 6 weeks, but chronic infections can also occur. *Giardia* has a very wide host range and is transmitted through the fecal-oral route, the sexual route, and contaminated water and food. Contaminated soil and fomites can also contain infective cysts.

#### 3.1.3.5 *Toxoplasma gondii*

*Toxoplasma gondii* belongs to the phylum Apicomplexa and family Sarcocystidae. They are obligate intracellular parasitic protozoa and parasites of birds and mammals. Felines are the only definitive host and the only animals that pass infective oocysts in their feces. Warm-blooded animals, including humans, are intermediate hosts that harbor tissue cysts in their bodies. It has a worldwide distribution. An estimated 15% to 85% of adults have a chronic infection, depending on where they live, with a documented higher prevalence of oocysts in the environment in moist and hot regions than in dry or cold regions. A documented outbreak of acute toxoplasmosis due to contaminated drinking water affected 110 individuals in Vancouver in 1995. Toxoplasma is transmitted by consuming contaminated water, food, or milk; by inhalation of aerosols containing oocysts; and by contact with sand or soil contaminated by cat feces. Other less relevant forms of transmission are also possible (e.g., transplacental, blood, mechanical). *Toxoplasma gondii* infection is usually nonpathogenic in immunocompetent adults. In acute infection, mild symptoms can arise, including fever, rash, headache, lymphadenopathy, organomegaly (liver and/or spleen), weight loss, weakness, pneumonia, and myalgia.

#### 3.1.3.6 *Balantidium coli*—Less Relevant

*Balantidium coli* is the largest and the only ciliated protozoan parasite of humans. *B. coli* is a member of the class Ciliata, order Trichomonatida and family Balantidiidae. The lifecycle of *B. coli* consists of two stages: a ciliated trophozoite and an environmentally resistant cyst. Cysts are the infectious stage and are shed in feces. The most common sites of *B. coli* infection ($> 95\%$) are the terminal ileum and colon. *B. coli* infections are similar to amoebiasis in that a large fraction of the population are asymptomatic carriers. However, dysenteric symptoms can be observed in patients with malnutrition, alcoholism, or suppressed immune systems. Acute disease is characterized by severe diarrhea and mucoid bloody stool, nausea, vomiting, and headache. It has a 30% mortality rate. *B. coli* infections are uncommon and only have a worldwide prevalence of 0.02% to 1%. It is more common in warm climates such as the tropics and subtropics, and regions with improper WASH conditions. The major risk factor for humans is close contact with pigs or pig feces, which are the primary reservoirs. Transmission occurs most commonly through the fecal-oral route by ingestion of contaminated water or food, or through direct contact with animals. Rats can be carriers of this pathogen; however, it is not clear if transmission to humans is possible. Mechanical vectors have also been reported.

#### 3.1.3.7 Blastocystis hominis

*Blastocystis hominis* is a single-celled parasite that infects humans and some animals. It has two lifecycle stages, and the cyst is the infective stage shed in feces. Blastocystis is highly controversial because it either can be regarded as a commensal or a pathogen; however, studies report that 50% to 80% of individuals infected with Blastocystis will show illness symptoms. Blastocystis is ubiquitous with a global distribution. It is the most common parasitic infection in the United States, where it infected approximately 23% of the total population during year 2000. In resource-limited areas, infection rates as high as 100% have been observed. People become infected through the fecal-oral route, ingestion of contaminated food or water, and poor hygiene.

#### 3.1.3.8 Cystoisospora belli (Formerly Isospora)

*Cystoisospora belli* is a coccidian parasite that infects the epithelial cells of the small intestine and is the least common of the three intestinal coccidia that infect humans. Like *Cryptosporidium*, it has two lifecycle stages: the resistant oocyst stage shed in feces and the sporozoite stage that infects other cells within the same host. It is common worldwide, especially in tropical and subtropical areas. Transmission is fecal-oral from one human to another and by ingestion of oocysts contaminating food or water or through contact with fomites. Infection occurs in immunodepressed individuals, and outbreaks have been reported in institutionalized groups in the United States. Infection causes acute, non-bloody diarrhea with crampy abdominal pain, which can last for weeks and result in malabsorption and weight loss. In immunodepressed patients, and in infants and children, diarrhea can be severe.

### 3.1.4 Fungi—Microsporidia

The microsporidia are a group of unicellular, intracellular parasites closely related to fungi, although the nature of the relation to the kingdom Fungi is not clear. The taxonomic position of this group has been debated and revised repeatedly; historically, they were considered protozoa and often remain managed by diagnostic parasitology laboratories. Microsporidia are characterized by the production of resistant spores that vary in size (usually 1 to 4 µm for medically important species). Many domestic and wild animals may be naturally infected with various medically important microsporidia. Microsporidia are being increasingly recognized as opportunistic infectious agents worldwide. Efforts to characterize the global distribution of species and genotypes are ongoing. The clinical manifestations of microsporidiosis are very diverse, varying according to the causal species and route of infection. Disseminated infection can be fatal. Of all the manifestations of microsporidiosis, *Enterocytozoon bieneusi*–associated diarrhea is the most common (Han & Weiss, 2017).

### 3.1.5 Helminths

The significance of these parasites in developed parts of North America is relatively low, but they are more of a concern in WASH- and resource-limited locations and communities (e.g., the Alabama Blackbelt, rural and remote indigenous communities in Canada, etc.). Reintroduction of these parasites is possible due to immigration, relocation (e.g., along the U.S.–Mexico border and in seasonal workers), and international travel, but the likelihood of these causing a waterborne outbreak in high-resource settings is unlikely. The list of helminths includes the following: *Ascaris, Diphylloborthrium, Hookworms (Ancylostoma and Necator), Schistosomes, Hymenolepis, Strongyloides, Taenia*, and *Trichuris*. The CDC DPDx database provides detailed lifecycles for each of these parasites and outlines transmission routes, lifecycle stages, intermediate hosts (if any), and symptoms of infection.

## 3.2 Other Pathogens of Interest

### 3.2.1 Enteric

Enteric pathogens infect the gut (the enteric or intestinal) tract. All are bacteria, except Hepatitis E virus. Some of these bacteria are always or frequently present in the human gut, cause no illness, and are shed in human feces. Others are rarely or never present in the gut, unless ingested with contaminated water or food to cause infection. One, hepatitis E virus, is only present and shed from the gut when one gets a liver infection and the disease called infectious hepatitis (a liver disease; it is reviewed in Section 3.2.1.9).

#### 3.2.1.1 *Leptospira*—Common in Mice, Rodent Vector Control Issues

Leptospira bacteria occur worldwide and infect humans and other mammals, but especially mice, other rodents, and many domestic and wild animals. The source of human infection is animal urine or water, food, or soil contaminated with such urine, in which the bacteria can persist for days or weeks. Human infection occurs via breaks in the skin and contact with mucous membranes of the mouth, nose, and eyes. Infection can cause the disease leptospirosis, which produces a range of symptoms including high fever, headache, chills, vomiting, diarrhea, abdominal pain, rash, muscle aches, and red eyes. In some case it can progress to liver and kidney disease or meningitis (disease of the outer tissues of the brain and spine), which can be life-threatening. Infection and illness can be treated with antibiotics. Upon exposure, infection and illness may occur as soon as 2 days or after a few weeks. Disease occurs especially among farmers, animal handlers, and those who come in contact with wastewater and contaminated water and soil through work or recreation. Wearing protective clothing and PPE, such as gloves and impervious boots, reduces exposure risks. The bacteria can be controlled or eliminated by effective water and wastewater treatment processes, especially disinfection. Vector control of rodents and other animals also reduces exposure risks.

#### 3.2.1.2 *Aeromonas hydrophila* and Other Pathogenic Species

*Aeromonas hydrophila* is the most important Aeromonas species infecting humans, although some other species do as well. These bacteria occur globally, reside naturally in aquatic environments, and occur widely in water, soil, sediments, and many foods (meat, milk, and produce). They can proliferate and survive for long time periods in the environment, including in biofilms. Human exposure to and infection from *Aeromonas* in environmental media can occur by direct contact, especially from wounds, traumatic accidents and near-drowning, respiratory tract infections and pneumonia, and sometimes bloodstream infections. Ingestion of wastewater, contaminated water, sediments, and foods also can result in diarrhea but is less common. Although *Aeromonas* can occur in drinking water distribution systems that lack and are high in nutrients to produce biofilms, the role of Aeromonas in drinking water outbreaks is uncertain. Wastewater reclamation and water treatment, especially disinfection, is effective in reducing *Aeromonas hydrophila* and other species, but these bacteria can regrow under many conditions, especially in nutrient-rich waters.

### 3.2.1.3 *Klebsiella* Spp., Especially *Klebsiella pneumoniae*

*Klebsiella* species and especially *Klebsiella pneumoniae* are widespread in the environment and occur worldwide in industrial wastewaters, and on and in plants and plant products, including trees, woody vegetation, wood products and byproducts, and effluents from pulp and paper mills and textile finishing plants. Many people and animals harbor *Klebsiella* spp. in the gut and shed them in feces. *Klebsiella pneumophila* is widely present in hospitals and other health care facilities as an opportunistic pathogen that colonizes hospital patients and causes hospital-acquired infection, illness, and death. It is a major cause of pneumonia, urinary tract infections, and infections from wounds, post-surgical procedures, and catheters. Patients at greatest risk are those who already have other health conditions. Contaminated water, presence in and release from biofilms, and aerosols may also be potential sources of human exposure in health care facilities. *Klebsiella* species, including *Klebsiella pneumophila*, are fecal indicator bacteria belonging to the coliform and fecal (thermotolerant) groups widely used to assess wastewater and water quality. They can be controlled in wastewater reclamation and water treatment systems by conventional treatment, including disinfection.

### 3.2.1.4 *Clostridium difficile*

*Clostridioides* (formerly *Clostridium*) *difficile* is a bacterium that requires an absence of free oxygen to live (anaerobic conditions) and produces highly resistant spores that are highly persistent in the environment. This bacterium resides in the human and animal gut, is shed in feces, and occurs worldwide. C. *difficile* is a health concern because it causes diarrhea in people and animals. Most illness in people occurs in the elderly and especially in those who have recently taken or are taking antibiotics. The diarrhea can progress to more a severe intestinal disease called colitis (inflammation and tissue damage of the large intestine) and in some cases spread to the rest of the body if the wall of the large intestine is breached, which is potentially life-threatening. Animal reservoirs, including pets such as dogs and cats, and contaminated foods are also a possible risk for C. *difficile* exposures that could result in infection and illness. Although C. *difficile* is commonly present in various environmental media, including wastewater and contaminated water, there is little evidence that these media have been a cause of human illness. In WRRFs, C. *difficile* concentrations can be greatly reduced by effective treatment, including disinfection with oxidants such as free chlorine, peracetic/peroxyacetic acids, and ozone and in biosolids by composting.

#### 3.2.1.5 *Clostridium perfringens* (Primarily Foodborne and Detected as a Fecal Indicator)

Like *C. difficile, Clostridium perfringens* is a bacterium that requires an absence of free oxygen to live (anaerobic conditions) and produces highly resistant spores that are highly persistent in the environment. It resides in the human and animal gut, is shed in feces, and occurs worldwide. *C. perfringens* is widely present in the environment, including human and animal wastewater and biosolids, fecally contaminated waters, soils, and sediments. It also occurs in many foods and is a major foodborne pathogen when foods are stored improperly at higher temperatures that allow the bacteria to grow and produce toxins. *C. perfringens* is associated with various systemic and enteric (gut) diseases in both humans and animals, including gas gangrene, severe disease of damaged soft tissues, and deep wounds marked by tissue destruction with blackening and gas production, as well as food poisoning and nonfoodborne diarrhea, and enterocolitis. It is not widely associated with waterborne disease. Gas gangrene can be life-threatening. Because *C. perfringens* is present in wastewater, biosolids, and contaminated waters, it is widely used as a fecal indicator microorganism, especially for highly resistant pathogens. Like *C. difficile,* it is reduced by water and wastewater reclamation and treatment, including disinfection with strong oxidants, such as free chlorine, ozone, and peracetic/peroxyacetic acids.

#### 3.2.1.6 *Clostridium tetani* (Cause of Tetanus From Wound and Related Exposures)

Like *C. difficile* and *Clostridium perfringens*, *Clostridium tetani* is a bacterium that requires an absence of free oxygen to live (anaerobic conditions) and produces highly resistant spores that are highly persistent in the environment. This bacterium is widely present in the environment, including soils, dusts, sediments, outdoor surfaces of all kinds, objects such as tools, and wastewater and biosolids. The main source of human exposure is through wounds and other breaches in the skin, and the risk of infection and illness is greatest when tissue penetration is deep. Infection resulting from exposure can result in tetanus disease after several days, typically 10 days. The disease produces distinctive jaw cramping (called "lockjaw"), as well as sudden, involuntary muscle tightening (muscle spasms)—often in the stomach—jerking, seizures or visual staring, overall muscle pain and stiffness, difficulty swallowing, headache, fever and sweats, and abnormal blood pressure and heart rate. Untreated, tetanus is life-threatening. Fortunately, a preventative tetanus vaccine is available and is required for all people in North America. However, the duration of protection from the vaccine decreases over time and revaccination is recommended every 10 years or possibly sooner, especially when an exposure occurs due to injury or wounds and the person cannot remember when last vaccinated. Because *C. tetani* is so common in the environment and always poses exposure risks in many occupational settings, including WRRFs, preventive measures such as vaccination are required and precautions such as using PPE are also usually required for many workers.

#### *3.2.1.7 Helicobacter Pylori*

*Helicobacter pylori* is a spiral-shaped bacterium that produces infection of the gut, especially the stomach, in a large segment of the population worldwide. About one-third of Americans have it, and in countries with poor sanitation and hygiene, most people harbor it. It is widely present in the environment in food and water, as well as on surfaces, especially those harboring saliva or other bodily releases such as feces from infected people. Many people harboring *H. pylori* are asymptomatic and appear "normal." However, others develop signs and symptoms of illness, typically stomach aches and ulcers. As a worst case, the illness progresses over time to stomach cancer, which is life-threatening. People with *H. pylori* ulcers can be treated with a combination of antibiotics and an acid-reducing drug, although a permanent cure is often not possible, and infection and illness can return. *H. pylori* can be extensively reduced in wastewater and contaminated water by treatment processes, including disinfection.

#### *3.2.1.8 MRSA and Other Antibiotic Resistant Bacteria*

Methicillin-resistant *Staphylococcus aureus* (called MRSA for short) is a bacterium that is resistant to the antibiotic methicillin and similar antibiotics. Many other bacteria, including those that cause enteric infections and illnesses, have also become resistant to a number of different antibiotics, including resistance to the so-called "antibiotics of last resort" or the last lines of antibiotic defense. People who become infected with such highly antibiotic resistant bacteria are no longer treatable with antibiotics, and they may die from such infections as a result. *Staphylococcus aureus* is widely present in the environment, occurs in about one-third of all humans in nasal passages and on skin, and is present in animals, foods, water, and other environmental media. A large proportion of *Staphylococcus aureus* infections, between 13% and 74% worldwide, are MRSA. MRSA most often causes skin disorders appearing as pimples that in some cases can progress to boils and abscesses. In some people it causes serious lung infections (pneumonia). MRSA also can also cause food poisoning from improperly stored foods that grow high concentrations of bacteria-producing toxins (poisons). The greatest risks from MRSA are in hospitals and other health care facilities, where patients get skin and soft-tissue, heart, and bloodstream infections, often from catheters and wounds after surgery. Although MRSA and other antibiotic-resistant bacteria can be extensively reduced by wastewater reclamation and water treatment processes, including disinfection, some of these bacteria survive, remain present, and further proliferate.

#### *3.2.1.9 Hepatitis E*

Hepatitis E virus (HEV) occurs globally but is greater where water, sanitation and hygiene are inadequate, mostly in less developed countries. The virus infects people as well as other mammals, including pigs, poultry, rodents, deer, and rabbits. Transmission is by the fecal-oral route, and contaminated water and food are important sources of exposure. Transmission through direct contact with animals, such as pigs, as well as transfusions and organ transplants and mother-to-child contact also occurs. HEV infections may be asymptomatic, but infectious hepatitis, a severe liver disease, is a major risk and can be life-threatening, especially for pregnant women. The disease cannot be easily distinguished from other forms of hepatitis. HEV cannot be readily cultured, but its nucleic acid has been detected in wastewater, biosolids and contaminated waters. The virus can be extensively reduced by wastewater and drinking-water treatment processes, including chlorine disinfection. Good sanitation and hygiene and vector control can reduce risks from HEV at WRRFs.

### 3.2.2 Non-enteric

Other potential pathogens of health concern do not usually infect the enteric tract but typically cause respiratory illnesses. Some cause skin infections and genito-urinary tract infections, bloodstream infections, and liver disease (infectious hepatitis) and can progress to more severe life-threatening health effects, especially in people with weakened immune systems and other health conditions. Of those that are bacteria, they are often referred to as "opportunistic" pathogens causing "opportunistic infections."

#### *3.2.2.1 Staphylococcus aureus*

*Staphylococcus aureus* is a widely present bacterium globally and is present in about one-third of all humans in their nasal passages and on skin and is also present in animals, foods, water, and other environmental media. *Staphylococcus aureus* is an opportunistic pathogen and is transmitted principally through personal contact (e.g., hands and other touching) or from contaminated objects (fomites), nasal and respiratory secretions, eye discharges, genito-urinary secretions, wounds and burns, and contact with animals, especially in intensive livestock agriculture. Various *S. aureus* types and strains cause pneumonias, skin, soft tissue, eye and urinary tract infections, bacteremia, toxic and septic shock conditions, arthritis, and joint diseases, particularly in susceptible hosts and those in health care facilities. Methicillin-resistant *S. aureus* causes more serious human health effects than methicillin-sensitive *S. aureus* (MSSA), but both can be pathogenic. *S. aureus* also causes food poisoning when prepared foods are improperly stored, which allows the bacteria to multiply and produce toxins. *S. aureus* can be present in wastewater, biosolids, and water, and its concentrations can be reduced extensively by wastewater and water treatment systems and disinfection processes. However, it can reproduce in environmental waters and wastes, but despite this ability, it has not been a documented cause of drinking-waterborne outbreaks.

#### *3.2.2.2 Hepatitis B*

Hepatitis B is a liver disease caused by hepatitis B virus (HBV). The disease caused is infectious hepatitis, and there are several other viruses that also cause this. HBV is transmitted from person to person by sexual contact, and by exposure to contaminated blood from injection (especially among drug abusers), needle sticks, and transfusions. The disease symptoms can include fatigue, poor appetite, stomach pain, nausea, and jaundice (yellowing of the skin and whites of the eyes). Not all infected people become ill with these symptoms, and asymptomatic carriers of the virus may harbor it indefinitely. However, some infected people may later develop severe disease conditions such as cirrhosis (severe damage and degradation of liver tissues) and liver cancer. Infected people can shed HBV in bodily secretions and excretions, and the virus or at least its nucleic acid can be present in raw wastewater. However, HBV does not survive well in the environment, and wastewater reclamation processes are effective in further reducing its presence. There are effective vaccines against HBV, and these vaccines are recommended for all people, including wastewater professionals.

#### *3.2.2.3* Free-Living Amoeba—*Acanthamoeba culbertsoni* and *A. keratitis*

*Acanthamoeba* spp. are microscopic, free-living, single-cell amoeba parasites that can alter their shape by extending and retracting projections called pseudopods. *A. culbertsoni* and *A. keratitis* occur worldwide in the environment in water and soil. These acanthamoebas cause severe infections of the eye, skin, and central nervous system. These amoebas can cause human infection of the eyes by contact lens use, by skin cuts or wounds, or by being inhaled into the lungs. People often become exposed to *Acanthamoeba*, but very few will become ill from such exposures. These *Acanthamoeba* cause three kinds of diseases: (1) *Acanthamoeba keratitis*, an eye infection that can result in permanent visual impairment or blindness; (2) Granulomatous Amebic Encephalitis (GAE), an infection of the brain and spinal cord that typically occurs in persons who have a compromised immune systems, are chronically ill and debilitated, or are receiving chemotherapy or taking broad-spectrum antibiotics; and (3) disseminated infection, which occurs mostly in people with compromised immune systems and affects the skin, sinuses, lungs, and other organs. GAE and disseminated infections can be life-threatening.

These amoebas can be in wastewater and water and can be reduced by treatment processes and especially disinfection. However, they can regrow in water and wastewater and pose exposure risks. Measures to reduce risks of eye infections from *A. keratitis* are to avoid contact with contaminated water, store contact lenses in fresh solution at every use, and never wear contact lenses when sleeping. To minimize risks of GAE and disseminated infections, immunocompromised and other at-risk people should practice good personal hygiene, live in settings with good environmental sanitation, and avoid exposure to environmental media in which these *Acanthamoeba* may be present at high concentrations. Workers who may contact or otherwise be exposed to wastewater should always use PPE and practice good personal hygiene.

#### *3.2.2.4* Free-Living Amoeba—*Naegleria fowleri*

*Naegleria fowleri* is a free-living amoeba similar to the *Acanthamoeba* already described. It thrives by feeding on bacteria in warm fresh waters and is commonly found worldwide in lakes, rivers, hot springs, and soils (and sometimes groundwaters). Because of its ubiquitous presence, people who enter such waters for recreation or work and then get water into their nose may become infected when these amoeba then travel to the brain. Once in the brain, the amoeba infect and proliferate in brain tissue and cause a severe disease called primary amebic meningoencephalitis, which is almost always fatal. Although it is possible for *Naegleria fowleri* to be present in wastewater, there is little if any evidence for its presence there. However, other waters that may be present in wastewater reclamation facilities could be a source of exposure for this pathogen, such as stored, nutrient-rich waters and cooling tower/heat exchange waters. Workers who may come in contact with or otherwise be exposed to such waters or wastewaters should always use PPE and practice good personal hygiene.

#### *3.2.2.5 Legionella pneumophila*

*Legionella pneumophila* is a naturally occurring bacterium that is widely present in the environment, including in water, wastewater, soil, and sediments. It exists and multiplies in water and causes respiratory disease (legionellosis) resulting from airborne water exposures. It proliferates in waters at warm but not hot temperatures, such as in hot water systems and cooling tower waters. It also colonizes biofilms and is widely present in building waters and their plumbing fixtures, as well as in the waters of cooling towers/heat exchangers, engineered recreational waters such as pools, tubs and recreational sprayers, and garden sprinklers and produce misters in food markets. *Legionella pneumophila* disease results from exposure to airborne droplets and smaller aerosol particles that contain this bacterium. The resulting disease, called Legionnaire's disease, is a severe pneumonia that can be life-threatening. A less severe respiratory illness caused by airborne *Legionella* exposure is called Pontiac fever and is rarely fatal. *Legionella pneumophila* is likely to be present in the wastewaters and waters of water reclamation facilities. Therefore, prevention and control measures are needed that focus on the development and use of an integrated management plan to minimize the presence of this pathogen and prevent worker exposures (see Chapter 4). As with measure for other pathogens, workers who may contact or otherwise be exposed to waters or wastewaters and their airborne droplets and aerosols should always use PPE and practice good personal hygiene.

#### 3.2.2.6 *Pseudomonas aeruginosa*

*Pseudomonas aeruginosa* is a common environmental bacterium that occurs globally and can multiply and persist in sewage, water, biofilms, soil, plants, and foods. It is an opportunistic (or conditional) pathogen that primarily causes health effects in people with other health conditions. In such at-risk people, it causes urinary tract, ear, eye, and respiratory infections. Contact with recreational water and with nonsterile medical instruments are important exposure sources. In health care facilities, transmission to at-risk patients can also occur from contaminated mechanical respiratory ventilators or other health care devices that are or become contaminated, and by direct contact with wounds or burns. Many strains of *P. aeruginosa* are resistant to a range of antimicrobial agents, which increases their potential to cause health risks. *Pseudomonas aeruginosa* in water and wastewater are detectable as members of the heterotrophic plate count group. Although wastewater reclamation processes, including disinfection, can reduce its concentrations, it can regrow in wastewater and water. This pathogen is not considered a great risk in WRRFs, and risks to workers can be managed by wearing PPE and practicing good hygiene.

#### 3.2.2.7 *Mycobacterium Avium Complex*

Mycobacterium species of bacteria include *Mycobacterium tuberculosis* (the cause of the respiratory disease tuberculosis) as well as non-tubercular

Mycobacteria (NTM) such the pathogenic *Mycobacterium avium* complex of species, called MAC. Non-tubercular Mycobacteria including MAC are widely distributed in the environment and present in wastewater, water, biofilms, soils, sediments. NTMs like MACs grow and survive in wastewater, water, and biofilms and are resistant to disinfection and other treatments due to their waxy outer layer. They are often present in building water supplies, especially in hospitals and other health care facilities. MAC infections can be asymptomatic in some people, but in those with weakened immune systems they also cause diseases. Illnesses include severe respiratory disease when acquired by inhalation of airborne droplets and aerosols, gut illness such as diarrhea from consumption of contaminated water sources, as well as genito-urinary tract infections. If untreated, these illnesses can progress to more severe ones that are life-threatening. Several waterborne outbreaks have been reported. MACs and other NTMs can be detected by culture methods, but the methods are challenging due to inefficient and slow growth and the added tests needed to specifically identify MACs; therefore, they are not widely analyzed in water or wastewater. Although MACs can be reduced by wastewater and water treatment processes, they can also regrow and persist after treatment. Wearing PPE and practicing good hygiene reduces the health risks from exposure to them.

Key information about the various pathogens described above in Section 3.2 are summarized in Table 2.2.

### 3.2.3 Viruses That Receive Public Attention but That Have Limited Relevance in Wastewater

In 2016, the WHO released *An R&D Blueprint for Action to Prevent Epidemics* after the Western Africa Ebola outbreak of 2013. The document is a global strategy and preparedness plan that allows the rapid activation of research and development activities during epidemics. The aim of the R&D Blueprint was to fast-track the availability of effective tests, vaccines, and medicines. The document listed a group of priority diseases and their infectious agents that should be the focus of future work. The list includes Middle East respiratory syndrome coronavirus (MERS-CoV), Severe Acute Respiratory Syndrome (SARS), SARS-CoV-2 and other coronaviruses, Ebola virus disease and Marburg virus diseases, Lassa fever, Crimean-Congo haemorrhagic fever (CCHF), Nipah and henipaviral diseases, Rift Valley fever (RVF), and Zika. However, these pathogens have limited relevance to wastewater and have no documented cases of waterborne transmission. For more information, see the WHO website: https://www.who.int/publications/m/item/an-r-d-blueprint-for-action-to-prevent-epidemics---update-2017.

**TABLE 2.2** Characteristics of Pathogens of Concern

| Microbe Name | Microbe Type | Exposure Sources and Reservoirs | Environmental Growth?/Survival | Health Effects/ Disease |
|---|---|---|---|---|
| | | **Enteric Pathogens** | | |
| *Leptospira* | bacterium | Environmental/animal urine | No/moderate survival | Leptospirosis/ systemic disease |
| *Aeromonas hydrophila* | bacterium | Environmental/various media | Yes/moderate survival | Multiple human infection sites |
| *Klebsiella* spp. & *Klebsiella pneumoniae* | bacteria | Environmental/various media | Yes/moderate survival | Multiple human infection sites |
| *Clostridium difficile* | bacterium | Humans and animals | No/long survival | Multiple human infection sites |
| *Clostridium perfringens* | bacterium | Humans and animals | Yes/long survival | Multiple human infection sites |
| *Clostridium tetani* | bacterium | Environmental media | Yes/long survival | Wounds and other infections sites |
| *Helicobacter pylori* | bacterium | Humans and environmental media | No/short survival | Stomach |
| *MRSA and other antibiotic-resistant* | bacteria | Humans, animals, and environmental media | Yes/long survival | Multiple infection sites |
| *Hepatitis E* | virus | Humans and some animals | No/long survival | Gut, liver, and bloodstream |
| | | **Non-enteric Pathogens** | | |
| *Staphylococcus aureus* | bacterium | Humans, animals, and environmental media | Yes/long survival | Nose, skin, and respiratory tract |
| *Hepatitis B* | virus | Humans, blood, and blood products | No/short survival | Infectious hepatitis disease |
| *Free-living amoeba—Acanthamoeba culbertsoni and A. keratitis* | Amoebic protozoan parasite | Environmental media, especially water | Yes/long survival | Eye infections |
| *Free-living amoeba—Naegleria fowleri* | Amoebic protozoan parasite | Environmental media, especially water | Yes/long survival | Brain infections; deadly but low exposure risk |
| *Legionella pneumophila* | bacterium | Environmental media, especially water, air, and biofilms | Yes/long survival | Pneumonia from airborne exposure |
| *Pseudomonas aeruginosa* | bacterium | Environmental media | Yes/long survival | Multiple infection sites |
| *Mycobacterium avium* complex *(MAC)* | bacteria | Environmental media, esp. water and biofilms | Yes/long survival | Pneumonia and sometimes other infection sites |

# 4.0 RESOURCES FOR ADDITIONAL INFORMATION ON SPECIFIC PATHOGENS

## 4.1 Publicly Available

Centers for Disease Control and Prevention. (2019). *Parasites A-Z index*. DPDx—laboratory identification of parasites of public health concern. https://www.cdc.gov/dpdx/az.html

Centers for Disease Control and Prevention. (2021). *Viral diseases*. National Center for Immunization and Respiratory Diseases. https://www.cdc.gov/ncird/DVD.html

European Centre for Disease Prevention and Control. (n.d.). *A–Z disease list*. Retrieved February 3, 2023, from https://www.ecdc.europa.eu/en/all-topics

Government of Canada. (2022, June 7). *Pathogen safety data sheets*. https://www.canada.ca/en/public-health/services/laboratory-biosafety-biosecurity/pathogen-safety-data-sheets-risk-assessment.html

World Health Organization. (n.d.). *Fact sheets*. Retrieved February 3, 2023, from https://www.who.int/news-room/fact-sheets

## 4.2 Additional Resources

Ambekar, A., Bharadwaj, R., Joshi, S., Kagal, A., & Bal, A. (2004). Sero surveillance of leptospirosis among sewer workers in Pune. *Indian Journal of Public Health, 48*(1), 27–29.

Arvanitidou, M., Mamassi, P., & Vayona, A. (2004). Epidemiological evidence for vaccinating wastewater treatment plant workers against hepatitis A and hepatitis B virus. *European Journal of Epidemiology, 19*(3), 259–262.

Bauer, H., Fuerhacker, M., Zibuschka, F., Schmid, H., & Puxbaum, H. (2002). Bacteria and fungi in aerosols generated by two different types of wastewater treatment plants. *Water Research, 36*(16), 3965–3970.

Becker-Dreps, S., Bucardo, F., & Vinjé, J. (2019). Sapovirus: An important cause of acute gastroenteritis in children. *The Lancet. Child & Adolescent Health, 3*(11), 758.

Berg, G., Sullivan, G., & Venosa, A. (1988). Low-temperature stability of viruses in sludges. *Applied and Environmental Microbiology, 54*(3), 839–841.

Bibby, K., & Peccia, J. (2013). Identification of viral pathogen diversity in sewage sludge by metagenome analysis. *Environmental Science & Technology, 47*(4), 1945–1951.

Bonanni, P., Comodo, N., Pasqui, R., Vassalle, U., Farina, G., Nostro, A. L., Boddi, V., & Tiscione, E. (2000). Prevalence of hepatitis A virus infection in sewage plant workers of Central Italy: Is indication for vaccination justified? *Vaccine, 19*(7–8), 844–849.

Bosch, A., Pintó, R. M., & Guix, S. (2014). Human astroviruses. *Clinical Microbiology Reviews, 27*(4), 1048–1074.

Brisolara, K. F., Maal-Bared, R., Sobsey, M. D., Reimers, R. S., Rubin, A., Bastian, R. K., Gerba, C., Smith, J. E., Bibby, K., Kester, G., & Brown, S. (2021). Assessing and managing SARS-CoV–2 occupational health risk to workers handling residuals and biosolids. *Science of the Total Environment, 774*, 145732.

Brown, R. (1985). *Demonstration of acceptable systems for land disposal of sewage sludge*. cs-805189. U.S. Environmental Protection Agency.

Burton, N. C., & Trout, D. (1999). *Health Hazard Evaluation Report 98–0118–2748*. National Institute for Occupational Safety and Health.

Cadilhac, P., & Roudot-Thoraval, F. (1996). Seroprevalence of hepatitis A virus infection among sewage workers in the Parisian area, France. *European Journal of Epidemiology, 12*(3), 237–240.

Cai, L., & Zhang, T. (2013). Detecting human bacterial pathogens in wastewater treatment plants by a high-throughput shotgun sequencing technique. *Environmental Science & Technology, 47*(10), 5433–5441.

Clark, C. S. (1981). *Evaluation of the health risks associated with the treatment and disposal of municipal wastewater and sludge*. U.S. Environmental Protection Agency.

Clark, C. S. (1987). Potential and actual biological related health risks of wastewater industry employment. *Water Pollution Control Federation Journal, 59*(12), 999–1008.

Clark, C. S., Linnemann Jr., C. C., & Sekla, L. (1986). The use of serum antibody as a means to determine infections from exposure to wastewaters and refuse. *Critical Reviews in Environmental Science and Technology, 16*(4), 305–326.

De Serres, G., & Laliberte, D. (1997). Hepatitis A among workers from a waste water treatment plant during a small community outbreak. *Occupational and Environmental Medicine, 54*(1), 60–62.

De Serres, G., Levesque, B., Higgins, R., Major, M., Laliberte, D., Boulianne, N., & Duval, B. (1995). Need for vaccination of sewer workers against leptospirosis and hepatitis A. *Occupational and Environmental Medicine, 52*(8), 505–507.

Dorn, C. R., Reddy, C. S., Lamphere, D. N., Gaeuman, J. V., & Lanese, R. (1985). Municipal sewage sludge application on Ohio farms: Health effects. *Environmental Research, 38*(2), 332–359.

Dudley, D. J., Guentzel, M. N., Ibarra, M., Moore, B., & Sagik, B. (1980). Enumeration of potentially pathogenic bacteria from sewage sludges. *Applied and Environmental Microbiology, 39*(1), 118–126.

Dumontet, S., Dinel, H., & Baloda, S. (1999). Pathogen reduction in sewage sludge by composting and other biological treatments: A review. *Biological Agriculture & Horticulture, 16*(4), 409–430.

Epstein, E. (2002). *Land application of sewage sludge and biosolids.* CRC Press.

Foladori, P., Cutrupi, F., Segata, N., Manara, S., Pinto, F., Malpei, F., Bruni, L., & La Rosa, G. (2020). SARS-CoV–2 from faeces to wastewater treatment: What do we know? A review. *Science of the Total Environment 743*, 140444.

Fracchia, L., Pietronave, S., Rinaldi, M., & Martinotti, M. G. (2006). Site-related airborne biological hazard and seasonal variations in two wastewater treatment plants. *Water Research, 40*(10), 1985–1994.

Friis, L., Engstrand, L., & Edling, C. (1996). Prevalence of Helicobacter pylori infection among sewage workers. *Scandinavian Journal of Work, Environment & Health*, 364–368.

Garner, F., Ellis, S., & Lacey, J. (1954). The size distribution and entrainment of droplets. *Transactions of the American Institute of Chemical Engineers, 32*, 222–235.

Górny, R. L. (2020). Microbial aerosols: Sources, properties, health effects, exposure assessment—A review. *KONA Powder and Particle Journal*, 2020005.

Haas, C. N., Rycroft, T., Bibby, K., & Casson, L. (2017). Risks from ebolavirus discharge from hospitals to sewer workers. *Water Environment Research, 89*(4), 357–368.

Han, B., & Weiss, L. M. (2017). Microsporidia: Obligate intracellular pathogens within the fungal kingdom. *Microbiology Spectrum, 5*(2), 3.

Harrison, E. Z., & Oakes, S. R. (2003). Investigation of alleged health incidents associated with land application of sewage sludges. *New Solutions: A Journal of Environmental and Occupational Health Policy, 12*(4), 387–408.

Heijnen, L., & Medema, G. (2011). Surveillance of influenza A and the pandemic influenza A (H1N1) 2009 in sewage and surface water in the Netherlands. *Journal of Water and Health, 9*(3), 434–442.

Heng, B., Goh, K., Doraisingham, S., & Quek, G. (1994). Prevalence of hepatitis A virus infection among sewage workers in Singapore. *Epidemiology & Infection, 113*(1), 121–128.

Jeggli, S., Steiner, D., Joller, H., Tschopp, A., Steffen, R., & Hotz, P. (2004). Hepatitis E, Helicobacter pylori, and gastrointestinal symptoms in workers exposed to waste water. *Occupational and Environmental Medicine, 61*(7), 622–627.

Jenkins, S. R., Armstrong, C. W., & Monti, M. M. (2007). *Health effects of biosolids applied to land: Available scientific evidence*. Virginia Department of Health.

Keene, W. C., Maring, H., Maben, J. R., Kieber, D. J., Pszenny, A. A., Dahl, E. E., Izaguirre, M. A., Davis, A. J., Long, M. S., & Zhou, X. (2007). Chemical and physical characteristics of nascent aerosols produced by bursting bubbles at a model air-sea interface. Journal *of Geophysical Research: Atmospheres, 112*(D21).

Khuder, S. A., Arthur, T., Bisesi, M. S., & Schaub, E. A. (1998). Prevalence of infectious diseases and associated symptoms in wastewater treatment workers. *American Journal of Industrial Medicine, 33*(6), 571–577.

Khuder, S., Milz, S. A., Bisesi, M., Vincent, R., McNulty, W., & Czajkowski, K. (2007). Health survey of residents living near farm fields permitted to receive biosolids. *Archives of Environmental & Occupational Health, 62*(1), 5–11.

Krämer, A., Akmatov, M., & Kretzschmar, M. (2009). Principles of infectious disease epidemiology. *Modern Infectious Disease Epidemiology*, 85–99.

Kraut, A., Lilis, R., Marcus, M., Valciukas, J. A., Wolff, M. S., & Landrigan, P. J. (1988). Neurotoxic effects of solvent exposure on sewage treatment workers. *Archives of Environmental Health: An International Journal, 43*(4), 263–268.

Kusnetsov, J., Neuvonen, L.-K., Korpio, T., Uldum, S. A., Mentula, S., Putus, T., Minh, N. N. T., & Martimo, K.-P. (2010). Two Legionnaires' disease cases associated with industrial waste water treatment plants: A case report. *BMC Infectious Diseases, 10*(1), 1–8.

LeChevallier, M. W., Mansfield, T. J., & Gibson, J. M. (2020). Protecting wastewater workers from disease risks: Personal protective equipment guidelines. *Water Environment Research, 92*(4), 524–533.

Levin, M., Froom, P., Trajber, I., Lahat, N., Askenazi, S., & Lerman, Y. (2000). Risk of hepatitis A virus infection among sewage workers in Israel. *Archives of Environmental Health: An International Journal, 55*(1), 7–10.

Lewis, D. L., Gattie, D. K., Novak, M. E., Sanchez, S., & Pumphrey, C. (2002). Interactions of pathogens and irritant chemicals in land-applied sewage sludges (biosolids). *BMC Public Health, 2*(1), 1–8.

Li, J., Zhou, L., Zhang, X., Xu, C., Dong, L., & Yao, M. (2016). Bioaerosol emissions and detection of airborne antibiotic resistance genes from a wastewater treatment plant. *Atmospheric Environment, 124*, 404–412.

Linnemann Jr., C. C., Jaffa, R., Gartside, P. S., Scarpino, P. V., & Clark, C. S. (1984). Risk of infection associated with a wastewater spray irrigation system used for farming. *Journal of Occupational Medicine*, 41–44.

Liu, Y. X., Qin, Y., Chen, T., Lu, M., Qian, X., Guo, X., & Bai, Y. (2020). A practical guide to amplicon and metagenomic analysis of microbiome data. *Protein & Cell, 12*, 315.

Lundholm, M., & Rylander, R. (1983). Work related symptoms among sewage workers. *Occupational and Environmental Medicine, 40*(3), 325–329.

Markt, R., Endler, L., Amman, F., Schedl, A., Penz, T., Büchel-Marxer, M., Grünbacher, D., Mayr, M., Peer, E., & Pedrazzini, M. (2022). Detection and abundance of SARS-CoV–2 in wastewater in Liechtenstein, and the estimation of prevalence and impact of the B. 1.1. 7 variant. *Journal of Water and Health, 20*(1), 114–125.

Mirskaya, E., & Agranovski, I. E. (2018). Sources and mechanisms of bioaerosol generation in occupational environments. *Critical Reviews in Microbiology, 44*(6), 739–758.

Muzaini, K., Yasin, S. M., Ismail Z., & Ishak, A. R. (2021). Systematic review of potential occupational respiratory hazards exposure among sewage workers. Front Public Health. 2021; 9: 646790. Published online 2021 Mar 8. doi: 10.3389/fpubh.2021.646790. PMCID: PMC7982603. PMID: 33763402

National Research Council. (2002). *Biosolids applied to land: Advancing standards and practices.* Committee on Toxicants and Pathogens in Biosolids Applied to Land, Board on Environmental Studies and Toxicology, Division on Earth and Life Studies. National Research Council.

Pascual, L., Pérez-Luz, S., Yáñez, M. A., Santamaría, A., Gibert, K., Salgot, M., Apraiz, D., & Catalán, V. (2003). Bioaerosol emission from wastewater treatment plants. *Aerobiologia, 19*(3), 261–270.

Pitt, T., & Barer, M. (2012). Classification, identification and typing of micro-organisms. *Medical Microbiology*, 24.

Prieto, J., Kilpatrick, C., & Randle, J. (2014). Infection prevention: Principles of safe practice in healthcare. In *Biofilms in infection prevention and control* (pp. 19–36). Elsevier.

Randall, K., Ewing, E. T., Marr, L. C., Jimenez, J., & Bourouiba, L. (2021). How did we get here: What are droplets and aerosols and how far do they go? A historical perspective on the transmission of respiratory infectious diseases. *Interface Focus, 11*(6), 20210049.

Rivadulla, E., & Romalde, J. L. (2020). A comprehensive review on human Aichi virus. *Virologica Sinica, 35*(5), 501–516.

Sánchez-Monedero, M., Aguilar, M., Fenoll, R., & Roig, A. (2008). Effect of the aeration system on the levels of airborne microorganisms generated at wastewater treatment plants. *Water Research, 42*(14), 3739–3744.

Schöniger-Hekele, M., Petermann, D., Weber, B., & Müller, C. (2007). Tropheryma whipplei in the environment: Survey of sewage plant influxes and sewage plant workers. *Applied and Environmental Microbiology, 73*(6), 2033–2035.

Shuval, H. I., Fattal, B., Jakubowski, W., Pahren, H. R., & Wax, Y. (1983). *Retrospective epidemiological study of disease associated with wastewater utilization.* U.S. Environmental Protection Agency, Health Effects Research Laboratory.

Sidhu, J. P., Hodgers, L., Ahmed, W., Chong, M., & Toze, S. (2012). Prevalence of human pathogens and indicators in stormwater runoff in Brisbane, Australia. *Water Research, 46*(20), 6652–6660.

Sidhu, J. P., & Toze, S. G. (2009). Human pathogens and their indicators in biosolids: A literature review. *Environment International, 35*(1), 187–201.

Snyder, C. (2003). Problem formulation for human health risk assessments of pathogens in land-applied biosolids. U.S. Environmental Protection Agency, Office of Research and Development, National Center for Environmental Assessment.

Thorn, J., & Kerekes, E. (2001). Health effects among employees in sewage treatment plants: A literature survey. *American Journal of Industrial Medicine, 40*(2), 170–179.

Tschopp, A., Joller, H., Jeggli, S., Widmeier, S., Steffen, R., Hifiker, S., & Hotz, P. (2009). Hepatitis E, Helicobater and peptic ulcers in workers exposed to sewage: A prospective cohort study. *Occupational and Environmental Medicine, 66*, 45–50.

Turovskiy, I. S., & Mathai, P. (2006). *Wastewater sludge processing.* John Wiley & Sons.

U.S. Environmental Protection Agency. (n.d.) *Municipal wastewater.* https://www.epa.gov/npdes/municipal-wastewater

U.S. Environmental Protection Agency. (1994). *A plain English guide to the EPA part 503 biosolids rule.* U.S. Environmental Protection Agency, Office of Wastewater Management.

U.S. Environmental Protection Agency. (2004). *Report to congress: Impacts and control of CSOs and SSOs*. U.S. Environmental Protection Agency, Office of Water.

Van Hooste, W., Charlier, A.-M., Rotsaert, P., Bulterys, S., Moens, G., van Sprundel, M., & De Schryver, A. (2010). Work-related Helicobacter pylori infection among sewage workers in municipal wastewater treatment plants in Belgium. *Occupational and Environmental Medicine, 67*(2), 91–97.

Venczel, L., Brown, S., Frumkin, H., Simmonds-Diaz, J., Deitchman, S., & Bell, B. P. (2003). Prevalence of hepatitis A virus infection among sewage workers in Georgia. *American Journal of Industrial Medicine, 43*(2), 172–178.

Viau, E., & Peccia, J. (2009). Survey of wastewater indicators and human pathogen genomes in biosolids produced by class A and class B stabilization treatments. *Applied and Environmental Microbiology, 75*(1), 164–174.

Virginia Department of Health. (2019). *Report to the Virginia General Assembly: Land applied biosolids epidemiological pilot study and pathogen content analysis*. Office of Environmental Health Services and Virginia Department of Health.

Warneck, P. (1999). *Chemistry of the natural atmosphere*. Elsevier.

Water Environment Federation. (2020). *Protecting wastewater professionals from Covid-19 and other biological hazards*.

Weirich, S. R., Silverstein, J., & Rajagopalan, B. (2011). Effect of average flow and capacity utilization on effluent water quality from US municipal wastewater treatment facilities. *Water Research, 45*(14), 4279–4286.

Weldon, M., VanEgdom, M. J., Hendricks, K. A., Regner, G., Bell, B. P., & Sehulster, L. M. (2000). Prevalence of antibody to hepatitis A virus in drinking water workers and wastewater workers in Texas from 1996 to 1997. *Journal of Occupational and Environmental Medicine,* 821–826.

Wong, K., Onan, B. M., & Xagoraraki, I. (2010). Quantification of enteric viruses, pathogen indicators, and Salmonella bacteria in class B anaerobically digested biosolids by culture and molecular methods. *Applied and Environmental Microbiology, 76*(19), 6441–6448.

Woodcock, A. (1955). Bursting bubbles and air pollution. *Sewage and Industrial Wastes*, 1189–1192.

World Bank Group, International Labour Office, WaterAid, & the World Health Organization. (2019). *Health, safety and dignity of sanitation workers*. The World Bank.

World Health Organization. (2016). *An R&D blue print for action to prevent epidemics.* https://www.who.int/publications/m/item/an-r-d-blueprint-for-action-to-prevent-epidemics

Xu, F., & Li, Y. (2017). Biomass digestion. In M. A. Abraham (Ed.), *Encyclopedia of sustainable technologies* (pp. 197–204). Elsevier.

Zabinski, J. W., Pieper, K. J., & Gibson, J. M. (2018). A Bayesian belief network model assessing the risk to wastewater workers of contracting Ebola virus disease during an outbreak. *Risk Analysis, 38*(2), 376–391.

# Chapter 3

# Current State of the Knowledge on Treatment Efficacy

*Kari Brisolara; Philip Block, PhD; Chuck Gerba; Bob Reimers; Jenny Reina; Stan Shmia; Jay Swift; & Yue Xu, PhD*

# 1.0 INTRODUCTION—OVERALL PATHOGENS REMOVALS

The past 2 centuries have seen significant achievements in the control of infectious diseases, which were previously the leading cause of death worldwide (Schlipköter & Flahault, 2010). In the 19th century, industrialization resulted in an increase in populations moving from rural to urban environments and a rise in immigration, both of which led to overcrowding in inadequate housing served by minimal to nonexistent public water supplies and sanitation systems. Not surprisingly, these conditions resulted in repeated outbreaks of cholera, dysentery, tuberculosis, typhoid fever, influenza, yellow fever, and malaria. It was around that time that public health professionals used the discovery that microorganisms were the causes of many serious diseases (e.g., cholera, tuberculosis) to develop infectious disease prevention strategies. These strategies focused on sanitation, water and food safety regulations, and improved nutrition and living standards (Centers for Disease Control and Prevention [CDC], 1999). One of the most notable improvements was the chlorination of drinking water in the early 1900s, which became a widespread public health practice, further decreasing the incidence of waterborne diseases. Between 1930 and 1950, governments also introduced sewage disposal, water treatment, food safety, organized solid waste disposal, and public education about hygienic practices (CDC, 1999).

Conventional wastewater treatment consists of a combination of physical, chemical, and biological processes to remove solids, organic matter, and nutrients from domestic, industrial, and other wastewaters (Asano & Tchobanoglous, 1987). The general terms used to describe different degrees of treatment (in order of increasing treatment level) are *preliminary* (or *pretreatment*), *primary*, *secondary*, *tertiary*, and *advanced* wastewater treatment. In preliminary treatment (e.g., coarse screening, grit removal), coarse solids and other large materials often found in raw wastewater are removed to enhance the operation and maintenance of subsequent treatment units. Primary treatment (e.g., flocculation, coagulation, sedimentation) removes the settleable organic and inorganic solids. Secondary treatment removes the remaining organics and suspended solids from primary effluent. In most cases, it removes the biodegradable dissolved and colloidal organic matter using aerobic biological treatment processes (e.g., activated sludge, trickling filters). Disinfection to inactivate pathogens follows the last treatment step and is meant to remove the risk of exposure to these biological hazards to protect public and occupational health and provide safer products from wastewater treatment processes (e.g., treated effluent, biosolids).

Disinfection can occur at a population level (e.g., water resource recovery facilities or WRRFs) or at a smaller scale (e.g., hospitals, cruise ships, army bases). Disinfection becomes especially important in the context of water reuse in the circular water economy. However, given the complexity of the potable and nonpotable water reuse topic, we will not be discussing this specifically and recommend the reader refer to the Water Environment Federation (WEF) *Water Reuse Roadmap* or relevant manuals of practice. Solids removed from the primary and secondary treatment processes are often combined and stabilized (reduction of pathogens including enteric viruses, pathogenic bacteria, and viable helminth ova) using either aerobic or anaerobic digestion, air drying, composting, or lime stabilization.

Regardless of the setting, during an infectious disease outbreak, water professionals can use the concept of disinfection hierarchy and an understanding of disinfectant modes of action to garner an understanding of the likelihood that the WRRF's disinfection process may be effective against the microorganism in question. The disinfection hierarchy describes the descending order of susceptibility of various classes of microorganisms to antimicrobial chemicals. Figure 3.1 illustrates the microbiological order from least susceptible class of microorganisms (i.e., prions and spores are the hardest to disinfect) to the most susceptible class of microorganisms (i.e., enveloped viruses are the easiest to disinfect) (CDC, 2008).

To help WRRF staff evaluate their disinfection options during infectious disease outbreaks, we provide some insight into specific commonly used disinfectants. For each disinfectant, we will cover chemistry, mode of action, disinfection efficacy (i.e., inactivation as assessed by log removals where possible) and pros and cons of usage.

## 2.0 WASTEWATER TREATMENT TO PROTECT PUBLIC HEALTH

### 2.1 Effluent—UV

Ultraviolet (UV) light as a disinfection method has become very popular in recent years because of its efficacy and reliability in inactivating key

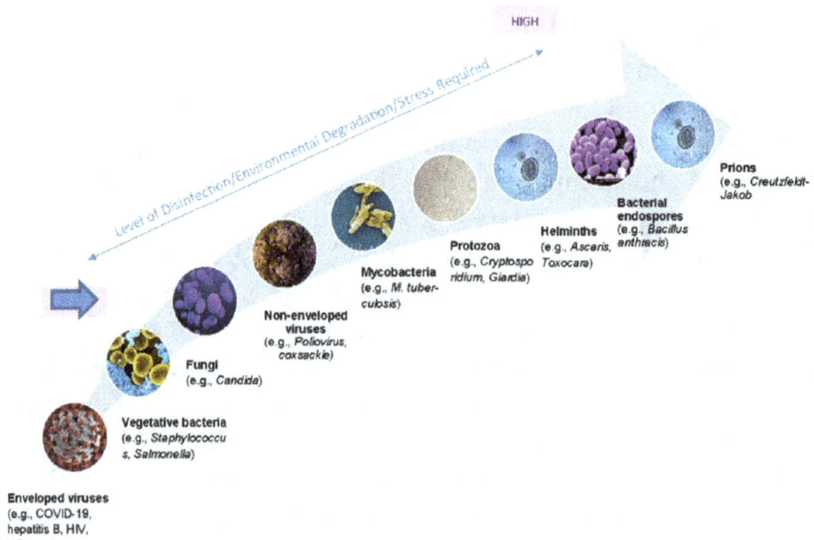


**FIGURE 3.1** Decreasing Order of Resistance of Microorganisms to Disinfection and Sterilization (Adapted from CDC, 2008)

pathogens, as well as its safety from chlorination disinfection byproduct creation. Additionally, costs of utilizing UV systems have decreased due to system design improvements. A UV disinfection system transfers electromagnetic energy from an ultraviolet lamp to an organism's genetic material, thus halting the pathogen's ability to reproduce.

Effectiveness of a UV system is dependent on flowrates and wastewater characteristics. One of the key characteristics is ultraviolet transmittance (UVT). Water with low UVT absorbs more UV light and therefore requires higher energy for a desired inactivation and results in higher capital, operating and maintenance costs.

A UV disinfection system consists of a power supply, lamps, reactor chambers, cleaning equipment, flow control, and controls and instrumentation. Reactor chambers (either open or enclosed channels, including in-pipe reactors) are set up with UV lamps in a horizontal or vertical configuration. Lamps are typically separated from the effluent by a quartz sleeve. Effluent flows through the reactor chambers and is exposed to UV light. Dose can be estimated through measurement of transmittance, UV intensity and flowrate. In an open-channel system, effluent weir control is necessary to maintain the submergence of the lamps.

A key maintenance task in a UV system is cleaning the surface of the quartz sleeves. Sleeves require periodic cleaning due to the tendency of constituents in wastewater to foul the surface of the lamp. Cleaning requirements are determined based on the reduction in measured intensity.

### *2.1.1 Chemistry*

Ultraviolet irradiation is a physical disinfection process and, as such, has several fundamental characteristics that distinguish it from chemical disinfection processes. Ultraviolet irradiation achieves disinfection by inducing photo-biochemical changes within microorganisms. Ultraviolet systems offer "chemical-free disinfection" in that chemicals are not mixed into the effluent to effect chemical reactions; however, UV lamps use chemicals including mercury as a UV source. Mercury inside the lamp is excited by applying a voltage to the gas mixture, by which the electrons in the mercury atoms are "excited" promoted to higher energy levels above the ground state. Transition of the electrons back to ground state results in the emission of electromagnetic energy (photons) at wavelengths characteristic of the mercury. These lamps are commonly referred to as Low Pressure (monochromatic, single wavelength) and Medium Pressure (polychromatic, multiple wavelengths) mercury vapor UV lamps. Low pressure UV lamps are further divided into conventional and high output types. Note that mercury lamps must be handled carefully and disposed of properly due to their hazardous nature.

### 2.1.2 *Mode of Action*

The germicidal UV range refers to the range the wavelengths of UV radiation that induce photochemical reactions within cells, which are effective in UV inactivation. Nucleic acid molecules absorb UV photons between 200 and 300 nm, with peak absorption at 265 nm, which is close to the peak emission of low-pressure lamps. Absorption does not occur above wavelengths of 300 nm and UV wavelengths below 200 nm cannot effectively transmit through water. Photochemical reactions in nucleic acid block DNA replication due to nucleic acid alterations, and ultimately result in reproductive cell death, hence loss of pathogenicity. Microorganism nucleic acids are the most important absorbers of light energy in the wavelength of 240 to 280 nm.

For medium pressure (polychromatic) lamps, most of the germicidal action of UV light is due to nucleic acid absorption; proteins, however, can also be involved in inactivation of microorganisms by UV. Microbial cells contain considerable amount of protein and nucleic acid. A microbial cell such as *Escherichia coli* contains 60% to 90% water that is transparent up to 300 nm. The next most common constituent is protein, which accounts for 50% of cell dry weight; nucleic acid makes up to 10% to 20% of dry weight, and carbohydrates and lipids are less than 10% of dry weight. Microorganism nucleic acids absorb in the range of 240 to 280 nm, 10 to 20 times higher per weight compared to protein. The absorption spectra of nucleic acids are very similar; however, the absorption spectra of proteins vary due to variation in the amino acid contents. Proteins usually show an absorption peak around 280 nm with a minimum around 240 to 250 nm, whereas DNA shows a minimum around 230 nm, and proteins show high absorption below 230 nm due to the peptide bonds.

### 2.1.3 *Inactivation/Log Removals*

The dose (also known as fluence) of UV light is measured as the product of intensity and exposure time, as milliwatt-seconds per square centimeter (mW-s/cm$^2$), which is equal to 1 millijoule per square centimeter (mJ/cm$^2$).

$$\underset{(\text{mJ/cm}^2)}{\text{Dose (energy/area)}} = \underset{(\text{mw/cm}^2)}{\text{UV Intensity (energy/time/area)}} \times \underset{(\text{sec})}{\text{Time}} \qquad (3.1)$$

Various dose requirements have been published for specific microorganisms and classes of microorganisms.

The Long Term 2 Enhanced Surface Water Treatment Rule details the required UV doses for up to 4 log inactivation of the protozoa *Cryptosporidium* and *Giardia* and for viruses as a group, as indicated in Table 3.1 (U.S. Environmental Protection Agency [U.S. EPA], 2006). Although there

**TABLE 3.1** UV Dose Requirements (in mJ/cm$^2$) for 1 to 4 Log Credit Disinfection of *Giardia, Cryptosporidium,* and Viruses per U.S. EPA Criteria (Adapted from U.S. EPA, 2006b)

| Microorganism | 1 log | 2 log | 3 log | 4 log |
|---|---|---|---|---|
| Giardia | 2.1 | 5.2 | 11 | 22 |
| Cryptosporidium | 2.5 | 5.8 | 12 | 22 |
| Viruses | 58 | 100 | 143 | 186 |

are no specific doses required by regulation for wastewater treatment, UV disinfection can typically meet fecal and total coliform (such as 100 mpn/100 mL) wastewater effluent discharge requirements at doses up to 25 mJ/cm$^2$. For water reuse applications, doses around 100 mJ/cm$^2$ are typically used. Details for specific microorganism classes are discussed below.

Table 3.2 is adapted from Masjoudi et al. (2021) and summarizes some UV sensitivity data for select bacteria, protozoa, viruses, and other microorganisms. It is important to note that some of the data provided (e.g.,

**TABLE 3.2** UV Dose Requirements (in mJ/cm$^2$) for 1 to 4 Log Credit Inactivation of Bacteria and Viruses (Adapted from Masjoudi et al., 2021)

| | Fluence (UV Dose) (mJ/cm$^2$) for a Given Log Reduction | | | |
|---|---|---|---|---|
| Organism | 1 | 2 | 3 | 4 |
| **Bacteria** | | | | |
| Bacillus anthracis (Sterne, “anthrax”) | 28 | 37 | 52 | |
| Enterococcus faecalis (ATCC 27285) | 3.4 | 9.1 | 18 | |
| E. coli (ATCC 11229) | 3.0 | 4.8 | 6.7 | 8.4 |
| Fecal coliforms | 6 | 9 | 13 | 22 |
| Legionella pneumophila (ATCC 33152) | 1.6 | 3.2 | 4.8 | 6.4 |
| Salmonella typhimurium (ATCC 6359) | 2.6 | 4.5 | 5.8 | 7 |
| Vibrio cholerae | 0.8 | 1.4 | 2.3 | 3.9 |
| **Viruses** | | | | |
| Adenoviruses (Type 1) | 35 | 69 | 103 | 138 |
| SARS-CoV-2 (“Coronavirus”) | 3 | 6.7 | 11 | 15 |
| Murine norovirus (NCIMB10108) | 10 | 15 | 22 | 27 |
| Poliovirus (Type 1 LSc2ab) | N/A | 5.6 | 11 | 17 |
| Rotavirus (SA-11) | 8 | 15 | 27 | 38 |

norovirus) are not based on cell culture but on molecular assays. Given that molecular methods detect genetic material but do not provide information on viral infectivity UV irradiation, the information provided must be interpreted with care.

### 2.1.4 Pros and Cons

#### 2.1.4.1 Pros

- UV as a disinfection method is a reliable, proven technology for wastewater applications.
- UV involves no chemical handling/storage and does not have the bulk chemical handling risks associated with other disinfection methods.
- UV does not create disinfection byproducts.
- There is no need to quench the disinfectant (e.g., dechlorinate the effluent).

#### 2.1.4.2 Cons

- The UV bulbs contain mercury, which is hazardous and, therefore, the bulbs need to be disposed of appropriately. Though this can be overcome with the use of new UV-LED technology.
- UV has a relatively high energy consumption, especially for low quality (low UVT) effluent.
- UV has ongoing lamp and ballast replacement costs.
- UV disinfection does not leave a residual, so there may be regrowth downstream.

## 2.2 Effluent—Chlorine, Chloramine, and Chlorine Dioxide

The use of chlorination for disinfection in water treatment began as early as the 1900s as a means to prevent common waterborne illnesses (e.g., cholera and dysentery). Chlorine has an irritating odor even at very low concentrations and reacts easily with other substances, usually causing an exothermic reaction. As the quality of water will differ from location to location, the chlorination methodology will need to take the water being treated into consideration.

Chlorine is mostly used in wastewater applications to disinfect the treated effluent prior to final discharge. It also used at other points in the WRRF process, including to eliminate odors associated with hydrogen sulfide formation, mitigate bulking or foaming effects of filamentous organisms in the activated sludge process, control algae growth on clarifier weirs, disinfect tertiary filter media and membranes, and reduce bacterial regrowth and control algae for onsite nonpotable reuse options.

### 2.2.1 Chemistry

When chlorine is applied to wastewater, it creates an aqueous chlorine solution that reacts with organics and inorganics. The reaction between chlorine and water forms hypochlorous acid (HOCl) and hypochlorite ions ($OCl^-$). Hypochlorous acid is 100 times more effective than hypochlorite ions and exists in greater concentration at a pH range of 2 to 6, whereas at pH greater than 7.8, hypochlorite ions are the predominant form. At pH less than 2, aqueous chlorine is the predominant form (WEF, 1996).

Free chlorine is defined as the sum of molecular chloride ($Cl_2$, nonreactive), hypochlorous acid (HOCl, most reactive), and hypochlorite ion ($OCl^-$, minimally reactive).

In wastewater with a pH of 7.5, approximately 50% of the chlorine present will be as HOCl, and approximately 50% as $OCl^-$. A higher pH level will result in a greater percent of $OCl^-$ and reduced germicidal efficacy.

Free available chlorine is an important consideration for supplying of chlorine disinfection products. For instance, the concentration of chlorine in hypochlorite can be estimated using the electrochemical equivalent of $Cl_2$ and the following equations:

$$Cl_2 + 2^{e-} <=> 2Cl^- \tag{3.2}$$

$$OCl^- + 2^{e-} + 2H^+ <=> Cl^- + H_2O \tag{3.3}$$

Equation 3.2 shows that one mole of elemental chlorine can react with two electrons to form two moles of chloride. Similarly, Equation 3.3 shows that one mole of hypochlorite reacts with two electrons to also form chloride. Therefore, one mole of hypochlorite is known to be equivalent to one mole of elemental chlorine. This means that both hypochlorite and elemental chlorine have approximately 71 g of available chlorine per mole of chemical.

The chlorine product sodium hypochlorite contains one mole of hypochlorite per mole of solution, which is approximately 71 g of available chlorine. Considering the solution molecular weight of sodium hypochlorite, the pure solution will then contain 95.2% available chlorine.

Commercially available sodium hypochlorite is generally available as a solution that contains 5% to 20% available chlorine with the most common specification being 12.5% active. This percent concentration is different from the weight percent of available chlorine described above, and it is selected to minimize degradation effects (American Water Works Association, 2006).

#### 2.2.1.1 Chlorine Demand

Molecular chlorine ($Cl_2$) is an extremely reactive compound rapidly undergoing chemical reactions with inorganic and organic substrates present in wastewater before disinfection can occur. The amount of chlorine needed to react with these organic and inorganic constituents and inactivate the target pathogen is known as the chlorine demand. Chlorine demand is then the difference between the amount of chlorine applied to the wastewater and the amount of free residual chlorine after a given contact time.

Inorganic reducing substances often present in wastewater streams that can quickly react with chlorine include $Fe^{2+}$, $Mn^{2+}$, $NO_{2-}$ and $H_2S$, and other sulfite salts. The chlorine product resulting from the reaction of chlorine and inorganic compounds is a stable nongermicidal ion. A portion of chlorine can also react with ammonia-nitrogen and other nitrogen-derived compounds present in wastewater. The combination of ammonia-nitrogen and chlorine is known as chloramines, which have a germicidal effect that varies with the type of chloramine species formed (see Table 3.3). Organic chloramines are formed when chlorine reacts with ammino acids, proteins, and other organic nitrogen forms that have a high chlorine demand and are nongermicidal.

#### 2.2.1.2 Breakpoint Chlorination

Figure 3.2 shows the classic breakpoint chlorination curve that illustrates the reactions that occur when chlorine is applied to water. Initially, oxidizable inorganics (e.g., iron, manganese, sulfide, nitrite) consume the added chlorine, which is why there seems to be no residual. When additional chlorine is added, reactions continue and combined chlorine residual (also

**TABLE 3.3** Comparative Efficiency of Disinfectants to Achieve 99% Bacterial Inactivation in Oxidant Demand-Free Systems (Adapted from the U.S. Department of the Interior, 2013)

| | *E. coli* | | | Heterotrophic bacteria | | |
|---|---|---|---|---|---|---|
| Chlorine Form | pH | Temp (°C) | Contact Time (mg/min·L) | pH | Temp (°C) | Contact Time (mg/min·L) |
| Hypochlorous Acid (HOCl) | 6 | 5 | 0.04 | 7 | 1–2 | 0.08 ± 0.02 |
| Hypochlorite ($OCl^-$) | 10 | 5 | 0.92 | 8.5 | 1–2 | 3.3 ± 1 |
| Monochloramine ($NH_2Cl$) | 9 | 15 | 64 | 7 | 1–2 | 94 ± 7 |

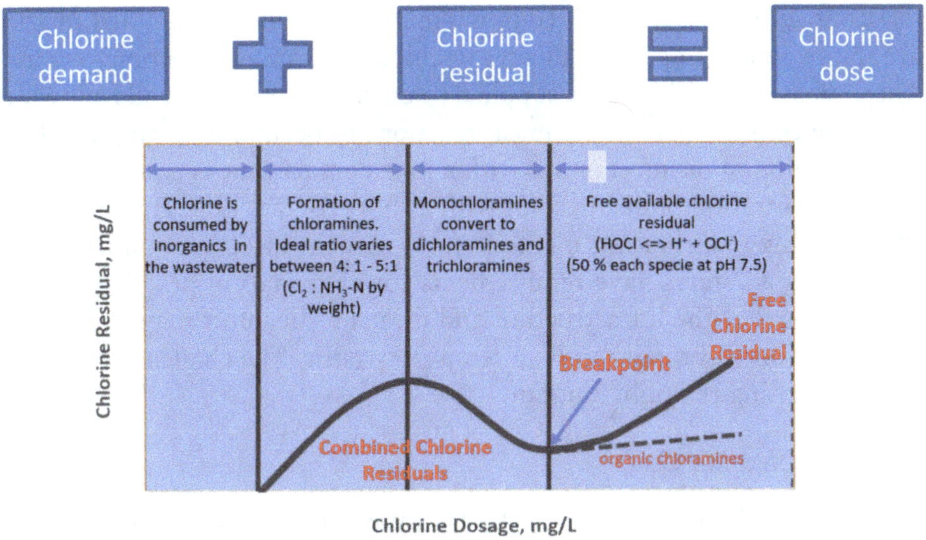


**FIGURE 3.2** Breakpoint Chlorination (Adapted from Spon, 2008)

known as chloramines) is formed. Application of more chlorine will lead to destruction of chloramines, which is followed by a breakpoint at which the amount of added chlorine is equal to the entire oxidant demand of the water (primary disinfection). After the breakpoint, any additional chlorine added will be in the form of residual free chlorine that is available to react with any organics, pathogens, or other contaminants in the water (secondary disinfection). Different water sources and wastewater types will require different doses of chlorine to reach breakpoint chlorination.

### *2.2.2 Mode of Action*

Chlorine dose and effectiveness are dependent on a number of water quality parameters, including, but not limited to

- concentration of oxidizable substances,
- particulate matter concentration (can reduce the effectiveness of the chlorine by increasing the demand),
- pH (lower pH results in more efficient use of chlorine),
- temperature (higher temperature is better),
- contact time (longer detention or contact time results in a higher degree of disinfection),
- contactor hydraulics,
- kinetics of inactivation, and
- particle sizes of the treated wastewater (Hejkal et al., 1981).

Contact time (CT, measured in mg/min·L) and chlorine residual (mg/L) are the two most important factors in chlorine-based disinfection systems. There must be sufficient dose supplied to disinfect the water while allowing adequate contact time for chemical reactions to occur. For example, a low concentration of disinfectantant with a long contact time is equivalent to a high concentration of disinfectant with a short contact time. An advantage of leveraging chlorination methods for emerging infectious outbreaks is that nearly all states have requirements for chlorination practices today. This means that the infrastructure and means to disinfect may already be on-site or can be brought in on a temporary basis. The challenge then just becomes dosing the right amount.

#### 2.2.2.1 Chlorine Gas

Chlorine gas ($Cl_2$) is considered 99.9% pure and has the unique property that, if properly contained (i.e., not open to atmosphere), it will remain at this concentration throughout its life in storage.

#### 2.2.2.2 Sodium Hypochlorite

Delivered or "bulk" sodium hypochlorite is typically delivered to treatment facilities at a concentration of 12.5% to 15% (although there are some higher concentration solutions starting to hit the market). All these solutions are, unlike chlorine gas, subject to the elements and will degrade over time, even when applying best practices in storage. Degradation rates vary and are affected by facility storage practices. This property of delivered sodium hypochlorite is important to understand when considering the CT values needed to achieve the proper log reductions for the infectious agents. On average, a 12.5% concentration solution, under standard storage solution conditions, may see about 15% degradation over a 30-day period of time.

The typical onsite sodium hypochlorite generation equation is generally expressed as follows:

$$NaCl + H_2O \rightarrow NaClO + H_2$$

There are three main types of on-site electrochlorination methods used today: open cell electrolyzer cells (typically $< 1\%$ generation), membrane electrolyzer cells (2.5% to 5% generation) and "wetted chlorine" electrolyzer cells (0.1% to 15% generation).

#### 2.2.2.3 Chloramine

Chloramine is formed by the chemical reaction of chlorine and ammonia, and the preferred form of chloramine is monochloramine, as it provides a long-lasting, stable disinfectant residual, with higher efficacy against biofilms microorganisms and minimal formation of disinfection byproducts when compared to free chlorine. However, partially nitrified effluents will require a higher chlorine dose because a portion of the chlorine will be needed to oxidize the nitrites. Typical equations for the different chloramine species are as follows:

$$NH_3 + HOCl <=> NH_2Cl + H_2O \text{ (monochloramine)}$$
$$NH_2Cl + HOCl <=> NHCl_2 + H_2O \text{ (dichloramine)}$$
$$NHCl_2 + HOCl <=> NCl_3 + H_2O \text{ (trichloramine)}$$

At the pH of most natural water (6.5 to 7.5), monochloramine and dichloramine exist together. Chlorine also reacts with organic nitrogen to form organochloramines with poor germicidal efficiency.

Literature suggests that the chlorine to ammonia ratio with the most efficient germicidal effect is 6 : 1. The 6 : 1 ratio may be adequate for secondary effluents with higher ammonia concentrations and when the reaction of combined chlorine with organic and inorganic constituents is expected to be slower. For fully nitrified or partial nitrified effluents, the 4 : 1 or 5 : 1 ratio allows better control of the disinfection efficiency.

### 2.2.3 Inactivation/Log Removals

When considering the log reduction for chlorine-based disinfection processes, CT value tables are used to determine the required CT value for a given level of inactivation, or log removal. Contact time values for 99% (2 log) and 99.99% (4 log) inactivation of bacteria and viruses for various disinfectant have been reported in technical references (Metcalf & Eddy, 2003) and are included in U.S. EPA (2020) and World Health Organization (WHO) guidelines. Additionally, recent scientific studies have evaluated CT values for enteric virus log removals using free chlorine and chloramines. Table 3.4 provides a summary of documented inactivation/log removals in the above mentioned refences.

Contact time tables based on measured temperature and pH for different microorganisms are commonly used, as these includes a wider range of water quality conditions. However, in most cases the available references do not consider kinetics of inactivation or chlorine decay rate. It should be noted that CT tables may also apply to specific treatment goals required under different regulatory frameworks. One example of this is the *EPA Disinfection Profiling and Benchmarking Technical Guidance Manual* (U.S. EPA, 2020), which includes CT tables for *Giardia* and viruses that are required under the Surface Water Treatment Rule. The CT tables in the

**TABLE 3.4** Contact Time Dose for Bacteria and Virus Inactivation by Chlorines

| Disinfectant | Reference | Unit | Bacteria Inactivation 2 log | Bacteria Inactivation 4 log | Viruses Inactivation 2 log | Viruses Inactivation 4 log |
|---|---|---|---|---|---|---|
| Chlorine | Metcalf & Eddy (2003) | mg/min·L | 0.1–0.2 | 10–12 | 2.5–3.5 | 6–7 |
| Free Chlorine | U.S. EPA (2020) | mg/min·L | 1–4 | | | 0.07–18.1 |
| Free Chlorine | White (1999) | | | | | 8–12 |
| Chloramines | White (1999) | mg/min·L | | | 214–1243 | 497–2883 |

*Note.* Temperature and pH have a significant effect on individual pathogen CT values. The U.S. EPA Surface Water Treatment Rule states that the required CT values for 4 log inactivation of viruses by free chlorine at a pH 6 to 9 are 12, 4, and 2 mg/min·L at 0.5, 15, and 25 °C, respectively.

EPA guidance manual that apply to chlorine disinfection include free chlorine, chlorine dioxide, and chloramines. A comparison of the CT values for these disinfectants illustrates that chloramines require significantly more contact time than chlorine to meet the requirement. It is also important to consider the contact chamber materials, downstream piping networks, and resistance of the specific infectious agent to the disinfectant (Buse et al., 2019). Synergistic effects of combined disinfectants have been tested using the mathematical model developed by Bernbaum and modified for disinfection kinetics by Kouame and Haas (1991), who showed that a combined synergistic interaction exists in the inactivation of *E. coli* when exposed simultaneously to free chlorine and monochloramine resulting in improved inactivation.

### 2.2.4 Pros and Cons

#### 2.2.4.1 Pros

1. Chlorination is a well-established technology.
2. Although in recent months some shortages in chlorine supply have been reported, the cost of chlorine-based disinfectants is still low compared to ozone and UV.
3. Chlorine disinfection is effective against a wide range organism.
4. Chlorine is a strong oxidant that is effective at oxidizing certain organic and inorganic compounds.

#### 2.2.4.2 Cons

1. Chlorine residual can have toxic effects to aquatic life.
2. Chlorine is highly reactive and corrosive.

3. Chlorine reacts with organic substances in the water to form disinfection byproducts.
4. The use of chlorine my lead to a modest increase of salinity in the water.
5. Some microorganisms have shown resistance to low doses of chlorine, including oocyst of Cryptosporidium parvum and cysts of *Giardia* spp.

## 2.3 Effluent—Peracetic Acid and Performic Acid

Peroxygen compounds have been used for disinfection and sterilization applications for many decades (Block, 1991). The most basic and ubiquitous of these compounds is hydrogen peroxide. However, due to its hydrophilic nature, hydrogen peroxide typically requires relatively high aqueous concentrations to provide effective biocidal behavior, is degraded by catalases and peroxidases manufactured within the microbial cell, and is susceptible to significant loss in waters (e.g., municipal wastewater) that contain carbonaceous oxygen demand. More lipophilic compounds that are not inactivated by catalase and peroxidase subsequently have been developed and successfully used for disinfection and sterilization, including the reduction of human pathogens in municipal and industrial wastewaters The most common of these are peracetic acid (PAA) and, more recently, performic acid (PFA).

Peracetic acid has been used as a wastewater disinfectant in Europe since the early 1990s and in the United States since the 2010s (Kitis, 2004; WEF, 2020). Peracetic acid is manufactured by combining concentrated acetic acid with hydrogen peroxide. The resultant reaction forms a relatively stable equilibrium consisting of peracetic acid, acetic acid, hydrogen peroxide, and water. Commercially available PAA solutions can have up to 1 year of shelf life without loss of assay under typical conditions found at most water treatment facilities. Stable PAA solutions are available from manufacturers in the range of 5 to 22 wt% PAA, with differing concentrations of acetic acid and hydrogen peroxide. Although not common, PAA can be generated on-site at the water treatment facility, delivering PAA in much lower concentration. Typical in-use effective concentrations range from 0.5 to 2 mg PAA/L wastewater (note that this concentration is based on mg active PAA, not mg of PAA equilibrium solution). There are examples of initial PAA dose concentrations higher than 2 mg/L for effectiveness (WEF, 2020), but these are often the exception.

In wastewater and in the receiving stream, PAA decomposes to water, acetic acid, oxygen, and water. The hydrogen peroxide within the PAA equilibrium solution likewise decomposes to water and oxygen. The aquatic toxicity of PAA is an order of magnitude less than that of chlorine (European Centre for Ecotoxicology and Toxicology of Chemicals, 2001) and generally does not need quenching at the facility outflow prior to release into the environment. The U.S. EPA regulatory registration of PAA for municipal wastewater is somewhat dependent on a specific manufacturers' product, but in general allows a PAA dose range up to 45 mg/L and sets a PAA residual at the outfall based on the effluent dilution factor (DF). The allowed residual is 1.0 mg PAA/L for DFs less than 12 and equal to 0.9 times the DF for DFs greater than 12. Within the United States, each state regulatory agency approves the use of PAA within their jurisdiction via their National Pollutant Discharge Elimination System (NPDES) permit requirements. As a result, each state set their own specifications of the allowable PAA residual at the outfall at or below the U.S. EPA–regulated limits. As of the time of the publication of this book, not every state has approved the use of PAA. For the states that have granted approval, the allowable PAA residual has ranged from 0.3 to 2 mg/L, with the majority of agencies setting the limit at 1 mg/L.

Measurement of PAA wastewater concentration is typically performed via grab samples utilizing the "DPD" method following Standards Method 4500-PAA (American Public Health Association [APHA] et al., 2023). In this method, the sample containing PAA is treated with an excess of potassium iodide (KI). The peracetic acid oxidizes the iodide to iodine. The iodine subsequently oxidizes the DPD (N,N-diethyl-p-phenylelnediamine) to a pink colored species. The pink color will be in direct proportion to the amount of peracetic acid in the sample and can be quantified via comparison to a color chart or measured via photometer. Peracetic acid test kits are commercially available. In addition, PAA can be measured in real time using commercially available submersible, amperometric probes, which operate on the same principles as amperometric probes used for the measurement of chlorine.

Performic acid has been used as a wastewater disinfectant in parts of Europe for the past several years, particularly in Denmark. At the time of the preparation of this book, PFA has not yet been approved in the United States for use. Performic acid is generated through the reaction of concentrated formic acid and hydrogen peroxide, generating an equilibrium solution containing PFA, hydrogen peroxide, formic acid, and water. The equilibrium solution is not stable enough for shipment and storage, and as a result, PFA needs to be generated on-site of the treatment facility. Performic acid rapidly decomposes in wastewater and in receiving streams to formic acid and water. The formic acid is readily biodegradable and itself rapidly decomposes to carbon dioxide. The residual hydrogen peroxide decomposes to water and oxygen. Due to its instability and rapid degradation, the environmental impact and aquatic toxicity of PFA will be minimal. As it is not yet approved for use in the United States, the U.S. EPA and state regulatory agencies have not yet publicly stated allowable PFA residual concentrations in the outflow from a treatment facility.

Peracids, such as PAA and PFA, are strong oxidants and are capable of oxidizing proteins in the microbial cell walls, the cytoplasm, and intracellular components (Block, 1991; WEF, 2020). As an example, PAA will react with the sulphydryl (-SH) and disfulfide (-S-S) and double bonds within enzymes and proteins. The mode of action of peracids in disinfection of microbes is not well understood. However, these oxidative reactions are believed to disrupt the cell wall, interfere with the chemiosmotic function, and disrupt biochemical and metabolic oxidation (WEF, 2020). In addition, both peracids can form reactive oxygen species, such as superoxides, hydroxyl radicals, and peracetyl radicals (Block, 1991; Karpova et al., 2013) under the water quality conditions of typical wastewaters, and their formation can be enhanced when coupled with other technologies such as the application of UV radiation.

Unlike typical chlorine dosing applications, in which chlorine is dosed to the point where a free chlorine residual can be maintained, PAA and PFA will continue to decompose within the wastewater. The decomposition profiles generally described by the Haas-Finch model (Haas & Finch, 2001), which is a first-order rate equation with an additional term for an "instantaneous" demand. The instantaneous demand and first-order rate constant will be dependent up the quality of the wastewater, and is affected by concentrations of reduced metals, phosphates, nitrates, organic components, and other oxidizable species (Henao, Cascio et al., 2018; Henao, Compagni et al., 2018). As a result, peracid decomposition rates will be specific to a given facility wastewater and may have a temporal change during the day or from season to season. Initial peracid decomposition rate may be measured in the laboratory via standard jar-testing procedures using site-specific wastewater samples or in the field using pilot or full-scale methods (WEF, 2020). The microbial reduction efficacy of peracids is determined by the area under the PAA concentration versus contact time, known as the integrated contact time (ICT) (Manoli et al., 2019).

### 2.3.1 Inactivation/Removal

Reduction in microbial concentrations generally occurs during the first 10 to 18 minutes of contact time with PAA (Koivunen & Heinonen-Tanski, 2005a, 2005b). Greater microbial inactivation can be achieved with longer contact times. Peracetic acid can function over a wide range of temperatures (Stampi et al., 2002) and is effective over a wide pH range (Baldry et al., 1991; Kitis, 2004), although PAA performance has been found to be better at lower pH. Above a pH of 9, the predominant PAA species is the dissociated acid ($pK_a$ =

8.2), which results in decreased disinfection efficiency in wastewater (Baldry & French, 1989; Sanchez-Ruiz et al., 1995). Dunkin et al. (2018) demonstrated peak PAA reduction of infectivity for human norovirus, murine norovirus, and MS2 bacteriophage occurring between a pH of 6.5 and 7.5. Peracetic acid has been shown to be very effective against bacteria in wastewaters (Kitis, 2004). Generally, requirements needed to meet regulated bacterial concentrations in the outflow require only 2 to 3 log reduction. As a result, typical PAA dose concentration × time (Ct) values needed to meet permit requirements for coliforms, *E. coli*, and enterococci are in the range of 30–120 mg/min·L.

Literature on PAA inactivation of viruses (Kitis, 2004) is more limited. F-specific coliphage, such as MS2, has been shown to be more resistant to PAA disinfection than enteric bacteria (Koivunen & Heinonen-Tanski, 2005a; Lefevre, 1992; Veschetti et al., 2003), and this may be due to different disinfection mechanisms as compared to bacteria. It has also been observed that poliovirus, echovirus, and coxsackie virus were more resistant to PAA disinfection in demineralized water than MS2 (Baldry et al., 1991). Dunkin et al. (Dunkin, Weng, Schwab et al., 2017; Dunkin, Weng, Coulter et al., 2017) recently published data on PAA disinfection efficacy on murine norovirus and human norovirus GI and GII. Their study shows that PAA can inactivate up to 4 logs of murine norovirus with a Ct of less than 95 mg/min·L, and with typical municipal PAA Cts of 30 to 60 mg/min·L, up to 1.5 log reduction was achievable for human norovirus GI (although at these Cts, minimal reduction in human norovirus GII were observed). Their studies show that the results for PAA are similar to those achieved with chloramines.

Performic acid also has shown to be effective in reducing microbial concentrations. For *E. coli,* 2 log reduction was observed for a Ct of 102 mg/min·L (Chhetri et al., 2015) in a combined sewer overflow and 3 log reduction with a Ct a low as 30 mg/min·L (Luukkonen, 2015) for secondary treated wastewater. Similarly for *Enterococcus*, a Ct of 102 mg/min·L resulted in a 3.3 log reduction in combined sewer overflow samples (Chhetri et al., 2015), and 45 mg/min·L achieved a 2.5 log reduction in secondary effluent (Luukkonen, 2015). A 2.7 log reduction in somatic coliphage was observed in combined sewer overflow streams with PFA concentrations between 12 and 24 mg/L (Tondera et al., 2016).

### 2.3.2 Pros and Cons

#### 2.3.2.1 Pros

- In comparison to chlorine disinfection, both PAA and PFA have advantages in that they are short-lived in the environment due to their rapid decomposition, do not typically require a quenching step, generally have low aquatic toxicity, and do not generate halogenated disinfection byproducts (Karpova et al., 2013; WEF, 2020).

- In many cases, the use of peracids can be cost-competitive to other disinfection technologies, especially due to the fact that their dosing systems have low capital cost requirements, especially compared to UV and ozone systems (WEF, 2020).
- Existing chlorination systems can be readily retrofitted to a peracid system.

#### *2.3.2.2 Cons*

Although PAA, and to some extent PFA, have been used as a wastewater disinfectant in Europe for decades, the use of these peracids in the United States have a much more limited history. Peracetic acid has been sold commercially in the United States since 2012, whereas PFA has not been approved by the U.S. EPA for use as of 2021. Even though the U.S. EPA has approved the use of PAA for wastewater applications as early as 2006, NPDES permits allowing the use of PAA as a wastewater disinfectant are reviewed on a state-by-state basis, and as of 2021, approximately only 18 states granted such modifications. Although the federal government has provided guidance on residual PAA concentrations at the outfall based on approved label specifications, each state has set its own requirements on allowable PAA residuals. These range from as low as 0.3 to 2.0 mg/L, with some states requiring facility by facility review. As a result, depending on the state, one of the greatest hurdles to the adoption of peracid disinfection is the regulatory pathway in gaining approval for use.

## 2.4 Effluent—Ozone

Like many common disinfecting agents used in wastewater treatment, ozone (chemical symbol $O_3$) is a powerful oxidant. Ozone is an unstable form of oxygen, consisting of three oxygen atoms. This inherent molecular instability gives ozone its strong oxidizing potential.

When ozone decomposes in water, it produces hydrogen peroxy ($HO_2\cdot$) and hydroxyl ($OH\cdot$) radicals, both of which have very strong oxidizing capabilities. These oxidants destroy the pathogens present in the water by breaking down their cell wall outer membranes (U.S. EPA, 2009). In addition to destroying living pathogens, ozone's oxidation capability is also used in wastewater and water reuse to remove iron and manganese, aid in reducing total organic carbon (TOC), and to reduce chemical oxygen demand (COD).

When ozone is combined with added hydrogen peroxide ($H_2O_2$), hydroxyl radicals ($OH\cdot$) are more quickly formed referred to as an advanced oxidation process (AOP). Conventional ozone treatments are used in wastewater treatment facility applications to handle a variety of water-related issues. The ozone AOP process is often used to treat contaminants of emerging concern (CECs), such as 1.4-Dioxane, pharmaceuticals, and pesticides.

Ozone treatment can produce unwanted disinfection byproducts including aldehydes, bromate, and N-nitroso dimethylamine (NDMA). The most common concern when using ozone is bromate formation. Bromate formation is a risk when the levels of bromide present in the source water and exceed 100 µg/L (100 ppb). The potential for bromate formation can be reduced by lowering the pH, minimizing the ozone residual and contact time, or using an AOP (Hart-Terry & Swift, 2020).

Because ozone is an unstable molecule, it must be generated onsite with a supply of pure oxygen. Larger systems use liquid oxygen, whereas smaller systems use oxygen generators, which sieve out the nitrogen present in air. In the ozone generator, the oxygen passes through a dielectric where an electric current is introduced. The current splits the $O_2$ molecules into individual oxygen atoms, approximately 10% to 14% of those atoms recombine to form $O_3$. This process produces heat; therefore, most ozone generators are water-cooled. The produced ozone is then introduced into the treated water through either a fine bubble diffusion system or a side stream injector. Excess ozone, or ozone that may degas from the system is plumbed to an ozone destruct system for removal (Hart-Terry & Swift, 2020).

### 2.4.1 Mode of Action

Ozone, either dosed conventionally or with hydrogen peroxide to produce hydroxyl radicals, yields a very high oxidation potential, as indicated in Figure 3.3. Ozone and the hydroxyl radical have the highest oxidation potential, much greater than the chlorine related chemistries often used in wastewater treatment.

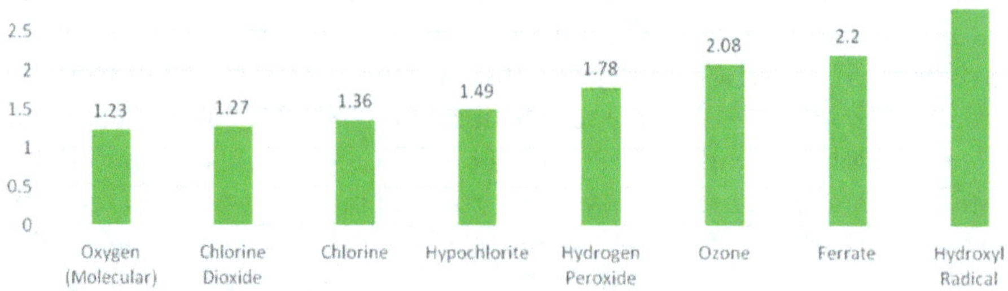


**FIGURE 3.3** Oxidation Potential (Adapted from Li et al., 2010)

### 2.4.2 Inactivation/Log Removal

Ozone and ozone AOP are broad, nonspecific oxidants and disinfectants, targeting a variety of pathogens, contaminants, and water-related issues (Metcalf & Eddy, 2003). Conventional ozone applications include taste and odor, iron and manganese removal, TOC reduction, and disinfection. The application of ozone depends on the dose and contact time (Wang et al., 2018). For example, a typical dose for iron oxidation is 0.5 ppm of ozone per ppm of iron and 5 minutes of contact time. For organic contamination, doses often exceed 5 to 10 ppm of ozone per ppm of contaminant.

Typical targets for ozone disinfection in WRRFs include *E. coli, Salmonella, Shigella, Cryptosporidium parvum*, and *Giardia lamblia*, as well as some helminths, enterovirus, and rotavirus (U.S. EPA, 2009). Doses range from 1 to 10 ppm per log reduction of the target organism depending on the organism's ozone response, with contact time ranging from just a few seconds up to several minutes.

When considering an AOP application, the ratio of ozone to hydrogen peroxide must be considered as well. Hydrogen peroxide dosages for typical ozone applications can range from 0.5 to 1 ppm of hydrogen peroxide for each ppm of ozone. In many instances, piloting of either ozone or ozone AOP is performed.

### 2.4.3 Pros and Cons

Treating wastewater and water reuse applications with ozone or ozone AOP has many advantages. Current environmental trends such as global warming, droughts, and local water shortages are leading many parts of the world to use raw water supplies that require more extensive treatments than in the past. Ozone and ozone AOP are commonly used in wastewater reuse systems, as well as drinking water facilities facing an increase in raw water organics due to algal blooms and other contaminants. Ozone is also effective at removing trace organics compounds and several contaminants of emerging concern. Ozone AOP has proven effective against 1,4-Dioxane as well as pesticides and personal care products such as pharmaceuticals.

Ozone must be generated on-site and therefore does not require the shipping of hazardous chemicals. Ozone decomposes rapidly, though it produces a short-lived residual, so quenching, common with chlorine applications, is not needed. Ozone also increases the dissolved oxygen, which can eliminate the need for post-aeration. Ozone also does not produce the regulated, carcinogenic disinfection byproducts commonly associated with traditional chlorine disinfection.

Water ozonation may produce bromate, an unwanted disinfection byproduct. Bromate is produced when sodium bromide is present in the raw water. The U.S. EPA established a bromate maximum contaminant level of 0.01 mg/L (U.S. EPA, 2010). Influent bromide as low as 0.03 mg/L may cause bromate formation issues after ozonation. However, when ozone is combined with hydrogen peroxide (ozone AOP), a lower amount of ozone is often needed for treatment and, therefore, the potential for bromate formation decreases.

Ozone can be more expensive than conventional disinfection technologies, from both a capital and operating expense. Initial investments in capital equipment are required to generate and dose ozone on-site. Ongoing operating expenses include electricity and often a liquid oxygen supply. Operating an ozone treatment system is also more complex than conventional treatment, requiring a more highly trained workforce than other technologies.

## 2.5 Biosolids

### 2.5.1 Class A and Class B

The U.S. EPA has mandated requirements to help ensure that biosolids are processed, handled, and treated in a manner that minimizes potential risk to human health. Based on the treatment methods, the U.S. EPA has divided biosolids into "class A" and "class B" designations (40 CFR 503; see Table 3.5). The treatment processes proposed for class A biosolids eliminate pathogens, including viruses, whereas treatment requirements for class B biosolids do not completely eliminate pathogens in the biosolids. As a result, the use of class B biosolids may require a permit from EPA with restrictions on land application, crop harvesting, and public access to allow time for pathogen degradation.

### 2.5.2 Overview of Treatment Types With Pros/Cons

#### 2.5.2.1 Anaerobic Digestion

Anaerobic digestion is a natural biological process that uses bacteria to break down organic matter into carbon dioxide ($CO_2$), methane ($CH_4$), and ammonia ($NH_3$) in the absence of oxygen. It takes place in digesters, which are enclosed concrete or steel tanks and can receive solids from primary and secondary clarification as well as fats, oils, and grease. The process generates biogas, which can be recovered for heat, power generation, and other beneficial uses. Overall, anaerobic digestion takes approximately 40% less time than aerobic processes until temperatures reach about 50 to 55 °C (122 to 131 °F).

The rate of pathogen inactivation typically increases as the solids content of anaerobic biosolids increases. Once inside the digester, microorganisms do not have access to a continuous food supply. When they run out of

**TABLE 3.5** Biosolids Treatment: Classes B and A

| | |
|---|---|
| **Class B biosolids treated through a process to significantly reduce pathogens (PRSP)** | |
| Aerobic digestion | 60 days at 15 °C (59 °F) to 40 days at 20 °C (68 °F) |
| Anaerobic digestion | 60 days at 20 °C (68 °F) to 15 days at 35–55°C (95–131 °F) |
| Alkaline (or lime) stabilization | pH ≥ 12 for at least 2 hours of contact |
| Mesophilic composting | Temperature of 40 °C (104 °F) or greater maintained for 5 days; for 4 hours during the 5-day period, temperature must exceed 55 °C (131 °F) |
| Air drying | Minimum of 3 months with at least 2 months of average ambient temperature above 0 °C (32 °F) |
| **Class A biosolids treated through a process to further reduce pathogens (PFRP)** | |
| Heat treatment (liquid sludge) | 180 °C (356 °F) for 30 minutes |
| High-temperature composting | Within vessel or static aerated pile: 55 °C (131 °F) or greater for 3 consecutive days<br>Windrow: 55 °C (131 °F) or greater for 15 consecutive days; minimum of 5 turnings |

organic matter, microorganisms use up their internally stored energy supplies to stay alive in a process called endogenous respiration. These starving microbes eventually die and become food for other microorganisms. Over an extended period, the overall biomass is significantly reduced. The process requires the maintenance of a steady pH, organic loading rate, mixing, and temperature. Sensitivity is also noted to elevated levels of sulfides, volatile acids, alkali/alkalines, heavy metals (e.g., copper; inhibitory at a soluble concentration above 0.5 mg/L), and ammonia. Ammonia is unique in that 50 to 200 mg/L is beneficial, 1500 to 3000 mg/L is inhibitory at pH above 7.4, and concentrations of > 3000 mg/L are toxic. See Table 3.6.

### *2.5.2.2 Aerobic Digestion*

Aerobic digestion uses bacteria to break down organic matter into carbon dioxide ($CO_2$), water ($H_2O$), and ammonia ($NH_3$) in the presence of oxygen and often used at small- to medium-sized WRRFs. It is not usually paired with primary clarifiers because this increases the cost and energy demands of supplying oxygen to the aerobic digester. It is highly reliant on temperature and detention time and is typically operated in the mesophilic temperature

**TABLE 3.6** Pros and Cons of Anaerobic Digestion (Adapted from U.S. EPA, 2003; WEF, 2023)

| Pros | Cons | Pathogen Efficacy (log reduction) |
|---|---|---|
| • Good volatile suspended solids destruction (40%–60%)<br>• Net operational cost can be low if gas (methane) is used.<br>• Broad applicability<br>• Biosolids suitable for agricultural use<br>• Reduces total sludge mass<br>• Low net energy requirements | • Requires skilled operators<br>• May experience foaming<br>• Methane formers are slow growing; hence, "acid digester" sometimes occurs.<br>• Recovers slowly from upset<br>• Supernatant strong in ammonia and phosphorus<br>• Cleaning is difficult (scum and grit).<br>• Can generate nuisance odors resulting from anaerobic nature of process<br>• High initial cost<br>• Potential for struvite (mineral deposit)<br>• Safety issues concerned with flammable gas | • Bacteria<br>○ 0.5–4.0<br>• Viruses<br>○ 0.5–2.0<br>• Parasites (protozoa and helminths)<br>○ 0.5 |

range, from 20 to 40 °C (68 °F to 104 °F). Below 20 °C (68 °F), the rate of digestion is significantly slower and operation in the thermophilic zone of more than 40 °C (104 °F) has become increasingly common. Microorganisms break down organic matter faster at higher temperatures. Operating digesters at higher, thermophilic temperatures has several advantages, including better dewatering characteristics, faster sludge stabilization, improved reduction of organic matter, and often better pathogen inactivation. See Table 3.7.

Autothermal thermophilic aerobic digestion is a modification of aerobic digestion that does not use an external heat source but instead relies on heat released by microorganisms during digestion to reach and sustain thermophilic operating temperatures. It requires higher solids concentration (4% volatile solids) and takes place in an insulated reactor with good mixing and an efficient aeration device to prevent heat loss. See Table 3.8.

### *2.5.2.3 Alkaline Stabilization*

Alkaline stabilization reduces pathogen concentrations and improves biosolids characteristics by adding alkaline chemicals to raise the pH. Pathogen inactivation results from a combination of high pH, temperature, gaseous

**TABLE 3.7** Pros and Cons of Aerobic Digestion (adapted from U.S. EPA, 2003; WEF, 2023)

| Pros | Cons | Pathogen Efficacy (log reduction) |
|---|---|---|
| • Low initial cost, particularly for small facilities<br>• Supernatant less objectionable than anaerobic<br>• Simple operational control<br>• Broad applicability<br>• If properly designed, does not generate nuisance odors<br>• Reduces total sludge mass | • High energy cost<br>• Generally lower volatile suspended solids destruction than anaerobic<br>• Reduced pH and alkalinity<br>• Potential for pathogen spread through aerosol drift<br>• Biosolids typically are difficult to dewater by mechanical means.<br>• Cold temperatures adversely affect performance.<br>• May experience foaming | • Bacteria<br>○ 0.5–4.0<br>• Viruses<br>○ 0.5–2.0<br>• Parasites (protozoa and helminths)<br>○ 0.5 |

ammonia, salts, and dryness. Traditionally, the process relies on quicklime (CaO) or hydrated lime [$Ca(OH)_2$] to raise the pH. It can be used to meet Class A or B requirements; to meet Class A stabilization, elevated pH is combined with elevated temperatures. These temperature increases are achieved by: (1) Overdosing with lime, which produces heat when it reacts with water; (2) Using a supplemental heat source; (3) Advanced technologies have emerged that vary the alkaline chemical used, dose and mix ratios, or add special equipment or steps (e.g., supplemental drying). See Table 3.9.

**TABLE 3.8** Pros and Cons of Autothermal Thermophilic Aerobic Digestion (Adapted from WEF, 2023)

| Pros | Cons |
|---|---|
| • Reduced hydraulic retention compared with conventional aerobic digestion<br>• Volume reduction<br>• Excess heat can be used for building heat.<br>• Pasteurization of the sludge pathogen reduction | • High energy costs<br>• Potential of foaming<br>• Requires skilled operators<br>• Potential for odors<br>• Requires 18%–30% dewatered solids |

**TABLE 3.9** Pros and Cons of Alkaline Stabilization (Adapted from U.S. EPA, 2003; WEF, 2023)

| Pros | Cons | Pathogen Efficacy (log reduction) |
| --- | --- | --- |
| • Low capital cost<br>• Easy operation<br>• Good as interim or emergency stabilization method | • Biosolids not always appropriate for land application<br>• Chemical intensive<br>• Overall cost very site specific related to product-management costs<br>• Volume of biosolids to be managed is increased.<br>• pH drop after treatment can lead to odors and biological growth in product. | • Bacteria<br>  ○ 0.5–4.0<br>• Viruses<br>  ○ 4.0<br>• Parasites (protozoa and helminths)<br>  ○ 0.5 |

### *2.5.2.4 Composting*

Microorganisms generate heat as they break down organic matter into carbon dioxide ($CO_2$), water ($H_2O$), and compost under controlled aerobic conditions. A proper ratio of carbon to nitrogen (25 : 1 to 35 : 1) creates optimum conditions for microbes to grow, degrade organic matter, and build up the temperatures needed for pathogen destruction. A bulking agent (e.g., woodchips or shredded yard waste) provides supplemental carbon and increases porosity so that oxygen can get into the composting mass. Specific temperature and length of time sludge is composted determines whether the end product qualifies as a class A or class B biosolids. The process occurs in open, enclosed, static (nonmixed), or dynamic (mixed) systems. In static piles, a lack of uniform sludge consistency can lead to pockets with lower temperatures and higher pathogen survival. Controlling pathogen reduction in the final product can be difficult, presenting some challenges for small WRRFs. See Table 3.10.

### *2.5.2.5 Thermal Drying*

The process of thermal drying typically follows thickening, stabilization, and dewatering to remove as much moisture as possible prior to treatment. The process uses heat to evaporate water retained in biosolids to below 8%, creating a final product that is 92% to 98% biosolids, thereby increasing pathogen destruction. Drying biosolids to about 95% solids can reduce bacterial pathogens by 0.5 to 4 log; storage at controlled temperatures can further improve pathogen destruction.

**TABLE 3.10** Pros and Cons of Composting (Adapted from U.S. EPA, 2003; WEF, 2023)

| Pros | Cons | Pathogen Efficacy (log reduction) |
|---|---|---|
| • High-quality, potentially saleable product suitable for agricultural use<br>• Can be combined with other processes<br>• Low initial cost (static pile and windrow) | • Requires bulking agent<br>• Requires either forced air (power) or turning (labor)<br>• Potential for pathogen spread through dust<br>• High operational cost: can be power, labor, or chemical intensive, or all three<br>• May require significant land area<br>• Potential odors | • Bacteria<br>○ 2.0–4.0<br>• Viruses<br>○ 2.0–4.0<br>• Parasites (protozoa and helminths)<br>○ 2.0–4.0 |

#### *2.5.2.6 Pasteurization*

Pasteurization subjects sludge to heat treatment at 70 °C (158 °F) for 30 minutes. The process is very energy intensive though highly effective at eliminating helminths eggs as well as bacterial and viral pathogens.

## 2.6 Hospital Effluent

### *2.6.1 Hazards Associated With Medical Waste*

*Medical waste* is defined as any waste or reusable material, other than a culture or stock of an infectious substance, that contains an infectious substance and is generated in the diagnosis, treatment, or immunization of human beings or animals. It is in the solid and hazardous waste classification. Medical waste falls under the Resource Conservation and Recovery Act as enacted in 1976, and these regulations have been dynamic depending on the evolution of the microbial infectious. Infective waste contains pathogens with sufficient virulence and quantity so that exposure to the waste by a susceptible host could result in an infectious disease. These include such things as sharps (needles, scalpels, etc.); pathological wastes; cultures/stocks of infectious agents; dialysis unit waste; human blood and blood products; contaminated autopsy, surgery, or patient care equipment; laboratory wastes; animal carcasses; body parts; and bedding.

Microbial resistance to disinfectants has been investigated and it is possible to list the major groups of microorganisms from most to least resistant as follows: bacterial spores > mycobacteria > parasites > hydrophilic

**TABLE 3.11** Pros and Cons of Thermal Drying (Adapted from U.S. EPA, 2006a)

| Pros | Cons |
|---|---|
| • Small footprint<br>• Can be designed to accept a variety of feed material<br>• Greatly reduces solids volume<br>• No additives or amendments needed | • Substantial capital investment<br>• Energy intensive<br>• Generates dust that presents health risks and fire/explosion hazards<br>• Complex operation<br>• Can produce odors |

viruses > lipophilic viruses > vegetative fungi and fungal spores > vegetative bacteria. Most parasites, such as *Acanthamoeba castellanii, Giardia*, and *Cryptosporidium* spp., are significantly resistant to disinfection and are usually rated between mycobacteria and viruses.

Adequate hospital rooms, labs, surgical suits, and laundry disinfection should reduce infectivity to patients, nurses, doctors, and other workers, with greatest attention to disinfection of highly resistant infectious microbes such as C. *diff*, antibiotic resistant microbes. The level of microbial inactivation required in hospitals is significant, as noted in Table 3.12. The primary infective agents are C. *diff* and *Staphylococcus aureus*. The degree of treatment necessary is at Level III.

The waste generated as a result of providing health care is typically not more dangerous than regular household waste. However, it does have certain streams within that waste that do represent a higher risk to health including infectious waste which represents 15% to 25% of total health care waste, among which are chemical or pharmaceutical waste (3%), human organ/tissue waste (1%), sharps waste (1%), and radioactive and cytotoxic waste or broken thermometers (less than 1%) (WHO, 2004). The disinfection of wastewater effluents to the sanitary sewers is controlled by municipality pretreatment discharge criteria approved by the U.S. EPA. Potential disinfection processes are elucidated in Table 3.13.

### *2.6.2 Approved Methods to Disinfect Infectious Wastes*

Health care facilities annually generate in excess of 600 million pounds of regulated medical waste (Nussbaum, 2009) disposed of within the health care setting in red bags and thus called "red bag waste." Under state laws, these wastes are decontaminated in one of six ways: incineration (see Prentice, 2012); thermal treatment, such as microwave technologies (see Farneman &

**TABLE 3.12** Establishing Levels of Microbial Inactivation (I–IV) (Adapted from State and Territorial Association on Alternate Treatment Technologies, 1994)

| Degree of Inactivation | Microbes Effected |
|---|---|
| Level I | Inactivation of vegetative bacteria, fungi, and lipophilic virus |
| Level II | Inactivation of vegetative bacteria, fungi, all viruses, and mycobacteria |
| Level III | Inactivation of vegetative bacteria, fungi, all viruses, mycobacteria at $10^6$ and *B. sterarothermophilus* spores at $> 10^4$; or *B. subtilis* spores at $> 10^4$ with chemical treatment |
| Level IV | Inactivation of vegetative bacteria, fungi, all viruses and mycobacteria and *B. sterarothermophilus* spores at $> 10^6$ |

**TABLE 3.13** Disinfection Effluent Pretreatment for Hospital Discharge Into Municipal Waste Systems

| Process | Effective Against | Ineffective Against | Byproduct Formation | Cost Cents/ 1000 gallons |
|---|---|---|---|---|
| Chlorine | Bacteria and viruses | Spores and Protozoa | Haloacetic acid, trihalomethanes, and other halogenide disinfection byproducts (over 700 chlorinated organics) | 10 |
| Ozone | Bacteria and viruses | Spores and Protozoa | Ozone produces bromated byproducts and oxygen | 15 |
| Chlorine Dioxide | Bacteria, protozoa, and viruses | Spores | Does not produce chlorinated organics | 15 |
| Peracetic Acid | Bacteria but has concerns for some | Viruses, coliphages, and protozoa | Acetic acid and water | 11 |
| Ferrate | Bacteria, protozoa, and viruses | Spores | Ferric hydroxide | 15 |
| Ultraviolet Irradiation (UV) | Bacteria, viruses, coliphages, and protozoa | Spores | No residual | 18 |

Welch, 2013); steam sterilization, such as autoclaving (see Smith & Bryant, 2013); plasma pyrolysis (see Foret, 2016); electropyrolysis (see Jensen & Houze, 2013); and chemical mechanical systems (see Mosenson et al., 2002).

These waste treatment methods have significant energy footprints, either in fuels or electricity, create hazardous air pollutants such as dioxins, and generate significant greenhouse gases. Many cannot be used on site, thus incurring significant transportation costs, and adding to urban congestion. The siting of offsite hazardous waste treatment facilities, to which red bag waste is often transported, frequently imposes environmental injustice (Brulle & Pellow, 2016). Advances in maceration technology and disinfection formulation allow for low-cost, small footprint, package red bag waste destruction that can be disposed of in local municipal solid waste facilities. This disclosure uses a maceration technology mated with a novel disinfection formulation that has the advantage of destroying bacteria, spore formers, fungi/yeasts, protozoa, helminths/helminth eggs, and viruses, spanning multiple kingdoms. The following are approved methods for disinfecting infectious wastes.

### *2.6.2.1 Incineration (Thermal Destruction)*

The common thermal destruction technology is incineration and has been developed with air pollution controls over the last 30 years. This is due to the observation of heavy metals, dioxins, furans, and hydrochloric acid releases as these issues arise. For example, the air pollution control technologies required to incinerate red bag wastes is equal to greater than the cost of incinerator. For medical waste, incineration combusts waste at high temperature, and currently, these medical waste incinerators have a two-chamber design along with air-pollution-control equipment. The controlled of combustion gets near-complete combustion by controlling appropriate time, temperature, and air turbulence. Incineration feeds the medical wastes into a primary chamber and exposed to lower temperatures 800 to 900 °C under low oxygen conditions, causing pyrolysis. The pyrolysis gases then are passed into a second chamber where they are burned at a higher temperature (around 1000 °C) resulting in the formation of carbon dioxide and water. In small-scale incinerators, mid-temperature incinerator waste is heated to approximately 80 °C in the secondary chamber with a residence time of at least 2 seconds, to control dioxin and furan formation. At small hospitals and universities, the incinerator only treats body parts and animal bodies. The environmental impacts of red bag waste incineration requiring control are listed below:

- Residual ash 3% to 4% must be disposed in a sanitary landfill to prevent soil and water contamination.

- Residual segregation policies prior to incineration practically eliminates lead and mercury emissions along with particulate emissions of arsenic, cadmium, and chromium.
- Dioxins and furans along with hydrochloric acid are controlled with the primary chamber coming from the combustion of halogenated organic materials at below 800 °C requiring combustion at above 800 °C and a residence time of greater than 2 seconds.

### *2.6.2.2 Steam Sterilization (Autoclave)*

Steam sterilization is nontoxic, inexpensive, rapidly microbicidal, sporicidal, and rapidly heats and penetrates fabrics. Like all sterilization processes, steam sterilization has some deleterious effects on some materials, including corrosion and combustion of lubricants associated with dental handpieces, reduction in ability to transmit light associated with laryngoscopes, and increased hardening time (5.6-fold) with plaster-cast (Bucx et al., 2000; Gilbert & Phillips, 1984; Joslyn, 2001; Silverstone & Hill, 1999).

The basic principle of steam sterilization, as accomplished in an autoclave, is to expose each item to direct steam contact at the required temperature and pressure for the specified time. Thus, there are four parameters of steam sterilization: steam, pressure, temperature, and time. The ideal steam for sterilization is dry saturated steam and entrained water; the dryness fraction or proportion of completely dry steam should not fall below 97% (Association for the Advancement of Medical Instrumentation, 2002, 2006). Pressure serves to obtain the high temperatures necessary to quickly kill microorganisms. Specific temperatures must be obtained to ensure the microbicidal activity. The two common steam-sterilizing temperatures are 121 °C (250 °F) and 132 °C (270 °F). These temperatures (and other high temperatures) must be maintained for a minimal time to kill microorganisms (Agalloco et al., 1998). Recognized minimum exposure periods for sterilization of wrapped health care supplies are 30 minutes at 121 °C (250 °F) in a gravity-displacement sterilizer or 4 minutes at 132 °C (270 °F) in a pre-vacuum sterilizer.

Steam sterilizers are used in health care facilities to decontaminate microbiological waste and sharps containers (Lauer et al., 1982; Palenik & Cumberlander, 1993; Rutala et al., 1982), but additional exposure time is required in the gravity-displacement sterilizer for these items. Autoclaving is an efficient wet thermal disinfection process. Typically, autoclaves are used in hospitals for the sterilization of reusable medical equipment. They allow for the treatment of only limited quantities of waste and are therefore commonly used only for highly infectious waste, such as microbial cultures or sharps. It is recommended that all general hospitals, even those with limited resources, be equipped with autoclaves.

#### 2.6.2.3 Dry Heat Sterilization/Thermal Inactivation

This method should be used only for materials that might be damaged by moist heat or that are impenetrable to moist heat (e.g., powders, petroleum products, sharp instruments). The advantages for dry heat include the following: it is nontoxic and does not harm the environment; a dry heat cabinet is easy to install and has relatively low operating costs; it penetrates materials; and it is noncorrosive for metal and sharp instruments. The disadvantages for dry heat are the slow rate of heat penetration and microbial killing makes this a time-consuming method. In addition, the high temperatures are not suitable for most materials (Lagergren, 1998). The most common time-temperature relationships for sterilization with hot air sterilizers are 170 °C (340 °F) for 60 minutes, 160 °C (320 °F) for 120 minutes, and 150 °C (300 °F) for 150 minutes. *B. atrophaeus* spores should be used to monitor the sterilization process for dry heat because they are more resistant to dry heat than are *G. stearothermophilus* spores. The primary lethal process is considered to be oxidation of cell constituents. Neither wet nor dry autoclave can be considered as an alternative technology to incineration in clinical solid waste management.

#### 2.6.2.4 Chemical Disinfection

Chemical disinfection, used routinely in health care to kill microorganisms on medical equipment and on floors and walls, is now being extended to the treatment of health care waste. Chemicals are added to waste to kill or inactivate the pathogens it contains, a treatment that results in disinfection rather than sterilization. Chemical disinfection is most suitable for treating liquid waste such as blood, urine, stools, or hospital sewage. However, solid—and even highly hazardous—health care wastes, including microbiological cultures, sharps, and so on, may also be disinfected chemically, with the following limitations:

- Shredding and/or milling of waste is usually necessary before disinfection: the shredder is often the weak point in the treatment chain, being subject to frequent mechanical failures or breakdowns.
- Powerful disinfectants have been required, which are themselves also hazardous and should be used only by well-trained and adequately protected personnel.
- Disinfection efficiency depends on operational conditions.
- Only the surface of intact solid waste will be disinfected.

### *2.6.2.5 Irradiation Technologies (Gamma, Microwave, and E-beam)*

Irradiation disinfects waste by exposing it to ionization radiation (gamma rays and electron beams) that are fatal to bacteria. A radioactive isotope of cobalt is used. Some irradiation treatment systems use electron beams. Both gamma rays and electron beams can penetrate plastic bags used for waste collection, so the waste does not need to be removed from the bag before treatment. Irradiation does not change the appearance of the waste, so process designers often install mechanical grinding or shredding upstream. This also makes the waste pieces smaller, which tends to enhance the efficacy of the treatment (Reimers et al., 2000).

Unlike some other treatment methods, irradiation requires a dedicated facility appropriate to radiation technologies. There are no mobile treatment modules that use radiation. It is expensive to build an irradiation facility and operating precautions must be taken to protect workers from radiation. These are among the reasons this method is not widely used, especially when heat treatment methods are typically just as effective. The efficiency of irradiation as a sterilization process depends to large extent on the total energy delivered, but even waste surfaces facing the radiation source get more sterile than the waste on the shaded side. Odd-shaped waste pieces may not get adequate exposure to the radiation, if contaminated surfaces face away from the cobalt source. Heat treatment, by contrast, brings every piece of waste to an adequate temperature for sterilization if done correctly, as does shredding with turbulent disinfectant spray technology. The ionizing irradiation processes are used to sterilize body implants and food products. For example, the E-beam process is used to disinfect food products to our space station (Pillai & Reimers, 2010).

Microwave radiation (non-ionization irradiation) is used to treat wastewater sludge and as a heat source to treat medical waste. Microwave treatment units can be either an on-site installation or a mobile treatment vehicle. The processing usually includes front-end shredding of the waste, both to increase the efficacy of the microwave treatment and to reduce the volume of the end waste for disposal. If the waste is dry, water is added, and the wet waste is introduced to the microwave chamber (Reimers et al., 2000).

These units are industrial-sized microwave ovens. Typical operation is at 2450 Hz. This frequency is the rotational frequency for water. Whereas an autoclave provides heat from outside the waste, like a conventional kitchen oven, the microwave unit transmits energy as microwaves and that energy turns into heat inside the wet waste (Reimers et al., 2000).

Microwave disinfection works only when there is water in the waste because the radiation directly works on the water, not the solid components of the waste. For this reason, treatment units are often supplied with a humidifier. Processing time is determined by the manufacturer and experience of the operators, but typically requires about 20 minutes per batch. Mechanical treatment is often positioned upstream of the microwave to make sure the waste pieces are small. Smaller pieces enhance the heating action as microwaves are able to penetrate to where infectious microbes are. With enough power, the water is converted to steam and makes all the waste around 100 °C. The entire process takes place within a single vessel. Bacteriological and virological tests are periodically conducted to ensure the process is effective. A common bacteriological test examines eradication of *Bacillus subtilis*. A 99.99% reduction of spores is considered an acceptable level of inactivation.

Plasma treatment has also been developed and proposed as an alternative to incineration. However, despite some interest in using it for medical waste, it has not found widespread use. It is used for treatment for sterilization of medical treatment items put into the human body.

#### 2.6.2.6 *Emerging Technologies: Shredding Medical Waste Followed by Disinfectant*

Shredding systems have been developed to handle red-bag wastes at the hospital or in transit. This technology reduces the liability of red-bag waste transport to offsite treatment.

Both mechanical grinding and chemical processes are usually batch processes—which is similar to a washing machine at a laundromat. They can be run in fixed batch sizes and consistent processing times. Shredding is required to maximize disinfection by breaking up the medical waste into small particles, which enables the disinfectant to contact the infectious microbes. Obviously, the shredding and mixing with the disinfection increases the risk of infectious aerosol emissions and potential toxic chemical byproducts. The critical element in this process is the disinfectant. In the past, chlorine, or bleach at 5000 ppm has been used. Chemical disinfection, primarily through the use of chlorine compounds, inactivates microorganisms in medical waste and can oxidize hazardous chemical constituents. Chlorine bleach has been used for disinfecting processes for years to disinfect *E. coli*; it will not, however, inactivate parasites, protozoan oocysts, and bacterial spores such as *C. diff* and *Staphylococcus aureus*. In place of these corrosive chlorine-based disinfectants, these shredder systems have the potential to increase disinfection efficacy and extend mechanical lifespans by using stabilized chlorine dioxide, quats, and ammonia blend. This disinfecting blend can also reduce the odors associated with chlorination disinfectants and oxidize the organics more efficiently because stabilized chlorine dioxide has a higher oxidation potential than the oxidizing potential of chlorine. The U.S. EPA identifies chemical processing as a more appropriate to treat liquid medical waste than incineration (U.S. EPA, 2022).

#### *2.6.3 Summary of Pros and Cons of Treatment and Disposal Options*

See Table 3.14 for the advantages and disadvantages of treatment and disposal methods.

## 3.0 MONITORING WATER FOR PATHOGENS

### 3.1 Molecular Versus Culture-Based Methods

Monitoring pathogens in disinfected or treated wastewater effluent is often required to ascertain treatment and/or disinfection efficacy. Because of the numerous types of pathogens and their differences in structure, physiology, genomes, and growth requirements, there is currently no unified method for the monitoring of all relevant pathogens of interest, and we often rely on the detection of indicator bacteria. Various detection methods have been applied to the identification and/or quantification of pathogens by two major method types: culture-based methods and molecular-based methods.

#### *3.1.1 Culture-Based Methods*

Despite the relatively high cost, delay in time, and intensive labor, traditional culture-based methods are still considered the gold standard for the monitoring of microbial water quality by many regulators. Culture-based methods also provide direct conformation of organism viability and infectivity. The different culture-based methods used for bacteria include standard bacterial cell culture, multiple-tube fermentation (MTF), and membrane filtration. Culturing viruses and protozoa is more complicated and requires the use of the appropriate host cell lines or actual living hosts.

The MTF technique is widely applied for the enumeration of fecal coliform bacteria. This is a three-stage procedure, including the presumptive stage, confirmed stage, and completed test. The results of MTF are statistically expressed in terms of the most probable number (MPN). The MPN technique is still commonly used for the microbial risk assessment of wastewater (Cheng & Fallowfield, 2022; Pecson et al., 2022) and coastal water (Saingam et al., 2021). Colilert™ testing (IDEXX™) is one of the most common variations of the MPN test and is widely used in the water sector. It enumerates total coliforms and *E. coli* simultaneously. Its popularity with

**TABLE 3.14** Summary of the Advantages and Disadvantages of Treatment of Infectious Microbes

| Treatment/ Disposal Method | Pros | Cons |
|---|---|---|
| Single-chamber incineration | • Good disinfection efficiency<br>• Drastic reduction of weight and volume of waste<br>• The residues may be disposed of in landfills.<br>• No need for highly trained operators<br>• Relatively low investment and operating costs | • Significant emissions of atmospheric pollutants (especially dioxins)<br>• Need for periodic removal of slag and soot<br>• Inefficiency in destroying thermally resistant chemicals and drugs such as cytotoxic |
| Pyrolytic incineration | • Adequate for all infectious waste and most pharmaceutical and chemical waste | • Incomplete destruction of cytotoxic<br>• Relatively high investment and operating costs |
| Wet thermal treatment (autoclave) | • Environmentally sound<br>• Drastic reduction in waste volume<br>• Relatively low investment and operating costs | • Operation requires qualified technicians.<br>• Inadequate for anatomical, pharmaceutical, and chemical waste and waste that is not readily steam-permeable |
| Chemical disinfection | • Highly efficient disinfection under good operating conditions<br>• Some chemical disinfectants are relatively inexpensive.<br>• Drastic reduction in waste volume<br>• The advent of multi-stertors disinfecting blends has advanced this treatment mode. | • Requires highly qualified technicians for operation of the process<br>• Uses hazardous substances that require comprehensive safety measures (chlorine and other oxidizing disinfectants)<br>• Inadequate for pharmaceutical, chemical, and some types of infectious waste |
| Irradiation technologies (ionization or non-ionization) | • The waste does not have to be removed from the bag.<br>• Simple process using conveyor belts passing the bags through the plasma beams<br>• E-beam better because the plasma beam can be controlled<br>• The ionization radiation is good for sterilizing implants such as knee and hips.<br>• Microwave is simple and can be easily placed in labs and facilities. | • Because the treatment does not alter the form of the waste, it will require a grinder.<br>• There is a requirement of safety handling and processing for workers operating the equipment (especially for $Co^{60}$ and $Cesium^{137}$).<br>• Microwave requires water because the rotation frequency of the water molecule produces heat energy.<br>• Irradiation process requires at least three times the amount of energy. |
| Safe burying | • Low costs<br>• Relatively safe; access to site is restricted and where natural infiltration is limited. | • Safe only if access to site is limited and certain precautions are taken<br>• Generally, the disposal of red-bag waste is allowed only if the residuals have been sterilized. |

operators and WRRF staff lies in the test simplicity (one step, no confirmation step needed), lack of biohazards exposure in trays where growth occurs and inclusion of all needed, sterilized consumables.

Membrane filtration is a pathogen-detection method that consists of the filtration of microorganisms through a membrane with small pore sizes (such as 0.45 uM) and cultivation of membrane with bacteria on selective agar. The agar is typically elaborately formulated to allow the growth of the targeted bacteria while inhibiting the growth of other microorganisms. For example, xylose lysine deoxycholate (XLD) agar is used in the isolation of salmonella and shigella (Taylor & Schelhart, 1971), and charcoal-based selective medium is used for campylobacter (Karmali et al., 1986). Tryptose-sulfite-cycloserine agar is used for the enumeration of *Clostridium perfringens*. Bacterial colonies on the membranes are counted after incubation of certain time depending on the type of targeted bacteria.

For a detailed description of microbiological methods used in the analysis of water, wastewater, sludge, biosolids, and sediment, refer to *Standard Methods for the Examination of Water and Wastewater* (APHA et al., 2023).

### 3.1.2 Molecular-Based Methods

Molecular-based methods, which are cultivation independent, have been developed in recent decades to achieve more rapid, specific, and sensitive detection of pathogens compared to traditional cultivation-based methods. Rather than the detection and enumeration of bacterial cells, molecular-based methods detect specific, unique sequences of nucleic acids (i.e., the genetic material) of target pathogens.

Conventional polymerase chain reaction (PCR) is one of the earliest molecular-based techniques used for detection of waterborne pathogens. Polymerase chain reaction involves several cycles of denaturation of sample DNA, annealing of specific primers to the target sequence, and the extension of the target sequence as facilitated by a thermostable polymerase. In PCR, the target DNA is replicated in an exponential manner, enabling the amplification of target DNA from a few copies to millions of copies in couple of hours. The PCR amplicons can be subsequently displayed by gel electrophoresis to examine the presence of pathogens.

Real-time PCR, also called quantitative PCR (qPCR), is a type of PCR using florescent reporter to enable the real-time monitoring of DNA product during the reaction. Basically, there are two common fluorescent reporter types that can be used for the detection of PCR products in qPCR: the nonspecific fluorescent dyes (such as SYBR Green) and sequence specific hydrolysis probes (such as TaqMan probe). Different from conventional PCR, qPCR provides copy numbers of target genes within samples through the construction of standard curves (i.e., concentration of the target). Quantitative PCR methods have been applied widely for quantitative monitoring of various pathogens in water environments, such as *Escherichia coli*, *enterococci*, and *Bacteroides thetaiotaomicron* (Aslan et al., 2018), *Salmonella*, *Cryptosporidium*, and *Giardia* (Stokdyk et al., 2020). The EPA has also established qPCR methods or assays for monitoring fecal indicator bacteria (FIB) including *Enterococci* (U.S. EPA, 2012) and *E. coli* (Haugland et al., 2021).

Next-generation sequencing (NGS) is a type of sequencing technology that was developed after Sanger sequencing. It is an emerging and promising tool for monitoring pathogens in aquatic environments and developing a better understanding of microbial populations. Next-generation sequencing still requires the extraction of DNA from an environmental sample and running a PCR. However, it allows the analysis and identification of nucleotide sequences from the PCR amplicon (e.g., 16S rRNA gene for bacterial community profiling) at high throughput with low cost and time requirements, thus providing fast and specific detection of pathogens. Unlike PCR, where genus, family, or species-specific primers have to be chosen, NGS is not target specific. You do not have to know what pathogen you are looking for.

Polymerase chain reaction amplicon sequencing, which is the strategy of sequencing to obtain the DNA sequences from PCR amplicons of DNA samples, has been commonly used for the detection of pathogens. Polymerase chain reaction amplicon sequencing can be based on PCR amplicons of 16S rRNA, 18S rRNA, virulence genes, or other genes with interests. The 16S amplicon sequencing was used for the identification of possible bacterial pathogens in different water environments including wells, ponds, rivers, springs (Shrestha et al., 2017), and wastewater (Kumaraswamy et al., 2014). 18S rRNA-based NGS has been applied for detection of waterborne protozoan *Cryptosporidium* spp., *Giardia* spp., and *T. gondii* (DeMone et al., 2020). Next-generation sequencing targeting at virulence factors has been applied for pathogen detection to improve the sensitivity of PCR or qPCR-based detection of *Salmonella* (Saingam et al., 2018).

Metagenomics is the study of genetic material recovered directly from environmental samples, based on DNA sequencing and subsequent bioinformatic annotation of sequences. Metagenomic sequencing, which uses the Shotgun sequencing technology, is a method that involves randomly breaking up the metagenomes from samples into small DNA fragments and sequencing those fragments individually. Unlike the PCR amplicon-based sequencing that targets specific DNA pieces in the sample, metagenomic sequencing is a nontargeting sequencing that can theoretically obtain the entire DNA sequence information from the sample depending on the sequence depth used in the experiment. In recent years, metagenomic sequencing has been used for screening of pathogens and fecal indicators. By metagenomic sequencing, 34 pathogenic (and indicator) bacterial species, including *Pseudomonas* spp. and *Shigella* spp., were identified in the freshwater beach sands (Mohiuddin et al., 2017). Bacterial pathogens including *Mycobacterium* spp., and viral pathogens including hepatitis A and E, were detected in high quantity in WRRFs using metagenomic sequencing (Giwa et al., 2020). Metagenomic sequencing can also target on specific microbial population coupled with selective culturing (Li et al., 2022) and provide quantitative detection using internal DNA standards (Li et al., 2021).

There are also many other molecular-based methods, such as fluorescence in situ hybridization (FISH) and DNA microarray, that can be applied for pathogen detection. However, these are becoming less commonly used due to concerns of lower sensitivity or specificity. Fluorescence in situ hybridization is a cytogenetic technique based on hybridization between specific DNA sequences and oligonucleotide probes labeled fluorescent dyes. Fluorescence in situ hybridization has been applied for the detection of various pathogens, including *Legionella* spp. in cooling water (Declerck et al., 2003), *Salmonella* spp. in wastewater (Santiago et al., 2018), *E. coli* in seawater (Tang et al., 2005), *Arcobacter* and *Campylobacter* strains in water and sewage (Moreno et al., 2003), and multiple pathogens in drinking water (Wilhartitz et al., 2007). It has also been used for the identification of filamentous bacteria in activated sludge.

A DNA microarray is collection of microscopic DNA spots attached to a solid surface or slide that enables the simultaneous detection of hundreds of genes in a single assay via nucleic acid hybridization. DNA microarray technology was initially designed to investigate the transcriptional levels of RNA transcripts derived from thousands of genes within a genome, however, it has also been adopted for environmental monitoring of pathogens. A DNA microarray was developed and used for the identification of 11 pathogens in drinking water (Zhou et al., 2011). DNA microarray technology also applied for high-throughput screening of 941 pathogenic bacterial species/groups in groundwater (Inoue et al., 2015).

### 3.1.3 Limitations of Each Approach (Pros and Cons) in Relation With Public Health Decision-Making

The most important requirements required for reliable monitoring of pathogens includes accuracy, specificity, sensitivity, and cost-effectiveness. Environmental samples, especially complex samples such as wastewater, are likely to have high microbial diversity (i.e., potential competition), low concentrations of the target microorganism, and inhibitors that hinder some detection methods. These characteristics of environmental samples pose challenges to current detection methods. The various approaches, both culture-based and molecular-based for monitoring pathogens, all have pros and cons.

#### 3.1.3.1 Culture-Based Approaches

Culture-based methods are the basic tools used for pathogen detection around the world in clinical and environmental settings, because their reliability in efficiency, sensitivity, and range of applications has been proven. Conventional culturing methods are still regarded as the gold standard for confirmation of the obtained results. Because obtaining a positive culture result requires the microorganisms to grow, culture-based methods confirm the viability and infectivity of the organism grown. However, culture techniques are not time efficient or cost-effective. They require considerable amounts of laboratory resources, labor, and time for pathogen detections. In some cases, growing specific pathogens requires higher levels of biosafety containment levels and safety measures, in addition to specialized lab skills. Finally, one major limitation of the culture-based detection methods for pathogens is that they are less effective at detecting viable but nonculturable (VBNC) cells (Sohier et al., 2014).

#### 3.1.3.2 Molecular Approaches

Compared with conventional culture-based methods, many of the molecular-based approaches can provide more rapid, specific, cost-effective, and efficient detection of pathogens. Results by some molecular-based technology, such as qPCR, can be achieved very quickly within a couple of hours and many of these methods are being automated. Many pathogens and proposed new indicators are unculturable or lack the efficient culture methods (e.g., noroviruses), whereas molecular techniques can overcome those limitations and can be used for the detection of any pathogens with well-designed assays.

Molecular-based methods—such as mPCR, high-throughput quantitative PCR (HT-qPCR), microarray, and sequencing—enable the simultaneous detection of numerous pathogens in the same experiment. This high-throughput capacity is especially important when screening for potential pathogens is required in samples without prior knowledge of pathogens present.

Molecular-based methods can provide highly specific detections that can be used for the identification of specific strains, genotyping, and confirmation of culture-based technologies. Specifically designed PCR or qPCR primers or sequencing allows the identification of single-nucleotide polymorphisms in genomic DNA or pathogen variants. These specific identifications of pathogens support the phylogenetic characteristics and even source tracking of the pathogens to identify pollution sources in watersheds.

Despite these advantages, molecular methods still face challenges to achieve more accurate, sensitive, and robust results. PCR-based methods—including conventional PCR, qPCR, and HT-qPCR—may all suffer from the occurrence of PCR inhibition and nonspecific amplification in the absence of the target pathogen. Environmental samples usually contain many PCR inhibitors that interact with the template DNA and the polymerase and prevent the enzymatic reaction even at low concentrations (Ijzerman et al., 1997). Sewage sludge usually contains fats, proteins, polyphenols, and heavy metals, and wastewater contains polysaccharides and metal ions (e.g., iron, aluminum), which affect specificity of primers.

Another notorious limitation of PCR methods is the lack of ability to differentiate whether the target DNA is coming from viable or dead cells. Some researchers have used staining with membrane impermeable fluorescent dyes (e.g., propidium iodide, propidium monoazide [PMA], and ethidium monoazide [EMA]) to distinguish cells with damaged or intact membranes. Still, these methods can only determine membrane integrity but not activity or viability of the cells. Using PMA or EMA treatment with PCR is increasingly adapted by researchers for viability detection of pathogens and might become a common methodology given its advantages over traditional PCR (Zeng et al., 2016).

In addition, the majority of regulators are still uncomfortable with the use of molecular methods to determine water safety or treatment efficacy. An increasing number of standard methods (e.g., the U.S. EPA enterococci assay) are being adopted by different states, but adoption remains slow.

One of the major advantages of sequencing methods is the large amount of sequence information obtained. However, this advantage also increases the difficulties in interpreting data that requires high proficiency in bioinformatic analysis and lots of computing knowledge due to the large datasets. For quantitative microbial risk assessment (QMRA), the ability to accurately quantify biological risks/agents is required within the framework of QMRA (see Chapter 4). However, the next generation sequencing (NGS) methods can provide only relative abundances of target sequences (estimates) as opposed to absolute quantification. More research is still required before the application of NGS for quantification of pathogens is reliable.

## 3.2 Types of Indicators

There are three major classes of pathogens that are responsible for the majority of water-associated illnesses in humans: bacterial pathogens, viral pathogens, and protozoan pathogens. Parasitic protozoa have been identified as one of the leading etiological agents of diarrheal waterborne outbreaks (Lim & Nissapatorn, 2017). Viral agents have been implicated in waterborne disease outbreaks in recreational water (Gibson, 2014; McMinn et al., 2017; Sinclair et al., 2009), reclaimed water (Deng et al., 2019), and groundwater (Gibson, 2014) settings, and the public health impact of the transmission of viruses specifically in the United States and abroad has been documented extensively (Dias et al., 2017; Gibson, 2014; O'Brien & Xagararaki, 2019).

Unfortunately, directly monitoring waterborne pathogens is costly and time-consuming, requires a highly specialized set of expertise, and is not always feasible for certain pathogens. As a result, for the last 120 years, we have been relying on the use of fecal indicator organisms, with the most commonly used indicator organism being FIB. We assume that ideal indicators: (a) are suitable for all water types; (b) are present in water, wastewater, reclaimed water, and biosolids when pathogens are present; (c) are present in greater numbers than the pathogens; (d) have similar survival, disinfection, and removal characteristics as the pathogen of interest; (e) are unable to multiply in water; (f) are nonpathogenic; and (g) are detectable in low number quickly, reliably, and at low cost. Over time, we have come to realize that no such ideal indicator organism exists. We will discuss the indicators currently used for each of the three main classes of pathogens below along with the pros and cons of each.

### 3.2.1 Bacterial Indicators and Their Standard Methods

Bacteria are found throughout our environment with many species performing functions that are necessary to human life. The ones we will cover in this section are the pathogenic species that cause disease. Of note in water are those bacteria found in the feces of humans and other animals that can find their way into waterbodies. This includes the coliform group, streptococcus, campylobacter, and others. Those pathogenic water borne bacteria of primary concern include *Escherichia coli O157:H7, Campylobacter jejuni*, and *Salmonella*. These can all cause illness in humans.

To better monitor the presence of potential pathogenic bacteria, FIB have traditionally been used. These indicators are less expensive and easier to detect than the pathogenic bacteria noted above. The most common FIB include total coliform group, fecal coliform bacteria, *Escherichia coli*, enterococci, and *Clostridium perfringens*.

Total coliforms were at one time the standard indicator; however, due to their natural presence in soil and plant material, they are not fully reliable at indicating fecal contamination specifically. Total coliforms are defined as all aerobic and facultative anaerobic, Gram-negative, non-spore-forming, rod-shaped bacteria that ferment lactose with gas formation within 48 hours at 35 °C based on Standard Methods 9221 B (https://www.standardmethods.org/doi/10.2105/SMWW.2882.192, from APHA et al., 2023). This makes them unsuitable particularly for recreational and shellfishing waters. With drinking water, they do continue to serve as an indicator, as the presence of plant or soil contamination would also be of concern. Total coliforms including fecal coliform and *E. coli* are contained in the primary drinking water regulations of the Safe Drinking Water Act, which are legally enforceable primary standards and treatment techniques that apply to public water systems (U.S. EPA, 1986).

Fecal coliforms are a subset of total coliforms and are present in large quantities. Fecal coliforms are relatively good indicators of microbial contamination, but they can also regrow in the aquatic environment outside the host in tropical waters, sediment, and biofilms (Byappanahalli et al., 2012; Hodgson et al., 2017; Payment et al., 2003). They are susceptible to environmental conditions and stress, which convert them into a VBNC state that cannot be detected using culture-based techniques. Fecal coliforms cannot pinpoint the source of pollution (McMinn et al., 2017), as they occur in the intestinal tract of warm-blooded animals and humans alike.

*Escherichia coli* is a further subgroup of fecal coliform and is the preferred bacterial indicator today particularly for recreational water due to its ability to predict swimming-associated gastroenteritis. In 1986, *E. coli* replaced total and fecal coliforms and fecal streptococci as the recommended indicator bacteria. *E. coli* is typically the preferred indicator for freshwater monitoring, and enterococci are used for marine waters because of their salt tolerance (U.S. EPA, 2004).

*Clostridium perfringens* is a spore-forming bacterium that has been associated with an estimate of nearly 1 million cases of foodborne illness per year in the United States (CDC, 2022d). The *C. perfringens* spores are also unique to sewage contamination and environmentally stable, both of which make it a candidate for an indicator organism. Its persistence, particularly in tropical waters along with the notable absence of the spores in waters with no human sewage contamination suggest that *C. perfringens* may be superior to *E. coli* and fecal coliform (Fujioka & Shizumura, 1985; Lisle et al., 2004; Stelma, 2018).

### 3.2.2 Protozoan Indicators and Their Standard Methods

The two primary protozoa of concern in water are Cryptosporidium and Giardia. Of the 251 identified waterborne protozoan outbreaks from 2017 to 2020 worldwide, 76.5% (192) were attributed to Cryptosporidium and 19.1% (48) to Giardia (Ma et al., 2022).

### 3.2.3 Viral Indicators and Their Standard Methods

#### 3.2.3.1 Bacteriophages

For monitoring purposes, model viral indicators would be organisms that behave like waterborne viruses with respect to fate, transport, decay, removal, inactivation, and disinfection but are readily detectable by simple, rapid, and inexpensive methods. The literature suggests that bacteriophages in general and coliphages in specific appear to be well suited for this purpose (Havelaar et al., 1993; Jofre et al., 2016, 2021; McMinn et al., 2017). Bacteriophages or phages are viruses that infect bacteria. They are estimated to be the most abundant organisms on Earth with more than an estimated $10^{31}$ total individual phage on the planet (Ross et al., 2016). They consist of at least a nucleic acid molecule, the genome, a capsid, and in some cases additional structures (e.g., tails, spikes, lipids). The size, composition, morphology, properties, and structure of bacteriophages are comparable to that of human and animal viruses, thereby making them relatively good indicators for the survival, persistence, removal, and inactivation of waterborne viruses of human concern (Jofre et al., 2016; McMinn et al., 2017).

Bacteriophages only can replicate inside susceptible and metabolizing host bacteria that contain specific receptors providing some level of host specificity. This host specificity is wide-ranging and depends on the bacteriophage itself. It is generally believed that bacteriophages can only infect a narrow range of closely related bacteria. For example, bacteriophage Mu can infect species of *Escherichia coli, Citrobacter freundii, Shigella sonnei, Enterobacter,* and *Erwinia*; this is due to a combination of factors including specificity of phages' host-binding proteins, biochemical interactions during infection, presence of related prophages or particular plasmids (especially for phages adsorbing to pili), and bacterial phage-resistance mechanisms (Ross et al., 2016). Phage receptors have been described in different parts of bacteria, such as the capsule, cell wall, flagella, and pili. Phages that attach to the receptors located on the cell wall are the most common and are known as somatic phages (Jofre et al., 2016).

F-specific bacteriophages, also known as sexual coliphages or male-specific bacteriophages, infect bacteria through the sex pili, which are coded by the F plasmid that was first detected in E. coli K-12 (Jofre et al., 2016). F-specific RNA bacteriophages (a subgroup of F-specific bacteriophages) consist of a simple capsid with cubic symmetry and a diameter of between 21 nm and 30 nm. They contain a single-stranded RNA molecule as the genome.

Other types of phages have also been investigated for monitoring viral water quality. Bacteriophages infecting *Bacteroides* spp. have been extensively studied to investigate their potential for microbial source tracking (McMinn et al., 2017). *Bacteroides* spp. are Gram-negative, obligately anaerobic, pleomorphic rod-shaped bacteria that are commonly found in high concentrations in both human and other animal feces (Harwood et al., 2014). Some *Bacteroides* spp. host strains (HSP40, RYC2056, and RYC4023) are useful in discriminating phages from different animal fecal sources (Gómez-Doñate et al., 2011), but major challenges prevail with this indicator, such as cross-reactivity, geographical limitations, and low/varying recoveries from wastewater (Harwood et al., 2013, 2014; McMinn et al., 2017).

Bacteriophages are of particular interest in monitoring contamination of bivalves (Hodgson et al., 2017). Enterophages are bacteriophages that infect enterococci specifically. Intestinal enterophages have also been used as viral indicators and for source tracking, especially in tropical and subtropical environments (McMinn et al., 2017). *Enterococcus* spp. are facultatively anaerobic, Gram-positive cocci that are abundant in both human and animal feces. They are commonly used for the assessment of microbial water quality in recreational waters and marine environments. Some species (e.g., *E. faecalis* and *E. faecium)* have been identified as being frequently found in human intestines, whereas members of other species (*E. casseliflavus, E. mundtii*, and *E. gallinarum*) typically reside in nonhuman hosts (Byappanahalli et al., 2012). Enterophages have been shown to have survival times similar to human enteric viruses in both fresh and marine waters, but more studies need to be conducted to assess their use in temperate climates (McMinn et al., 2017).

Coliphages are viruses that can infect *E. coli* specifically. Coliphages, like other bacteriophages, are split into two categories based on the route of bacterial host infection. Somatic coliphages are lytic phages that attach to specific protein receptors on the outer host cell membrane (Kott et al., 1974), whereas male-specific (F+) coliphages (F-RNA and F-DNA) infect via the host's sex (F) pili. Male-specific phages have been extensively studied due to their similarity to pathogenic human enteric viruses in particular enteroviruses, caliciviruses, adenovirus, and hepatitis A and E virus (Hodgson et al., 2017).

Even though multiple candidate viral indicators exist, the two most commonly used bacteriophages for viral water-quality monitoring are somatic and male-specific (F+) coliphages. The US EPA Coliphage Experts Workshop (2016) reported that coliphages are more like human pathogenic viruses than the commonly used FIB. The workshop concluded that coliphages mimic the persistence of human pathogenic viruses and can therefore be useful in evaluating individual wastewater treatment processes, disinfection efficacy, and shellfish-harvesting waters (Griffith et al., 2016; Soller et al., 2016; U.S. EPA, 2016). For somatic coliphage, the methods are ISO-10705-2, which includes both the double agar layer (DAL) plaque assay method and the presence/absence test, and the two U.S. EPA methods: EPA Method 1601, which uses the presence/absence procedure, and EPA Method 1602, which uses the single agar layer (SAL) plaque assay for quantification. Results of quantification by plaque forming units (PFU) can be obtained within 6 hours. For F+ specific coliphages, ISO-10705-1 method includes both the

DAL assay and the presence/absence assay. The ISO procedure for the detection and enumeration of F-specific bacteriophages detects both F-specific and RNA F-specific bacteriophages. EPA Method 1601 uses the presence/absence method, and EPA method 1602 uses the SAL for the detection of F-specific bacteriophages that rely on *E. coli* Famp as the host strain. A simplified version of the ISO and the U.S. EPA protocols is found in *Standard Methods for the Examination of Water and Wastewater* (SM9224F; see APHA et al., 2023).

#### *3.2.3.2 Other (PMMoV, Adenovirus)*

In addition to the bacteriophages, alternative viral indicators researched and determined to have viability include pepper mild mottle virus (PMMoV) and Adenovirus. The relevance of a plant pathogen PMMoV as a fecal indicator was initiated due to work by Zhang et al. (2005) that identified the presence of PMMoV in high concentrations in humans ($10^9$ virions per gram dry weight of fecal matter). In addition, the ubiquity of PMMoV combined with its high persistence as compared to human enteric viruses make it a good candidate as an indicator. The morphological differences (rod-shaped versus the icosahedral shape of human enteric viruses) and its inconsistent occurrence have been noted as limitations to the use of PMMoV (Kitajima et al., 2018). Methodologies are the same as those used for enteric RNA viruses (RT-PCR and RT-qPCR; Kitajima et al., 2018).

Adenoviruses present another alternative with characteristics such as stability, persistence (in particular, resistance to ultraviolet), wide distribution, human specificity, and abundance in conjunction with the availability of easier methods of detection as compared to RNA viruses (Allard & Vantarakis, 2017; Rames et al., 2016).

## 4.0 LOW- AND HIGH-RESOURCE SETTINGS

### 4.1 Defining the Issue

Water is an essential public health resource. Secure access to clean water has been defined as a basic human right. These ideals highlight the competing uses and social values that revolve around water and clash with the current state of aging infrastructure and limited resources. As we move into a future with greater impacts from a changing climate such as new and reemerging pathogens, imbalances of flooding and drought, integrated management of limited resources, and consideration of social disparities in access and quality must be considered. It is important to note the significant disparities that exist in water systems not only internationally but also within the

United States. These disparities have historically been defined in terms of resources, finances, and population served. Recognition is emerging of the impacts associated with these disparate resources in terms of the public's health. In the United States, more than 97% of the nation's 157 000 public water systems are small systems, meaning they serve 10 000 or fewer people (U.S. EPA, 2022j). These small community systems face unique financial and operational challenges significantly different to those of larger municipalities with the same goal of providing safe drinking water to the public.

Though all levels are covered by the Safe Drinking Water Act (SDWA), disparities exist in the violations of this act and how these violations are resolved. In an analysis by the National Resources Defense Council (NRDC, 2019), nationwide violations of the SDWA that occurred from 2016 to 2019 were found to be correlated to multiple sociodemographic characteristics—especially race. At the county level, as the number of people of color, low-income people, non-native English speakers, and those with little access to transportation increased, the rate of drinking water violations also increased. Their data also indicated that race, ethnicity, and language spoken had the strongest relationship to slow and inadequate enforcement of the SDWA of any sociodemographic characteristic analyzed (McDonald & Jones, 2018; NRDC, 2019; Switzer & Teodoro, 2017). In the fourth quarter of 2021 alone, there were 151,239 violations impacting a population total of 118,804,588; of these, 10,844,041 people were impacted by violations related to the total coliform rule within the SDWA (U.S. EPA, 2022j). Within the violations from 2016 to 2019, small systems accounted for 85.2% of all violations and 80.1% of health-based violations of SDWA (NRDC, 2019).

In addition to financial concerns, emerging and reemerging pathogens can tax already-limited resources in the water industry. As climate change is felt through increasing health risks associated with exposure of populations and infrastructure to weather hazards and defined vulnerability (socioeconomic, systemic), monitoring and surveillance becomes even more essential. The associations with climate and waterborne pathogen risks have been documented through a multitude of research studies (Semenza, 2020). These studies have attempted to disentangle the complex relationship between pathogens and climate. For example, both drought and flooding have a positive association with diarrheal disease incidence. Decreasing water availability has been linked to increasing diarrhea incidence, likely due to increased water contamination via stagnation (Singh et al., 2001), and increased E. coli O157 linked to livestock using traditionally human drinking-water sources (Schijven et al., 2013). Increase in flooding has also been associated with increases in diarrheal diseases, for example, through the overload of water treatment systems, as has been the case with *Cryptosporidium* (Semenza &

Nichols, 2007). Rose et al. (2000) also noted that 20% to 40% of waterborne disease outbreaks from ground and surface water (1971 to 1994) were associated with extreme precipitation. The primary issue continues to be the heterogeneity of the studies, whether in design, health endpoints examined, how they defined precipitation, timeline, or observed effects (Guzman Herrador et al., 2015; Levy et al., 2016; Lo Iacono et al., 2017). From an epidemiological perspective, some pathogens (bacterial and protozoan more so than viral) have a greater tendency to express virulence at elevated temperatures (Guzman Herrador et al., 2015; Levy et al., 2016; Lo Iacono et al., 2017). For example, in some bacterial pathogens of mammals, including *Shigella* species, temperature increases can signal successful infection of its host and lead to expression of bacterial virulence genes encoding type III secretion systems and other important virulence regulators (Shapiro & Cowen, 2012)

## 4.2 Technologies and Solutions

The water industry has used the multiple-barrier approach as the most effective way to protect public health and reduce the risk of contamination. This approach works through using a series of technical and managerial barriers for each stage from source to monitoring. This same approach can also work for small systems and any systems with limited resources. Consideration is given at each phase of water access to determine ways to limit pathogens along with other pollutants. The steps are typically as follows:

- *Source water*: Barriers at this phase include both selecting the best source of water supply and then protecting that source from potential contamination from external sources. Depending on the source, consideration of local geology, topography, land use, and activities both within and outside the watershed are important. The U.S. EPA provides resources for the development of a plan to protect source water that includes both assessment and protection phases. It is important to inventory and characterize the potential threats, develop an action plan that includes strategies and needed resources for the prevention of contamination, and then finally the implementation of that plan including evaluation and monitoring (U.S. EPA, 2023c).
- *Treatment*: The treatment of the water is the most commonly associated barrier. The various levels including primary (physical), secondary (biological), and tertiary (disinfection) are implemented to varying degrees based on the quality of the water source, cost-effectiveness, population served, and access. There are a variety of options for communities to consider such as centralized treatment facility, community

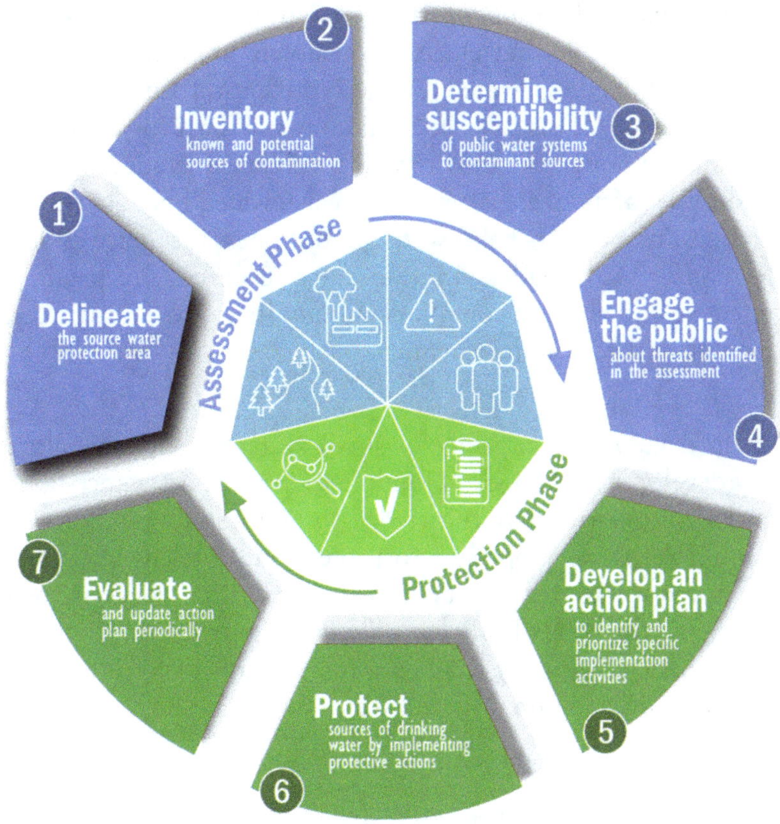


**FIGURE 3.4** Components of a Source Water Protection Program (U.S. EPA, 2023c).

wells for drinking water and/or community small septic systems, and individual or point-of-use wells or septic systems for treatment. These all come with their own set of regulations and technologies along with cost/benefit considerations.

- *Storage and distribution*: For drinking water, storage of treated water until it is used is an important point for barriers. Constructing, operating, and maintaining well-engineered storage facilities and distribution systems can protect public health by preventing pathogen contamination post-treatment from both external and internal sources. These systems represent the majority of the physical infrastructure for water systems and serve as the final barrier against direct contamination from microbial growth or corrosion. Some of the areas for consideration include microbial growth, biofilms, sediments, disinfectants/disinfection byproducts, water age, storage facility management, pressure management, water losses, main breaks, corrosion, permeation, leaching, cross-connections, and backflow.

- *Monitoring and public information*: Within the SDWA is the Public Notification Rule that ensures consumers are alerted to any issues with their drinking water (Public Notification Rule, 65 FR 25982, May 4, 2000). This sets the minimum standard for communication but should not be the only time the public is included in discussions about the safety of their water and wastewater. Engaging and serving community partners helps water practitioners recognize and build on local efforts and leadership to leverage resources and promote more effective and efficient communication with improved social, environmental, and economic benefits.

To help develop the capacity of small systems, states have access to "capacity development" program funding to provide assistance both financially and with institutional knowledge to help reliably and consistently use the multiple-barrier approach to protection. Capacity development is a process for water systems to acquire and maintain adequate technical, managerial, and financial (TMF) capacity (see Figure 3.5). Technical, managerial, and financial capacity enables water systems to have the capability to consistently provide safe drinking water to the public.

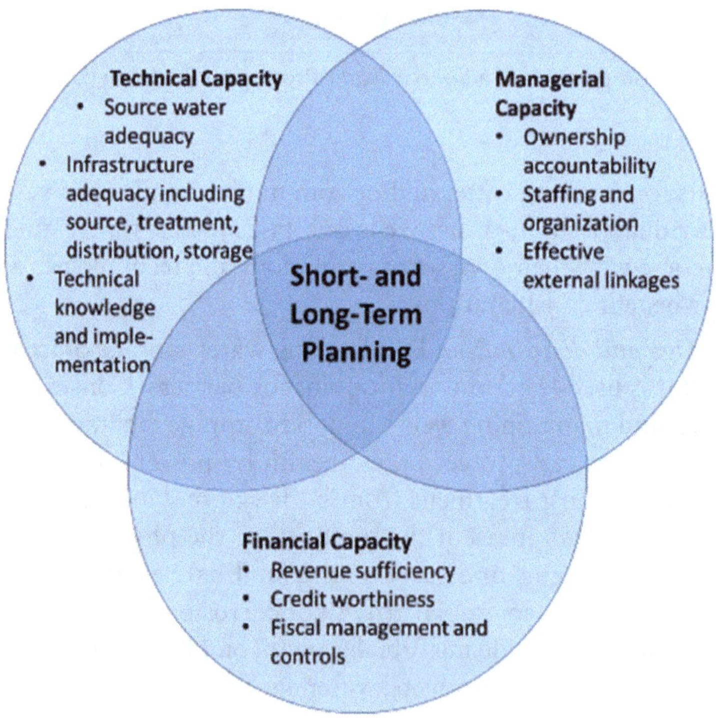


**FIGURE 3.5** Planning for Capacity Development (U.S. EPA, 2022j)

Capacity development is a fundamental component of the 1996 Safe Drinking Water Act amendments. The SDWA amendments provide a framework for states and water systems to work together to protect public health. Every state has developed a capacity development program to assist public water systems in building TMF capacity. Local officials and consumers play an important role in helping small water systems meet regulatory requirements and protect public health. Besides protecting public health, communities that support their water systems are making long-term investments in sustainable communities and economic well-being.

## 5.0 RESOURCES

### 5.1 Publicly Available

Centers for Disease Control. (2022a, February 10). Drinking water. https://www.cdc.gov/healthywater/drinking/index.html

Centers for Disease Control. (2022b, March 29). Waterborne disease & outbreak surveillance reporting. https://www.cdc.gov/healthywater/surveillance/index.html

Centers for Disease Control. (2022c, May 26). Water contamination and diseases. https://www.cdc.gov/healthywater/drinking/public/water_diseases.html

U.S. Food and Drug Administration (2018, September 22). FDA regulates the safety of bottled water beverages including flavored water and nutrient-added water beverages. https://www.fda.gov/food/buy-store-serve-safe-food/fda-regulates-safety-bottled-water-beverages-including-flavored-water-and-nutrient-added-water

U.S. Environmental Protection Agency. (2022a, March 11). A plain English guide to the EPA part 503 biosolids rule. https://www.epa.gov/biosolids/plain-english-guide-epa-part-503-biosolids-rule

U.S. Environmental Protection Agency. (2022b, April 18). Safe Drinking Water Act: Consumer confidence reports (CCR). https://www.epa.gov/ccr

U.S. Environmental Protection Agency. (2022c, May 26). Private drinking water wells. https://www.epa.gov/privatewells

U.S. Environmental Protection Agency. (2022d, June 15). Ground water and drinking water. https://www.epa.gov/ground-water-and-drinking-water

U.S. Environmental Protection Agency. (2022e, July 6). About NPDES. https://www.epa.gov/npdes/about-npdes

U.S. Environmental Protection Agency. (2022f, August 12). Municipal wastewater. https://www.epa.gov/npdes/municipal-wastewater

U.S. Environmental Protection Agency. (2022g, August 29). Small and rural wastewater systems. https://www.epa.gov/small-and-rural-wastewater-systems

U.S. Environmental Protection Agency. (2022h, September 26). Septic systems (decentralized/onsite systems). https://www.epa.gov/septic

U.S. Environmental Protection Agency. (2022i). *Medical Waste*. https://www.epa.gov/rcra/medical-waste#Treatment%20and%20Disposal%20of%20Medical%20Waste

U.S. Environmental Protection Agency. (2023a, January 6). Health and environmental agencies of US states and territories. https://www.epa.gov/aboutepa/health-and-environmental-agencies-us-states-and-territories

U.S. Environmental Protection Agency. (2023b, January 9). National primary drinking water regulations. https://www.epa.gov/ground-water-and-drinking-water/national-primary-drinking-water-regulations

U.S. Geological Survey. (n.d.). Water resources: Science. Retrieved January 26, 2023, from https://www.usgs.gov/mission-areas/water-resources/science

## 5.2 Require Access

Agalloco, J. P., Akers, J. E., & Madsen, R. E. (1998). Moist heat sterilization—myths and realities. *PDA Journal of Pharmaceutical Science and Technology, 52*, 346–350.

Allard, A., & Vantarakis, A. (2017). Adenoviruses. In J. B. Rose & B. Jiménez-Cisneros (Eds.), *Water and sanitation for the 21st century: Health and microbiological aspects of excreta and wastewater management* (Global Water Pathogen Project), J. S. Meschke & R. Girones (Eds.), *Part 3: Specific excreted pathogens: Environmental and epidemiology aspects*). Michigan State University; UNESCO. https://doi.org/10.14321/waterpathogens.11

American Public Health Association, American Water Works Association, & Water Environment Federation. (2023). *Standard methods for the examination of water and wastewater* (24th ed.). American Public Health Association.

American Water Works Association. (2006). *Water chlorination/chloramination practices and principles* (AWWA Manual M20, 2nd ed.).

Asano, T., & Tchobanoglous, G. (1987). *Municipal wastewater treatment and effluent utilization for irrigation*. FAO.

Aslan, A., Cole, Z., Bhattacharya, A., & Oyibo, O. (2018). Presence of antibiotic-resistant Escherichia coli in wastewater treatment plant effluents utilized as water reuse for irrigation. *Water, 10*(6), 805.

Association for the Advancement of Medical Instrumentation. (2002). *Steam sterilization and sterility assurance in health care facilities.* (ANSI/AAMI ST46).

Association for the Advancement of Medical Instrumentation. (2006). *Comprehensive guide to steam sterilization and sterility assurance in health care facilities.* (ANSI/AAMI ST79).

Baldry, M., & French, M. (1989). Disinfection of sewage effluent with peracetic acid. *Water Science and Technology, 21*, 203–206.

Baldry, M. G. C., French, M. S., & Slater, D. (1991). The activity of peracetic acid on sewage indicator bacteria and viruses. *Water Science and Technology, 24*, 353–357.

Block., S. S. (1991). *Disinfection, sterilization, and preservation.* Lea & Febiger.

Brulle, R. J., & Pellow, D. N. (2016). Environmental justice: Human health and environmental inequalities. *Annual Review of Public Health, 27*, 103–124. doi:10.1146/annurev.publhealth.27.021405.102124

Bucx, M. J., Veldman, D. J., Beenhakker, M. M., & Koster, R. (2000). The effect of steam sterilization at 134 degrees C on light intensity provided by fibrelight Macintoch laryngoscopes. *Anaesthesia, 55*, 185–186.

Buse, H., Morris, B., Struewing, I., & Szabo, J. (2019). Chlorine and monochloramine disinfection of Legionella pneumophila colonizing copper and polyvinyl chloride drinking water biofilms. *Applied and Environmental Microbiology, 85*(7). https://doi.org/10.1128/AEM.02956-18

Byappanahalli, M. N., Nevers, M. B., Korajkic, A., Staley, Z. R., & Harwood, V. J. (2012). Enterococci in the environment. *Microbiology and Molecular Biology Reviews, 76*(4), 685–706. https://doi.org/10.1128/MMBR.00023-12

Centers for Disease Control and Prevention. (1999). Achievements in public health, 1900–1999: Control of infectious diseases. *Morbidity and Mortality Weekly Report, 48*(29), 621–629.

Centers for Disease Control and Prevention. (2008). *Guideline for disinfection and sterilization in healthcare facilities.* https://www.cdc.gov/infectioncontrol/guidelines/disinfection/

Centers for Disease Control and Prevention. (2022d). *Prevent illness from* C. perfringens. https://www.cdc.gov/foodsafety/diseases/clostridium-perfringens.html

Cheng, N. N., & Fallowfield, H. J. (2022). Comparison of disinfection in intermittently mixed (6 am–6 pm) and continuously mixed high rate algal ponds treating domestic wastewater in winter. *Environmental Science: Water Research & Technology, 8*(4), 771–780.

Chhetri, R., Flagstad, R., Munch, E., Horning, C., Berner, J., Kolte-Olsen, A., Thornberg, D., & Andersen, H. (2015). Full scale evaluation of combined sewer overflows disinfection using perfomic acid in a sea-outfall pipe. *Chemical Engineering Journal, 270*, 133–139.

Declerck, P., Verelst, L., Duvivier, L., Van Damme, A., & Ollevier, F. (2003). A detection method for Legionella spp in (cooling) water: Fluorescent in situ hybridisation (FISH) on whole bacteria. *Water Science and Technology, 47*(3), 143–146.

DeMone, C., Hwang, M.-H., Feng, Z., McClure, J. T., Greenwood, S. J., Fung, R., Kim, M., Weese, J. S., & Shapiro, K. (2020). Application of next generation sequencing for detection of protozoan pathogens in shellfish. *Food and Waterborne Parasitology, 21*, e00096.

Deng, S., Yan, X., Zhu, Q., & Liao, C. (2019). The utilization of reclaimed water: Possible risks arising from waterborne contaminants. *Environmental Pollution, 254*(A), 113020. https://doi.org/10.1016/j.envpol.2019.113020

Dias, E., Ebdon, J., & Taylor, H. (2017). The application of bacteriophages as novel indicators of viral pathogens in wastewater treatment systems. *Water Research, 129*. https://doi.org/10.1016/j.watres.2017.11.022

Dunkin, N., Weng, S., Coulter, C., Jacangelo, J. G., & Schwab, K. J. (2017). Reduction of human norovirus GI, GII, and surrogates by peracetic acid and monochloramine in municipal secondary wastewater effluent. *Environmental Science & Technology, 51*(20), 11918–11927. https://doi.org/10.1021/acs.est.7b02954

Dunkin, N., Weng, S., Schwab, K. J., McQuarrie, J., Bell, K., & Jacangelo, J. G. (2017). Comparative inactivation of murine norovirus and MS2 bacteriophage by peracetic acid and monochloramine in municipal secondary wastewater effluent. *Environmental Science & Technology, 51*(5), 2972–2981. https://doi.org/10.1021/acs.est.6b05529

Dunkin, N., Coulter, C., Weng, S., Jacangelo, J., & Schwab, K. (2018). Effects of pH variability on peracetic acid reduction of human norovirus GI, GII RNA, and infectivity plus RNA reduction of selected surrogates. *Food and Environmental Virology*. doi.org/10.1007/s1256 0-018-9359-z

European Centre for Ecotoxicology and Toxicology of Chemicals, Joint Assessment of Commodity Chemicals programme. (2001). *JACC No. 40—Peracetic acid and its equilibrium solutions*.

Farneman, J. O., & Welch, K. L. (2013). *Recycling and material recovery system*. Micro Recovery Solutions, LLC. Patent 8,382,957.

Foret, T. (2016). *Method for operating a plasma arc torch having multiple operating modes*. Foret Plasma Labs, LLC. Patent 9,241,396.

Fujioka, R. S., & Shizumura, L. (1985). Clostridium perfringens a reliable indicator of stream water quality. *Journal of Water Pollution Control Federation, 57*, 986–992.

Gibson, K. E. (2014). Viral pathogens in water: Occurrence, public health impact, and available control strategies. *Current Opinion in Virology, 4*, 50–57. https://doi.org/10.1016/j.coviro.2013.12.005

Gilbert, J. A., & Phillips, H. O. (1984) The effect of steam sterilization on plaster casting material. *Clinical Orthopaedics and Related Research, 1984*, 241–244.

Giwa, A. S., Ali, N., Athar, M. A., & Wang, K. (2020). Dissecting microbial community structure in sewage treatment plant for pathogens' detection using metagenomic sequencing technology. *Archives of Microbiology, 202*(4), 825–833.

Gómez-Doñate, M., Payán, A., Cortés, I., Blanch, A. R., Lucena, F., Jofre, J., & Muniesa, M. (2011). Isolation of bacteriophage host strains of Bacteroides species suitable for tracking sources of animal faecal pollution in water. *Environmental Microbiology, 13*(6), 1622–1631. https://doi.org/10.1111/j.1462-2920.2011.02474.x

Griffith, J. F., Weisberg, S. B., Arnold, B. F., Cao, Y., Schiff, K. C., & Colford, J. M. (2016). Epidemiologic evaluation of multiple alternate microbial water quality monitoring indicators at three California beaches. *Water Research, 94*, 371–381. https://doi.org/10.1016/j.watres.2016.02.036

Guzman Herrador, B. R., De Blasio, B. F., MacDonald, E., Nichols, G., Sudre, B., Vold, L., Semenza, J. C., & Nygård, K. (2015). Analytical studies assessing the association between extreme precipitation or temperature and drinking water-related waterborne infections: A review. *Environmental Health, 14*(29). https://doi.org/10.1186/s12940-015-0014-y

Haas, C. N., & Finch, G. R. (2001). *Methodologies for the determination of disinfection effectiveness*. American Water Works Association Research Foundation and American Water Works Association.

Hart-Terry, J., & Swift, J. (2020). *Ozone disinfection*. The Water Environment Federation.

Harwood, V. J., Boehm, A. B., Sassoubre, L. M., Vijayavel, K., Stewart, J. R., Fong, T.-T., Caprais, M.-P., Converse, R. R., Diston, D., Ebdon, J., Fuhrman, J. A., Gourmelon, M., Gentry-Shields, J., Griffith, J. F., Kashian, D. R., Noble, R. T., Taylor, H., & Wicki, M. (2013). Performance of viruses and bacteriophages for fecal source determination in a multi-laboratory, comparative study. *Water Research, 47*(18), 6929–6943. https://doi.org/10.1016/j.watres.2013.04.064

Harwood, V. J., Staley, C., Badgley, B. D., Borges, K., & Korajkic, A. (2014). Microbial source tracking markers for detection of fecal contamination in environmental waters: Relationships between pathogens and human health outcomes. *FEMS Microbiology Reviews, 38*(1), 1–40. https://doi.org/10.1111/1574-6976.12031

Haugland, R., Oshima, K., Sivaganesan, M., Dufour, A., Varma, M., Siefring, S., Nappier, S., Schnitker, B., & Briggs, S. (2021). Large-scale comparison of E. coli levels determined by culture and a qPCR method (EPA Draft Method C) in Michigan towards the implementation of rapid, multi-site beach testing. *Journal of Microbiological Methods, 184*, 106186.

Havelaar, A. H., van Olphen, M., & Drost, Y. C. (1993). F-specific RNA bacteriophages are adequate model organisms for enteric viruses in fresh water. *Applied Environmental Microbiology, 59*(9), 2956–2962. doi:10.1128/aem.59.9.2956-2962.1993

Hejkal, T. W., Wellings, F. M., Lewis, A. L., & LaRock, P. A. (1981). Distribution of viruses associated with particles in waste water. *Applied Environmental Microbiology*, 628–634. https://doi.org/10.1128/aem.41.3.628-634.1981

Henao, L. D., Cascio, M., Turolla, A., & Antonelli, M. (2018). Effect of suspended solids on peracetic acid decay and bacterial inactivation kinetics: Experimental assessment and definition of predictive models. *Science of the Total Environment, 643*, 936–945. https://doi.org/10.1016/j.scitotenv.2018.06.219

Heano, L. D., Compagni, R. D., Turolla, A., & Antonelli, M. (2018). Influence of inorganic and organic compounds on the decay of peracetic acid in wastewater disinfection. *Chemical Engineering Journal, 337*, 133–142. https://doi.org/10.1016/j.cej.2017.12.074

Hodgson, K. R., Torok, V. A., & Turnbull, A. R. (2017). Bacteriophages as enteric viral indicators in bivalve mollusc management. *Food Microbiology, 65*, 284–293. https://doi.org/10.1016/j.fm.2017.03.003

Ijzerman, M. M., Dahling, D. R., & Fout, G. S. (1997). A method to remove environmental inhibitors prior to the detection of waterborne enteric viruses by reverse transcription-polymerase chain reaction. *Journal of Virological Methods, 63*(1–2), 145–153.

Inoue, D., Hinoura, T., Suzuki, N., Pang, J., Malla, R., Shrestha, S., Chapagain, S. K., Matsuzawa, H., Nakamura, T., & Tanaka, Y. (2015). High-throughput DNA microarray detection of pathogenic bacteria in shallow well groundwater in the Kathmandu Valley, Nepal. *Current Microbiology, 70*(1), 43–50.

Jensen, R. C., & Houze, G. C. (2013). *System and method for processing material to generate syngas*. Responsible Energy Inc. Patent 8,574,325.

Jofre, J., Lucena, F., & Blanch, A. R. (2021). Coliphages as a complementary tool to improve the management of urban wastewater treatments and minimize health risks in receiving waters. *Water, 13*, 1110. https://doi.org/10.3390/w13081110

Jofre, J., Lucena, F., Blanch, A. R., & Muniesa, M. (2016). Coliphages as model organisms in the characterization and management of water resources. *Water, 8*(5), 199. https://doi.org/10.3390/w8050199

Joslyn, L. (2001). Sterilization by heat. In S. S. Block (Ed.), *Disinfection, sterilization, and preservation* (pp. 695–728). Lippincott Williams & Wilkins.

Karmali, M. A., Simor, A., Roscoe, M., Fleming, P., Smith, S., & Lane, J. (1986). Evaluation of a blood-free, charcoal-based, selective medium for the isolation of Campylobacter organisms from feces. *Journal of Clinical Microbiology, 23*(3), 456–459.

Karpova, T., Pekonen, P., Gramstad, R., Ojstedt, U., Laborda, S., Heinonen-Tanski, H., Chavez, A., & Jimenez, B. (2013). Performic acid for advanced wastewater disinfection. *Water Science and Technology, 68*, 2090–2096.

Kitajima, M., Sassi, H. P., & Torrey, J. R. (2018). Pepper mild mottle virus as a water quality indicator. *npj Clean Water, 1*, 19. https://doi.org/10.1038/s41545-018-0019-5

Kitis, M. (2004). Disinfection of wastewater with peracetic acid: A review. *Environment International, 30*, 47–55.

Koivunen, J., & Heinonen-Tanski, H. (2005a). Inactivation of enteric microorganisms with chemical disinfectants, UV irradiation and combined chemical/UV treatments. *Water Research, 39*, 1519–1526.

Koivunen, J., & Heinonen-Tanski, H. (2005b). Peracetic acid (PAA) disinfection of primary, secondary and tertiary treated municipal wastewaters. *Water Research, 39*, 4445–4453.

Kott, Y., Roze, N., Sperber, S., & Betzer, N. (1974). Bacteriophages as viral pollution indicators. *Water Research, 8*, 165–171.

Kouame, Y., & Haas, C. N. (1991). Inactivation of E. coli by combined action of free chlorine and monochloramine. *Water Research, 25*(9), 1027–1032.

Kumaraswamy, R., Amha, Y. M., Anwar, M. Z., Henschel, A., Rodríguez, J., & Ahmad, F. (2014). Molecular analysis for screening human bacterial pathogens in municipal wastewater treatment and reuse. *Environmental Science & Technology, 48*(19), 11610–11619.

Lagergren, E. R. (1998). Recent advances in sterilization. *Journal of Infection Control (Asia), 1*, 11–13.

Lauer, J. L., Battles, D. R., & Vesley, D. (1982). Decontaminating infectious laboratory waste by autoclaving. *Applied and Environmental Microbiology, 44*, 690–694.

LeFevre, F., Audic, J., & Ferrand, F. (1992). Peracetic acid disinfection of secondary effluents discharged off coastal seawater. *Water Science and Technology, 25*, 155–164.

Levy, K., Woster, A., Goldstein, R., & Carlton, E. (2016). Untangling the impacts of climate change on waterborne diseases: A systematic review of relationships between diarrheal diseases and temperature, rainfall, flooding, and drought. *Environmental Science & Technology, 50*(10), 4905–4922. https://doi.org/10.1021/acs.est.5b06186

Li, B., Li, X., Wang, B., & Yan, T. (2022). A metagenomic approach for characterizing antibiotic resistance genes in specific bacterial populations: Demonstration with Escherichia coli in cattle manure. *Applied and Environmental Microbiology, 88*(7): e02554–02621.

Li, B., Li, X., & Yan, T. (2021). A quantitative metagenomic sequencing approach for high-throughput gene quantification and demonstration with antibiotic resistance genes. *Applied and Environmental Microbiology, 87*(16): e00871–00821.

Li, W., Zhou, Q., & Hua, T. (2010). Removal of organic matter from landfill leachate by advanced oxidation processes: A review. *International Journal of Chemical Engineering, 2010*.

Lim, Y. A., & Nissapatorn, V. (2017). Transmission of waterborne parasites in the Association of Southeast Asian Nations (ASEAN): Overview and direction forward. *Food and Waterborne Parasitology, 8*, 75–83.

Lisle, J. T., Smith, J. J., Edwards, D. D., & McFeters, G. A. (2004). Occurrence of microbial indicators and Clostridium perfringens in wastewater, water column samples, sediments, drinking water, and Weddell Seal feces collected at McMurdo Station, Antarctica. *Applied and Environmental Microbiology*. https://doi.org/10.1128/AEM.70.12.7269-7276.2004

Lo Iacono, G., Armstrong, B., Fleming, L., Elson, R., Kovats, S., Vardoulakis, S., & Nichols, G. (2017). Challenges in developing methods for quantifying the effects of weather and climate on water-associated diseases: A systematic review. *PLOS Neglected Tropical Diseases*. https://doi.org/10.1371/journal.pntd.0005659

Luukkonen, T., Heyninck, T., Ramo, J., & Lassi, U. (2015). Comparison of organic peracids in wastewater. *Water Research, 85*, 275–285.

Ma, J. Y., Li, M. Y., Qi, Z. Z., Fu, M., Sun, T. F., Elsheikha, H. M., & Cong, W. (2022). Waterborne protozoan outbreaks: An update on the global, regional, and national prevalence from 2017 to 2020 and sources of contamination. *Science of the Total Environment, 806*(Pt 2), 150562. doi:10.1016/j.scitotenv.2021.150562

Manoli, K., Sarathy, S., Maffettone, R., & Santoro, D. (2019). Detailed modeling and advanced control for chemical disinfection of secondary effluent wastewater by peracetic acid. *Water Research, 153*, 251–262.

Masjoudi, M., Mohseni, M., & Bolton, J. R. (2021). Sensitivity of bacteria, protozoa, viruses, and other microorganisms to ultraviolet radiation. *Journal of Research of the National Institute of Standards, 126*, 126021. https://doi.org/10.6028/jres.126.021

McDonald, Y., & Jones, N. (2018). Drinking water violations and environmental justice in the United States, 2011–2015. *American Journal of Public Health, 108*(10), 1401–1407. https://doi.org/10.2105/AJPH.2018.304621

McMinn, B. R., Huff, E. M., Rhodes, E. R., & Korajkic, A. (2017). Concentration and quantification of somatic and F+ coliphages from recreational waters. *Journal of Virological Methods, 249*, 58–65. https://doi.org/10.1016/j.jviromet.2017.08.006

Metcalf & Eddy. (2003). *Wastewater engineering: Treatment and reuse* (4th ed.). McGraw Hill Education.

Mohiuddin, M. M., Salama, Y., Schellhorn, H. E., & Golding, G. B. (2017). Shotgun metagenomic sequencing reveals freshwater beach sands as reservoir of bacterial pathogens. *Water Research, 115*, 360–369.

Moreno, Y., Botella, S., Alonso, J. L., Ferrús, M. A., Hernández, M., & Hernández, J. (2003). Specific detection of Arcobacter and Campylobacter strains in water and sewage by PCR and fluorescent in situ hybridization. *Applied and Environmental Microbiology, 69*(2), 1181–1186.

Mosenson, B., Litinsky, Y., & Mark, I. (2002). *Apparatus for treating waste, particularly medical waste, to facilitate its disposition.* M.C.M. Environmental Technologies Ltd. Patent 6,494,391.

National Resources Defense Council. (2019). *Watered down justice.* R:19-09-A. https://www.nrdc.org/sites/default/files/watered-down-justice-report.pdf

Nussbaum, G. F. (2009). Alternative waste management strategies. *Perioperative Nursing Clinics, 3*(1), 63–72. https://www.ncbi.nlm.nih.gov/pmc/articles/PMC7147228/#bib2

O'Brien, E., & Xagoraraki, I. (2019). A water-focused one-health approach for early detection and prevention of viral outbreaks. *One Health, 7*, 100094. https://doi.org/10.1016/j.onehlt.2019.100094

Palenik, C. J., & Cumberlander, N. D. (1993). Effects of steam sterilization on the contents of sharps containers. *American Journal of Infection Control, 21*, 28–33.

Payment, P., Waite, M., & Dufour, A. (2003). Introducing parameters for the assessment of drinking water quality. In *Assessing Microbial Safety of Drinking Water* (pp. 47–77). World Health Organization.

Pecson, B. M., Darby, E., Danielson, R., Dearborn, Y., Di Giovanni, G., Jakubowski, W., Leddy, M., Lukasik, G., Mull, B., & Nelson, K. L. (2022). Distributions of waterborne pathogens in raw wastewater based on a 14-month, multi-site monitoring campaign. *Water Research, 213*, 118170.

Pillai, S. D., & Reimers, R. S. (2010). *Disinfecting and stabilizing biosolids using e-beam and chemical oxidants*. (Water Environment Research Foundation Report Number U4R06). WERF Publishing.

Prentice, J. A. (2012). *Combustion methods and fuels for the production of energy.* Advanced Combustion Energy Systems, Inc. Patent 8,132,416.

Rames, E., Roiko, A., Stratton, H., & Macdonald, J. (2016). Technical aspects of using human adenovirus as a viral water quality indicator. *Water Research, 96*, 308–326. https://doi.org/10.1016/j.watres.2016.03.042

Reimers, R. S., Holland, J. E., Bankston, W. S., Davis K. A., Shelton, J. J., Bankston, J. C., & Meckling, J. H. (2000). Application of applied fields in the disinfection of water and waste. In *WEF disinfection 2000: Disinfection of wastes in the new millennium*. Water Environment Federation.

Rose, J. B., Daeschner, S., Easterling, D. R., Curriero, F. C., Lele, S., & Patz, J. A. (2000). Climate and waterborne disease outbreaks. *Journal of the American Water Works Association, 92*, 77–87. https://doi.org/10.1002/j.1551-8833.2000.tb09006.x

Ross, A., Ward, S., & Hyman, P. (2016). More is better: Selecting for broad host range bacteriophages. *Frontiers in Microbiology, 7*.

Rutala, W. A., Stiegel, M. M., & Sarubbi, F. A., Jr. (1982). Decontamination of laboratory microbiological waste by steam sterilization. *Applied and Environmental Microbiology, 43*, 1311–1316.

Saingam, P., Li, B., Sung, S., & Yan, T. (2021). Immediate impact of Hurricane Lane on microbiological quality of coastal water in Hilo Bay, Hawaii. *Environmental Science & Technology, 55*(5), 2960–2967.

Saingam, P., Li, B., & Yan, T. (2018). Use of amplicon sequencing to improve sensitivity in PCR-based detection of microbial pathogen in environmental samples. *Journal of Microbiological Methods, 149*, 73–79.

Sanchez-Ruiz, C., Martınez-Royano, S., & Tejero-Monzon, I. (1995). An evaluation of the efficiency and impact of raw wastewater disinfection with peracetic acid. *Water Science and Technology, 32*, 159–166.

Santiago, P., Jiménez-Belenguer, A., García-Hernández, J., Estellés, R. M., Pérez, M. H., López, M. A. C., Ferrús, M. A., & Moreno, Y. (2018). High prevalence of Salmonella spp. in wastewater reused for irrigation assessed by molecular methods. *International Journal of Hygiene and Environmental Health, 221*(1), 95–101.

Schijven, J., Bouwknegt, M., de Roda Husman, A. M., Rutjes, S., Sudre, B., Suk, J. E., & Semenza, J. C. (2013). A decision support tool to compare waterborne and foodborne infection and/or illness risks associated with climate change. *Risk Analysis, 33*, 2154–2167. https://doi.org/10.1111/risa.12077

Schlipköter, U., & Flahault, A. (2010). Communicable diseases: Achievements and challenges for public health. *Public Health Reviews, 32*(1), 90–119.

Semenza, J. C. (2020). Cascading risks of waterborne diseases from climate change. *Nature Immunology, 21*, 484–487. https://doi.org/10.1038/s41590-020-0631-7

Semenza, J. C., & Nichols, G. (2007). Cryptosporidiosis surveillance and water-borne outbreaks in Europe. *Eurosurveillance, 12*(5), 711. https://doi.org/10.2807/esm.12.05.00711-en

Shapiro, R., & Cowen, L. (2012). Thermal control of microbial development and virulence: Molecular mechanisms of microbial temperature sensing. *mBio, 3*(5). https://doi.org/10.1128/mBio.00238-12

Shrestha, R. G., Tanaka, Y., Malla, B., Bhandari, D., Tandukar, S., Inoue, D., Sei, K., Sherchand, J. B., & Haramoto, E. (2017). Next-generation sequencing identification of pathogenic bacterial genes and their relationship with fecal indicator bacteria in different water sources in the Kathmandu Valley, Nepal. *Science of the Total Environment, 601*, 278–284.

Silverstone, S. E., & Hill, D. E. (1999). Evaluation of sterilization of dental handpieces by heating in synthetic compressor lubricant. *Journal of General Dentistry, 47*, 158–160.

Sinclair, R. G., Jones, E. L., & Gerba, C. P. (2009). Viruses in recreational water-borne disease outbreaks: A review. *Journal of Applied Microbiology, 107*(6), 1769–1780. https://doi.org/10.1111/j.1365-2672.2009.04367.x

Singh, R., Hales, S., de Wet, N., Raj, R., Hearnden, M., & Weinstein, P. (2001). The influence of climate variation and change on diarrheal disease in the Pacific Islands. *Environmental Health Perspectives, 109*. https://doi.org/10.1289/ehp.01109155

Smith, O. L., & Bryant, R. M. (2013). *Medical waste processor and processing method.* Oncore Technology, LLC. Patent Application 20130287628.

Sohier, D., Pavan, S., Riou, A., Combrisson, J., & Postollec, F. (2014). Evolution of microbiological analytical methods for dairy industry needs. *Frontiers in Microbiology, 5*, 16.

Soller, J. A., Eftim, S., Wade, T. J., Ichida, A. M., Clancy, J. L., Johnson, T. B., Schwab, K., Ramirez-Toro, G., Nappier, S., & Ravenscroft, J. E. (2016). Use of quantitative microbial risk assessment to improve interpretation of a recreational water epidemiological study. *Microbial Risk Analysis, 1*, 2–11. https://doi.org/10.1016/j.mran.2015.04.001

Spon, R. (2008). Do you really have a free chlorine residual? *Opflow, 34*(6), 24–27.

Stampi, S., De Luca, G., Onorato, M., Ambrogiani, E., & Zanetti, F. (2002). Peracetic acid as an alternative wastewater disinfectant to chlorine dioxide. *Journal of Applied Microbiology, 93*, 725–731.

State and Territorial Association on Alternate Treatment Technologies. (1994). *Technical assistance manual: State regulatory oversight of medical waste treatment technologies*. https://archive.epa.gov/epawaste/nonhaz/industrial/medical/web/pdf/1-6.pdf

Stelma, G. (2018). Use of bacterial spores in monitoring water quality and treatment. *Journal of Water and Health, 16*(4), 491–500. doi:10.2166/wh.2018.013

Stokdyk, J. P., Firnstahl, A. D., Walsh, J. F., Spencer, S. K., de Lambert, J. R., Anderson, A. C., Rezania, L.-I. W., Kieke Jr., B. A., & Borchardt, M. A. (2020). Viral, bacterial, and protozoan pathogens and fecal markers in wells supplying groundwater to public water systems in Minnesota, USA. *Water Research, 178*, 115814.

Switzer, D., & Teodoro, M. (2017). Class, race, ethnicity, and justice in safe drinking water compliance. *Social Science Quarterly, 99*(2), 524–535. https://doi.org/10.1111/ssqu.12397

Tang, Y. Z., Gin, K. Y. H., & Lim, T. H. (2005). High-temperature fluorescent in situ hybridization for detecting Escherichia coli in seawater samples, using rRNA-targeted oligonucleotide probes and flow cytometry. *Applied and Environmental Microbiology, 71*(12), 8157–8164.

Taylor, W. I., & Schelhart, D. (1971). Isolation of shigellae: VIII. Comparison of xylose lysine deoxycholate agar, hektoen enteric agar, salmonella-shigella agar, and eosin methylene blue agar with stool specimens. *Applied and Environmental Microbiology, 21*(1), 32–37.

Tondera, K., Klaer, K., Koch, C., Hamza, I., & Pinnekamp, J. (2016). Reducing pathogens in combined sewer overflows using perfomic acid. *International Journal of Hygiene and Environmental Health, 219*, 700–708.

U.S. Department of the Interior, Bureau of Reclamation. (2013). *Effect of chlorine vs. chloramine treatment techniques on materials degradation in reclamation infrastructure* (Tech Memo No. MERL-2013-57). Bureau of Reclamation, Technical Service Center. file:///Users/rasha/Downloads/

effect_of_chlorine_vs_chloramine_treatment_techniques_on_materials_degradation_in_reclamation_infrastructure.pdf

U.S. Environmental Protection Agency. (1986). *Safe Drinking Water Act Amendments of 1986*. Pub. L. 99–359; 100 Stat. 642. 1986-06-19.

U.S. Environmental Protection Agency. (2003). *Control of pathogens and vector attraction in sewage sludge* (EPA/625/R-92/013). Revised July 2003.

U.S. Environmental Protection Agency. (2004). *Implementation guidance for ambient water quality criteria for bacteria* (EPA-823-B-04-002).

U.S. Environmental Protection Agency. (2006a). *Biosolids technology fact sheet: Heat drying* (EPA 832-F-06-029).

U.S. Environmental Protection Agency. (2006b). *Ultraviolet disinfection guidance manual for the final long term 2 enhanced surface water treatment rule* (EPA 815-R-06-007).

U.S. Environmental Protection Agency. (2009). *Wastewater technology fact sheet ozone disinfection.*

U.S. Environmental Protection Agency. (2010). *Comprehensive disinfectants and disinfection byproducts rules (stage 1 and stage 2): Quick reference guide.*

U.S. Environmental Protection Agency. (2012). *Method 1611: Enterococci in water by TaqMan® quantitative polymerase chain reaction (qPCR) assay.*

U.S. Environmental Protection Agency. (2016). *Proceedings from the US Environmental Protection Agency (EPA) Coliphage Experts Workshop.* No. 822-R-17-003, p. 94. https://www.epa.gov/sites/production/files/2017-08/documents/coliphage-workshop-proceedings-2016-508.pdf

U.S. Environmental Protection Agency. (2020). *Disinfection profiling and benchmarking technical guidance manual* (EPA 815-R-20-003).

U.S. Environmental Protection Agency. (2022j, December 22). *Building the capacity of drinking water systems.* https://www.epa.gov/dwcapacity

U.S. Environmental Protection Agency. (2023c, January 10). *Source water protection.* https://www.epa.gov/sourcewaterprotection

Veschetti, E., Cutilli, D., Bonadonna, L., Briancesco, R., Martini, C., Cecchini, G., Anastasi, P., & Ottaviani, M. (2003). Pilot plant comparative study of peracetic acid and sodium hypochlorite wastewater disinfection. *Water Research, 37*, 78–94.

Wang, H., Sikora, P., Rutgersson, C., Lindh, M., Brodin, T., Bjorlenius, B., Larsson, D. G. J., & Norder, H. (2018). Differential removal of human pathogenic viruses from sewage by conventional and ozone treatments. *International Journal of Hygiene and Environmental Health, 221*, 479–488.

Water Environment Federation. (1996). *Wastewater disinfection.* Manual of Practice No. FD-10.

Water Environmental Federation. (2020). *Peracetic acid disinfection: Implementation considerations for water resource recovery facilities.*

Water Environment Federation. (2023). *Standards for design of water resource recovery facilities.* WEF 8.

White, G. C. (1999). *Handbook of chlorination and alternative disinfectants.* John Wiley & Sons, Inc.

World Health Organization. (2004). *Safe health care waste management: Policy paper.* World Health Organization. https://www.who.int/publications/i/item/WHO-SDE-WSH-4.12

Wilhartitz, I., Mach, R. L., Teira, E., Reinthaler, T., Herndl, G. J., & Farnleitner, A. H. (2007). Prokaryotic community analysis with CARD-FISH in comparison with FISH in ultra-oligotrophic ground- and drinking water. *Journal of Applied Microbiology, 103*(4), 871–881.

Zeng, D., Chen, Z., Jiang, Y., Xue, F., & Li, B. (2016). Advances and challenges in viability detection of foodborne pathogens. *Frontiers in Microbiology, 7*, 1833.

Zhang, T., Breitbart, M., Lee, W. H., Run, J. Q., Wei, C. L., Soh, S. W. L., Hibberd, M. L., Liu, E. T., Rohwer, F., & Ruan, Y. (2005). RNA viral community in human feces: Prevalence of plant pathogenic viruses. *PLOS Biology, 4*(1), e3. https://doi.org/10.1371/journal.pbio.0040003

Zhou, G., Wen, S., Liu, Y., Li, R., Zhong, X., Feng, L., Wang, L., & Cao, B. (2011). Development of a DNA microarray for detection and identification of Legionella pneumophila and ten other pathogens in drinking water. *International Journal of Food Microbiology, 145*(1), 293–300.

# Chapter 4

# Understanding Risk and Knowledge Gaps

*Mark Sobsey, Lee Gary, Bo Li, Prakit Saingam, & Samendra Sherchan*

## 1.0 ASSESSING RISK

In 1966, the American Risk and Insurance Association defined *risk* as the uncertainty of the outcome of an event when two or more possible outcomes exist. In the public and occupational health literature, however, *risk* is defined as the chance or probability that a person will be harmed or experience an adverse health effect if exposed to a hazard, with the hazard being the source of the potential harm. Risks to human health are inevitable and come from a wide range of sources in the natural and human-made environment, making the concept of risk difficult to grasp and assess. Risk is also the focal topic in the management of many activities, procedures, and technologies in any organization or industry (Fischhoff et al., 1984). This is especially true during infectious disease outbreaks, when the water sector and its employees are potentially at higher risk of exposure to pathogens. The objective of this chapter is to provide the reader with the literacy and understanding to critically examine and consider claims about risk from the media, scientists, politicians, public and occupational health experts, industry, or other sources. It provides the reader with some tools to identify and mitigate risk proactively.

### 1.1 Qualitative Comparisons

Risks to human health can be described and compared qualitatively and quantitatively according to their consequences and severity. Such risks include drowning, exposures to natural phenomena such as lighting strikes; experiencing hurricanes; floods and earthquakes and deaths; injuries and other adverse health effects and outcomes from falling, motor vehicles and other machinery-related accidents, gunshot wounds, animal bites, burns; and exposures to harmful chemicals, radiation, and disease-causing microorganisms, among others. Risk exposures can occur in many settings including at home and in various other community locations, including workplaces (occupational risks), schools, public gatherings and venues for religion (churches, mosques and temples), commercial and business activities such as consumer retail (markets, restaurants, bars, etc.), transportation activities and settings (motor vehicles, trains, subways, airplanes, etc.), government service locations, libraries, museums, and entertainment venues, such as concerts and movie theaters. That is, human health risks and adverse events can occur anywhere there are people. Adverse health effects are defined as any changes in body function or the structures of cells that can lead to disease or health problems.

People have the ability to reduce or minimize their health risks based on the potential risk-producing activities in which they engage and their risky behaviors, such as driving a motor vehicle (with or without wearing a seatbelt), engaging in unprotected sex with a stranger, and drinking or swimming in fecally contaminated water.

Risks to health can be reduced by wearing seatbelts and other personal protective equipment (PPE), avoiding or minimizing harmful chemical exposures, such as not smoking or ingesting other harmful chemicals (avoiding or minimizing drug and alcohol use), and avoiding or minimizing exposures to pesticides and other toxic chemicals.

Health risks from exposures to agents in air, water, and on land can sometimes go unnoticed, such as physical, chemical, radiation, and biological (such as microorganism) exposures that may not immediately become manifest in obvious adverse health effects. In this case, their adverse effects are inapparent and unnoticed or perhaps manifest only at a later time. Other exposures cause observable adverse health effects that are experienced immediately or very soon when exposed, are experienced by the affected person, and can be observed by others, such as traumatic injuries or illnesses of differing severity and duration. Some health effects from exposures may cause health effects that are reversible or from which recovery is eventually complete. Other health effects may result in long-lasting and even permanent health deficits or disabilities. It is notable that exposures to some enteric pathogens can not only cause adverse health effects soon after exposure, such as gastrointestinal illness (nausea, vomiting, diarrhea, etc.), but also cause other adverse health effects that arise later, such as reactive arthritis and neuromuscular disease such as paralysis or Guillain-Barre syndrome. Such physical, chemical, or biological adverse effects include loss of sight, hearing loss, loss of a limb, decreased mobility, chronic respiratory disease, neurological conditions, or other health decreases or deficiencies. Finally, other exposures can result in either immediate or subsequent death, the worst-case outcome. Hence, there is a broad range of adverse outcomes that can occur from exposures to risk-causing agents in the environment. The most important consideration in assessing health risks from environmental exposures to any agent is the magnitudes (severities) and durations of the adverse health risks, which can greatly differ.

For more information on the use of qualitative risk assessment in the water sector and ISO 45001 OH&S management systems, see Chapter 6.

## 1.2 Quantitative Assessments Using Quantitative Microbial Risk Assessment and Disability Adjusted Life Year

To address both severity and duration of health effects, a standard metric is used to grade or characterize the risks in a way that allows uniform and consistent risk comparisons across different hazards and exposures.

### *1.2.1 How the World Health Organization Uses DALYs to Quantify Risks*

The World Health Organization (WHO) and other health agencies and stakeholders have adopted a uniform measure for health effects of all causes called the disability adjusted life year (DALY). Disability adjusted life year are the metric by which deaths at different ages and disabilities of all kinds may be measured. One DALY can be quantified as one lost year of "healthy" life. The extent or burden of disease can be considered the measurement of the difference between current health status and an ideal situation in which everyone lives into old age, free of disease and disability. Disability adjusted life year for a disease or injury are calculated as the sum of the years of life lost due to premature mortality (years of life lost, or YLL) and the years lost due to disability (years lived with disability, or YLD) for incident cases of the disease or injury, as shown in Figure 4.1.

Years of life lost are calculated from the number of deaths at each age multiplied by a global standard life expectancy of the age at which death occurs. Years lived with disability for a particular causative agent in a particular time period is estimated as follows:

$$\text{YLD} = \text{number of incident cases in that period} \times \text{average duration of the disease} \times \text{disability weight}$$

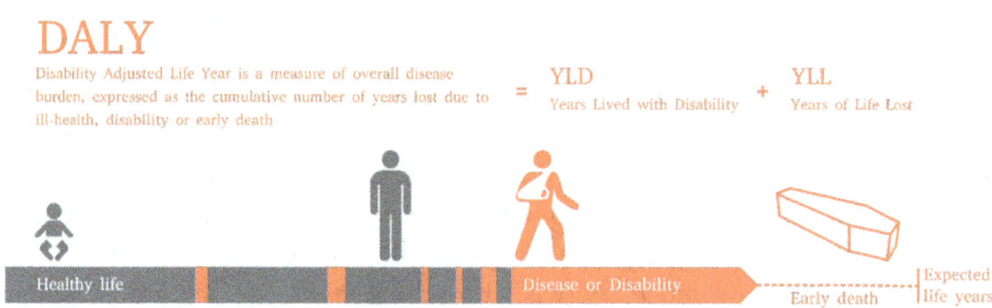


**FIGURE 4.1** Disability Adjusted Life Years as Shown Graphically Over a Lifetime
*Note.* From *Adjusted Life Year*, by Planemad, 2012, Wikipedia (https://en.wikipedia.org/wiki/Disability-adjusted_life_year). CC BY-SA 3.0.

The disability weight is based on the **severity** of the disease on a scale from 0 (perfect health) to 1 (death). The DALY estimates are established by an expert group on behalf of the WHO. Examples of various diseases as a share of total DALYs are shown in Figure 4.2. Note that respiratory infections including tuberculosis, respiratory diseases, and enteric diseases account for about 6.4%, 4.1%, and 3.8% (respectively) of the total DALY disease burden globally as of 2019 (Global Burden of Disease Collaborative Network, 2021).

It is important to keep in mind that people differ greatly in responses to health risks from environmental exposures to various agents because of their existing specific health conditions that influence their risks and often make them more susceptible. Factors that influence heath risk responses to various environmental exposures include age, gender, differences in genetic susceptibilities to infection by a pathogen (such as blood type differences in susceptibility to norovirus infections and gastrointestinal illness), and underlying health conditions such as asthma, chronic bowel disease, chronic respiratory disease, diabetes, heart disease, high blood pressure, obesity, and malnutrition. People who are at greater risk need to be mindful to take greater precautions because of the increased risk of adverse health effects.

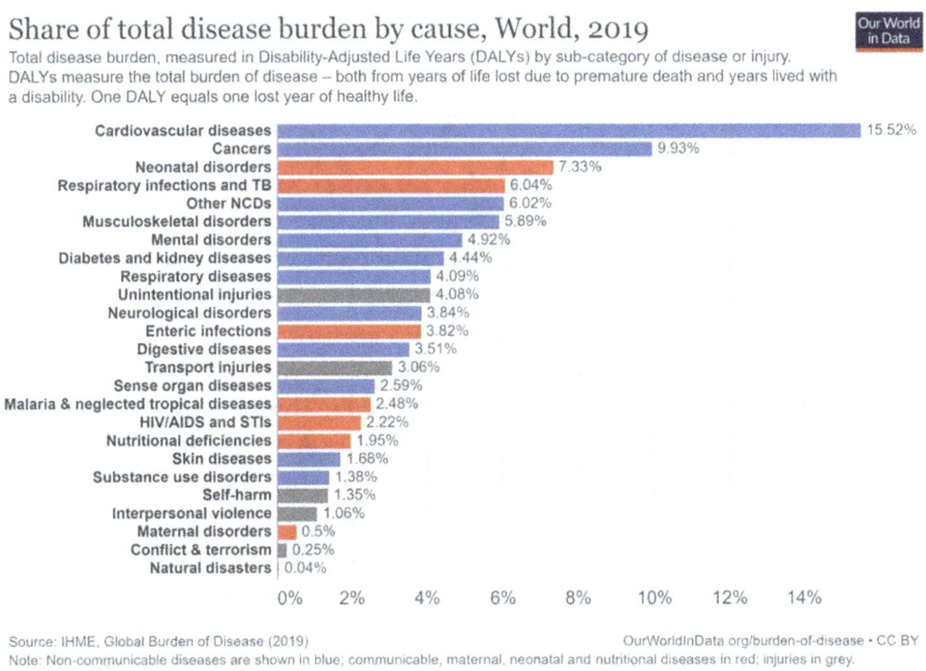


**FIGURE 4.2** Share of Global Disease Burden in 1990–2019 in DALYs Shown by Cause (http://ghdx.healthdata.org/gbd-results-tool)

### *1.2.2 What Is QMRA, and How Is It Conducted?*

Quantitative microbial risk assessment (QMRA) is a system and analytical tool to quantitatively estimate human health risks from exposures to pathogens (disease-causing microorganisms) via water, food, air, land-based sources, and other environmental routes. According to the U.S. Environmental Protection Agency (U.S. EPA; Whelan et al., 2017), "A Quantitative Microbial Risk Assessment (QMRA) is a modeling approach that integrates disparate data, including those linked to microbial fate/transport and exposure, and human health-effect relationships to characterize potential health impacts/risks from exposure to pathogenic microorganisms." It uses a structured mathematical approach to characterize and quantify human health risks from pathogen exposures in the environment. It is a different and more practical approach compared to epidemiology, which uses real-world field data on observable human health effects from exposures to microbes from specific sources (such as ingestion of contaminated drinking or recreational waters) to estimate health risks. Epidemiological data may come from either field investigations of waterborne disease outbreaks in which illness as a disease cluster or an epidemic is recognized, is attributed to a specific pathogen, if possible, and comes from a specific source of exposure, such as contaminated drinking water. Alternatively, epidemiological data can be obtained from carefully designed controlled studies that are usually prospective. Controlled prospective field studies attempt to detect and quantify resulting illnesses, such as gastroenteritis, in people who are observed to drink or swim in specific waters in which concentrations of pathogens or surrogate microbes in the exposure source are determined at the time human exposures occur, in comparison to a control group of similar people who are not exposed to the water. Data from waterborne outbreaks are difficult and often expensive to obtain, and the quality of the data gathered is often incomplete, uncertain, and subject to bias and high variability. Compared to epidemiological field studies, QMRA is practical, rapid, and economical and uses existing data.

Quantitative microbial risk assessment is typically described as a sequence of four steps, which will be introduced in this section and described in greater detail in Section 2.0 of this chapter. A useful introduction to QMRA basics can be found in this summary from the Food and Agriculture Organization: https://www.fao.org/3/y4722e/y4722e05.htm.

*Step 1. Hazard identification.* In this initial step the pathogen of interest or concern is identified, and the adverse human health effects, such as a disease or illness it causes, are also determined. Some specific pathogens cause very distinctive disease health effects such as infectious hepatitis, a liver disease (caused by hepatitis A and hepatitis E virus), or a disease of the nerves and muscles causing paralysis, like poliomyelitis (caused by polioviruses). Other pathogens infecting the gut may cause a range of signs and symptoms of illness of varying severity, such as gastroenteritis, nausea, and diarrhea, with or without high fever. Examples are viruses such as norovirus, bacteria such as salmonella or campylobacter, and protozoan parasites like *Cryptosporidium* and *Giardia.*

*Step 2. Exposure assessment.* In this step the pathways of exposure (water, air, food, etc.) are determined, preferably by direct measurement, or if need be, by modeling the pathogen exposure doses during defined exposure events. An exposure dose is the quantity of pathogens to which a person is exposed, usually at one time. The exposure dose is determined from the concentration of the pathogen in the exposure medium, such as water, and the amount of water that is ingested at one time or over a period of time, such as 1 day. For example, if the concentration of a pathogen in water is 2 organisms per liter of water, and a person drinks 1 liter of water per day, the dose is 2 organisms per day. When available, direct measurements of pathogen levels in exposure media as doses are preferred over modeled or assumed levels because they provide a more realistic estimate of the exposure dose.

*Step 3. Dose-response assessment.* Using existing pathogen-specific disease outbreak records, if available, or using data from controlled laboratory dosing studies in which human volunteers are given specific amounts of the pathogen as doses, the relationship between the exposure dose and the likelihood of a measurable health outcome, such as infection or illness are quantified. Quantitative dose-response relationships are already available for a variety of pathogens of interest for the route of exposure to it, such as ingestion or inhalation. These dose-response relationships are usually developed by experts having the skill to fit various mathematical models to the dose-response data and identify the best fit model for the available data. Quantitative dose-response relationships are often presented graphically in a dose response curve, as shown in Figure 4.3.

*Step 4. Risk characterization.* This step uses the pathogen exposure doses from Step 2 and the dose-response relationship from Step 3 to calculate the likelihood of the health outcome, such as infection, illness, or death, caused by a dose of pathogen acquired from a specific route of exposure. The calculated dose (or a dose of interest) from the exposure source is put into the dose-response model for a quantitative estimate of the health effects as a function of the dose. The result from the analysis gives the likelihood of the health effect in an exposed person from the exposure dose of the pathogen. Risk characterization is often done using mathematical models (e.g., Monte Carlo simulations), which have the ability to account for variability and uncertainty in estimating health risks. Such model simulations also allow for a *sensitivity analysis,* which makes it possible to determine

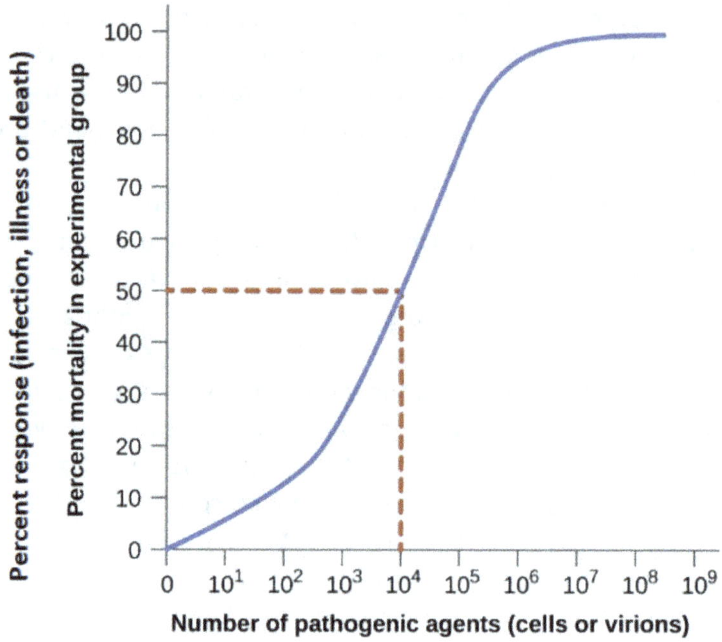


**FIGURE 4.3** Human Infectivity Dose-Response Relationships

the extent to which the calculated risk characterization value is affected by changes in the values of other key variables that can influence risk factors (e.g., exposure variables, health effects variables, different health effects outcomes, different host susceptibilities such as age and infirmity, etc.). Some of these other variables may have a greater effect on the risk characterization estimate than do others.

### *1.2.3 Approaches and Metrics Used to Evaluate and Quantify Risk*

Infection, illness, and death are different health outcomes from exposures with successively greater health effects. That is, the severity of health effects from exposures can range from no infection, to no observed health effect (*asymptomatic* or *inapparent infection*), to an observable *illness* or *disease* that may be mild or severe, to the worst-case outcome of death. This progression of outcomes from pathogen exposures can be viewed as a pyramid and is shown in Figure 4.4. For any particular pathogen, once a risk of illness is determined, a corresponding DALY value can be calculated based on the standard metric provided by the WHO. The DALY as a health effect resulting from an exposure was defined in Section 1.2.1.

In characterizing risk of a health outcome for an exposure—be it infection, illness, or death—the range of outcomes can be zero (no chance of infection from exposure, with no effect on the exposed person) or as high as

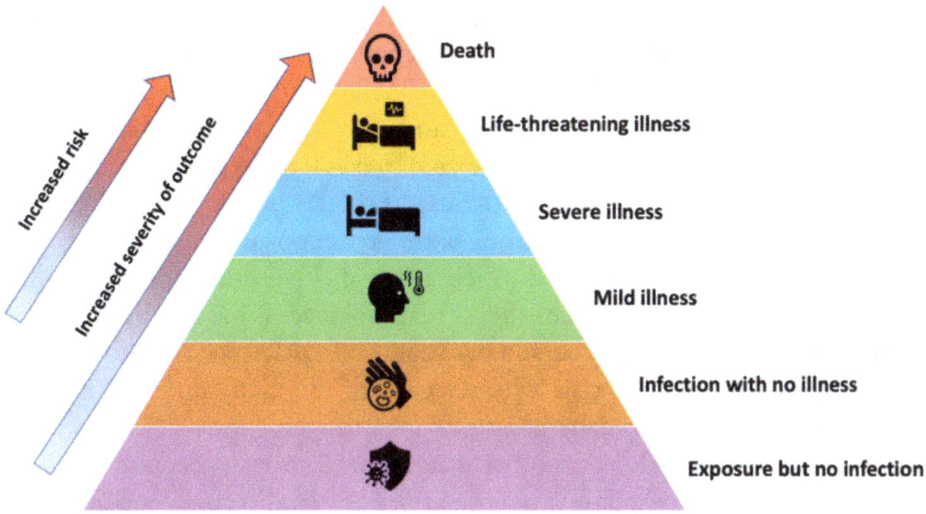


FIGURE 4.4 Health Outcome Pyramid for Pathogen Exposures. Bottom: No Effect. Top: Death

1 (100% chance of a health effect). A useful way to illustrate the different health risk outcomes is as a pyramid of possible effects, as shown above in Figure 4.4. Typically, of those exposed to a certain dose of a pathogen from an environmental exposure source, only some will become infected, for example, 60% of them. Furthermore, of those who become infected, only some of them become ill and have the disease, for example, 50% (that is, 30% of those exposed become ill or develop disease). Even fewer of those with disease will go on to die, perhaps 10% of those who became ill or 5% or 3% of those who initially become infected. The severity of illness can also differ from mild to severe to life-threatening, as shown in Figure 4.4. Therefore, risks can be quantified for each of the key health effects outcomes: infection, illness of varying severity, and death.

## 2.0 EXAMPLES OF RISK ANALYSES AND ASSOCIATED METRICS

Organizations such as the WHO, the U.S. EPA, the International Standards Organization (ISO), and the American Society of Heating, Refrigerating and Air-Conditioning Engineers (ASHRAE) regularly conduct quantitative risk assessments to inform decision making and protect public and occupational health. A quantitative risk assessment analysis typically provides quantitative data on the extent or magnitude of risk from environmental exposure. Such data can then be used to determine what actions, if any, need to be taken to manage the risk. In some cases, a QMRA may find that the risk is very low and considered tolerable (acceptable), in which case no further action needs to be taken. However, the QMRA may show that the risk is sufficiently unacceptable (intolerable), and therefore, actions would be recommended or required to reduce the risk to a tolerable level.

The WHO often recommends the use of the DALY metric for quantifying and managing health risks from pathogen exposures and more generally for all sources of health risk, such as toxic chemical exposures, motor vehicle accidents, gun violence, and sexual violence. It has established $10^{-6}$ DALYs (1 micro-DALY) per person per year (or 1 DALY per 1 million people per year) as the tolerable risk level for each of the many different adverse health effects experienced by people. For pathogen exposures from environmental sources such as drinking water, one approach recommended by the WHO is to first calculate a pathogen-specific health risk, such as illness based on infections per person per year, and then convert this to a DALY value per person per year. From such a QMRA, the magnitude of the risk can be calculated, and the acceptability of the risk can be determined and then judged, based on this standard metric of $10^{-6}$ DALYs per person per year. However, some QMRAs have been done using only pathogen-specific risks of health effects per defined time period, such as per exposure dose, per day, or per year.

Different sources of pathogen exposures can result in different health approaches to risk reduction. Below are different examples of how QMRA and epidemiological analyses led to pathogen risk reduction approaches and metrics for different exposure sources in different settings and contexts related to water and human fecal wastes.

## 2.1 Health Risk-Based Guidance for the Safe Use of Human Fecal Wastes for Agriculture

The WHO used QMRA to develop its health-risk-based guidance for the safe reuse of human fecal wastes for agriculture, with an emphasis on reducing pathogen exposure risks (WHO, 2006, 2016). It is important to note that because untreated human wastewater and other fecal wastes are still widely used for agriculture in many countries worldwide, such guidance was much needed. To achieve the WHO-recommended tolerable health-based target of $10^{-6}$ DALY per person per year for pathogen reductions from wastewater, it was calculated by QMRA that total pathogen reductions of 6 to 7 $\log_{10}$ would be required, based on the use of human rotavirus (a gut virus causing diarrhea) as the worst-case reference pathogen, for crops eaten raw, such as lettuce (an above ground food) and onions (a below ground food). Pathogen control measures were addressed for bacterial, viral, and protozoan pathogen contamination combined, without distinguishing between different types and species of pathogens. Because it is impossible to detect and quantify all pathogens in wastewater, achievement of tolerable risk levels in practice was based on quantifying concentrations of the fecal bacterium *E. coli* found in the gut (and possibly also helminth eggs (eggs of worms infecting the gut) in locations where they are sufficiently present to pose human health risks.

Because of the wide range of options and conditions for human fecal waste use in agriculture, WHO also provided guidance to support the development of country and local level Sanitation Safety Plans that create systems to treat and otherwise manage fecal wastes by reducing pathogen risks to acceptable levels for agricultural use, especially for irrigation of crops consumed by people. In practice, achievement of tolerable risk levels is based on determining concentrations of the fecal bacterium *E. coli* in the treated wastewater (and possibly also helminth eggs in locations where they are sufficiently present to pose human health risks). As an example, the quantified *E. coli* concentrations achieved by waste treatment and other management measures to reduce pathogens in different fecal wastes to tolerable risk levels for large-scale treatments systems are shown in Table 4.1 (WHO, 2006, 2016).

Note that *E. coli* concentrations are different for agricultural use of wastes, depending on two categories of use: restricted irrigation and unrestricted irrigation of crops eaten raw. A lower *E. coli* concentration is

**TABLE 4.1** WHO Guideline Values for Verification Monitoring in Large-Scale Treatment Systems of Graywater, Excreta, and Fecal Sludge for Use in Agriculture (WHO, 2006, 2016)

| Matrix Used in Agriculture | Helminths Eggs (number per gram total solids or per liter) | *E. coli* (number per 100 mL) |
|---|---|---|
| Treated feces and fecal sludge | < 1/g total solids | < 1000 g/total solids |
| Grey water for use in *Restricted irrigation* | < 1/L | $< 10^5$<br>Relaxed to $< 10^6$ when exposure is limited or regrowth is likely |
| *Unrestricted irrigation of crops eaten raw* | < 1/L | $< 10^3$<br>Relaxed to $< 10^4$ for high-growing leaf crops or drip irrigation |

specified for unrestricted use on crops eaten raw, compared to a higher concentration where irrigation is restricted by other management factors, such as a lengthy duration of storage time before crop harvest, to allow for pathogen die-off. Therefore, pathogen risk management can be adapted to site-specific conditions of waste treatment and other management measures to achieve a tolerable level of pathogen risk.

## 2.2 Health Risk-Based Microbial Standards for Treatment of Human Wastes Using Commercial Nonsewered Sanitation Systems On-Site

The QMRA was used to develop performance and testing standards for the microbial quality of the residual solids and liquids and the required $\log_{10}$ reduction values (LRVs) required for commercial nonsewered sanitation systems (NSSS) by the ISO and the American National Standards Institute (ANSI; ISO, 2018). This voluntary system is ISO standard 30500, which was developed using QMRA to specify LRVs and the microbial quality of the resulting fecal wastes residuals, both solids and liquids, for any commercial NSS system that seeks certification (approval). In addition, ISO 30500 requires the use of their specific protocol for independent testing and then certification of the performance of an NSSS. The protocol describes how to obtain the results from a specified series of controlled laboratory tests using fecal waste matrices consisting of human fecal matter and urine containing required concentrations of fecal bacteria, viruses, protozoan parasites, and helminth eggs, and usually achieved by spiking the wastes with these microorganisms, when subjected NSS treatment. It also specifies procedures for additional field testing of the NSS. Performance of the NSSS is based on specified LRVs of bacteria, viruses, protozoan parasites, and helminth eggs or in some cases surrogate fecal indicator organisms and well as the ability to achieve allowable concentrations of these microorganisms in the resulting solid and liquid residuals from treatment. These specifications are shown in Table 4.2 for residual waste solids and in Table 4.3 for residual waste liquids. The QMRA was developed by a team of experts (Jahane et al., 2016; Schoen et al., 2017).

## 2.3 Health Risk Analysis for the Microbial Quality of Drinking Water

The U.S. EPA has defined the acceptable (tolerable) risk of infection resulting from exposure to a specific pathogen in drinking water for a year to be 1 infection per 10 000 people or an individual person's risk of $10^{-4}$ per year. It applied QMRA to determine the risks of human consumption of drinking water to estimate the annual risks from various waterborne pathogens and

**TABLE 4.2** Pathogen (or Indicator) Maximum Concentrations and $Log_{10}$ Reductions of Pathogens in Treated Residual Solids for Commercial On-Site Nonsewered Sanitation Systems (ISO, 2018)

| Parameter (Pathogen Class) | Human Enteric Bacterial Pathogens | Human Enteric Viruses | Human Enteric Helminths | Human Enteric Protozoa |
|---|---|---|---|---|
| Surrogate | *E. coli*[b], measured in CFU or MPN | MS2 Coliphage, measured in PFU | *Ascaris suum* viable ova | viable *Clostridium perfringens* spores, measured in CFU |
| Max. concentration in solids [number/g (dry solids)] | 100 | 10 | < 1 | < 1 |
| Overall LRV for solid[a] | ≥ 6 | ≥ 7 | ≥ 4 | ≥ 6 |

[a]Log-reduction values (LRVs) were derived from a QMRA, as described by WHO (2016), assuming 1 g of fecal solids contains approximately the same range of reference pathogens as in 1 L of liquid effluent (for LRVs derived in Table 4.3).

[b]*E. coli* strain K011 (ATCC 55124) is selected because it is chloramphenicol resistant. Therefore, this antibiotic may be added to the plating medium to suppress the growth of other, interfering bacteria.

**TABLE 4.3** Pathogen (or Indicator) Maximum Concentrations and $Log_{10}$ Reductions of Pathogens in Treated Liquid for Commercial On-Site Nonsewered Sanitation Systems (ISO, 2018)

| Parameter (Pathogen Class) | Human Enteric Bacterial Pathogens | Human Enteric Viruses | Human Enteric Helminths | Human Enteric Protozoa |
|---|---|---|---|---|
| Surrogate | *E. coli*[b], measured in CFU or MPN | MS2 Coliphage, measured in PFU | *Ascaris suum* viable ova | viable *Clostridium perfringens* spores, measured in CFU |
| Max. concentration in liquids [number/L] | 100 | 10 | < 1 | < 1 |
| Overall LRV for liquid[a] | ≥ 6 | ≥ 7 | ≥ 4 | ≥ 6 |

[a]Log-reduction values (LRVs) were derived from a quantitative microbial risk assessment (QMRA) as described by WHO (2016).

[b]*E. coli* strain K011 (ATCC 55124) is selected because it is chloramphenicol resistant. Therefore, this antibiotic may be added to the plating medium to suppress the growth of other, interfering bacteria.

the QMRA procedure, as described initially in a scientific publication (Regli et al., 1991). Summary results for the analysis are shown in Table 4.4 below.

It is apparent from this analysis that the concentrations of the different microorganisms corresponding to a $10^{-4}$ annual risk of infection are quite different by as much as more than 300-fold. Further, the acceptable risk concentrations of these microorganisms are too low to be readily quantified in drinking water samples and would require specialized and cumbersome methods to concentrate them from the water for detection by specialized methods. Therefore, the U.S. EPA developed separate risk-based regulations for both surface and ground water sources of drinking water to address the major pathogens of concern. For the surface water treatment rule, the U.S. EPA identified the protozoan parasite *Cryptosporidium* as a worst case due to its chlorine resistance and small size. The rule requires sanitary surveys to identify vulnerable source waters and it requires treatment by filtration or other processes, such as UV light disinfection to reduce this pathogen by 2 $\log_{10.}$ It also requires protection such as covers on new finished water reservoirs as well as monitoring of finished waters for *E. coli* and coliform bacteria. For the ground water rule, the U.S. EPA also requires sanitary surveys to identify vulnerable sources under the influence of surface water, encourages ground water monitoring for *E. coli* and enterococci bacteria and coliphages as fecal indicator viruses, and requires ground water treatment (usually by disinfection) to achieve 99.99% (4 $\log_{10}$) virus inactivation as well as ongoing compliance monitoring of total coliforms and *E. coli* of finished water.

## 2.4 Epidemiological Health Risk-Based Approach to Address Pathogen Risks in Recreational Waters

The development of recommended levels of tolerable health risk for bathing waters by the U.S. EPA was established from field epidemiological studies

**TABLE 4.4** Acceptable Concentrations of Microorganisms From Dose Response Curves to Achieve Annual Risk of Infection of 1:10 000 per Year From Drinking Water (Regli et al., 1991)

| Microorganism | Acceptable Concentration in 10 000 000 liters[a] |
|---|---|
| Rotavirus | 2.2 |
| Poliovirus 3 | 2.6 |
| *Giardia* | 68 |
| Poliovirus 1 | 150 |
| Echovirus 12 | 680 |

[a]Assumes 2 L per day of consumption. In the case of *Giardia*, an exponential dose-response model was used.

rather than estimated from QMRA analysis. In prospective field studies gastrointestinal illness risks in volunteer bathers (compared to nonbathers as unexposed controls) were enumerated for those who bathed in waters that were concurrently analyzed for their concentrations of fecal indicator bacteria, specifically enterococci in marine waters and *E. coli* in fresh waters. From the results of these studies, two different tolerable levels of gastrointestinal illness risk and the corresponding concentration of fecal indicator bacteria were recommended, as shown in Table 4.5 (U.S. EPA, 2012). The decision on the acceptable level of risk would be up to the authority that was responsible for the beach and its water quality, such as a local, county, or state government.

## 2.5 Categorical Health Risk Analysis in Relation to Bacteriological Quality of Bathing Waters Based on Field Epidemiological Studies

The WHO recommendations for estimated tolerable health risks from bathing in recreational waters were also based on quantifiable concentrations of fecal indicator bacteria (WHO, 2021a). Like those of U.S. EPA, the WHO guidelines for acceptable risk of recreational bathing waters were established based on field epidemiological studies of bathing-associated gastrointestinal illness and the relationship of gastrointestinal illness, and the microbial quality of bathing water as determined for fecal indicator bacteria, specifically fecal enterococci bacteria. However, instead of setting an acceptable level of risk as a single outcome (safe or unsafe), WHO categorized bathing waters into a range of increasing gastrointestinal and respiratory illness risks and corresponding microbial water quality concentrations for fecal enterococci bacteria, using a four-category risk scale (see Figure 4.5). The lowest risk category, designated A, specified < 40 fecal enterococci per 100 mL as the upper 95% confidence level, which corresponded to < 1 case

**TABLE 4.5** U.S. EPA Recommendations for Estimated Tolerable Health Risks From Bathing in Recreational Waters Based on Quantifiable Concentrations of Fecal Indicator Bacteria for Two Different Health Risk Levels (U.S. EPA, 2012)

| Criteria Elements | Recommendation 1 Estimated Illness Rate 36/1000 | | Recommendation 2 Estimated Illness Rate 32/1000 | |
|---|---|---|---|---|
| Indicator | GM (cfu/100 mL) | STV (cfu/100 mL) | GM (cfu/100 mL) | STV (cfu/100 mL) |
| Enterococci (marine & fresh) | 35 | 130 | 30 | 110 |
| *E. coli* (fresh) | 126 | 410 | 100 | 320 |

| Intestinal enterococci (95th percentile value per 100 mL [rounded values]) | Basis of derivation | Estimated risk per exposure |
|---|---|---|
| ≤40<br>A | This range is below the NOAEL in most epidemiological studies.<br>Low risk or low probability of adverse effects. | • <1% GI illness risk.<br>• <0.3% AFRI risk.<br>• The upper 95th percentile value relates to an average probability of less than 1 case of gastroenteritis in every 100 exposures. The AFRI burden would be negligible. |
| 41–200<br>B | The 200/100 mL value is above the threshold of illness transmission reported in most epidemiological studies that have attempted to define a NOAEL or LOAEL for GI illness and AFRI. | • 1–5% GI illness risk.<br>• 0.3–1.9% AFRI risk.<br>• The upper 95th percentile value relates to an average probability of 1 case of gastroenteritis in 20 exposures. The AFRI illness rate at this upper value would be less than 19 per 1000 exposures, or less than approximately 1 in 50 exposures. |
| 201–500<br>C | This range represents a substantial elevation in the probability of all adverse health outcomes for which dose–response data are available. | • 5–10% GI illness risk.<br>• 1.9–3.9% AFRI risk.<br>• This range of 95th percentiles represents a probability of 1 in 10 to 1 in 20 of gastroenteritis for a single exposure. Exposures in this category also suggest a risk of AFRI of 19–39 per 1000 exposures, or approximately 1 in 50 to 1 in 25 exposures. |
| >500<br>D | Above this level, there may be significant risk of high levels of minor illness transmission. | • >10% GI illness risk.<br>• >3.9% AFRI risk.<br>• There is a greater than 10% chance of gastroenteritis per single exposure. The AFRI illness rate at the 95th percentile value of >500/100 mL would be greater than 39 per 1000 exposures, or greater than approximately 1 in 25 exposures. |

A–D: microbial water quality assessment categories (refer to section 4.3) used in the classification procedure; AFRI: acute febrile respiratory illness; GI: gastrointestinal; LOAEL: lowest-observed-adverse-effect level; NOAEL: no-observed-adverse-effect level.

**FIGURE 4.5** Guideline Values for Microbial Quality of Coastal and Freshwater Recreational Waters

of gastrointestinal illness per 100 people and a negligible risk of respiratory illness per bathing exposure event. The highest risk level, designated D, specified > 500 fecal enterococci per 100 mL as the upper 95% confidence level, which corresponded to > 10 cases of gastrointestinal illness per 100 people and > 39 cases of respiratory illness per 1000 people per bathing exposure event. Risk levels of B and C are intermediate between these lowest and highest risk categories.

## 2.6 Assessing and Managing Health Risks from *Legionella* in Water

*Legionella* spp. is a naturally occurring environmental opportunistic pathogen that exists and multiplies in water. The approach to addressing, preventing, and minimizing the risks from *Legionella* spp. bacteria and the disease they cause, called legionellosis, and related respiratory illness due to airborne exposures from ventilation systems such as cooling towers and water systems in buildings and other structures, is through the development and use of comprehensive and site-specific risk management plans, as described by ASHRAE (2018). *Legionella* spp. bacteria are ubiquitous and present in a wide range of environmental media, including water and wastewater, and have caused illnesses and outbreaks from airborne exposures to droplets and aerosols from drinking water, recreational water, and various waters in workplaces, including water resource recovery facilities (WRRFs). Risk management plans in general consist of a series steps, as shown in Figure 4.6.

These steps for *Legionella* spp. management are to fully understand and map water and wastewater systems and their flows; identify potentially hazardous locations and conditions, such as stagnant lines and vessels and dead-ends in them; implement and maintain control measures such as disinfection and control of biofilms and keeping hot water hot (50 to 60 °C) and cold water cold (20 to 25 °C); identify and eliminate or minimize human exposures to areas or sources potentially harboring *Legionella* spp., applying disinfection, biofilm control, and other sanitization measures to reduce *Legionella* spp. presence; and regularly determine bacteriological water quality to assure that concentrations of *Legionella* spp. or an indicator bacteria group, typically heterotrophic plate count bacteria, are within acceptable limits, and invoke appropriate corrective measures if they are not. In workplaces, such as WRRFs, regularly monitor and assess worker health to identify possible cases of legionellosis so that infected workers can be identified and then isolated and possibly exposed workers can be told to quarantine.

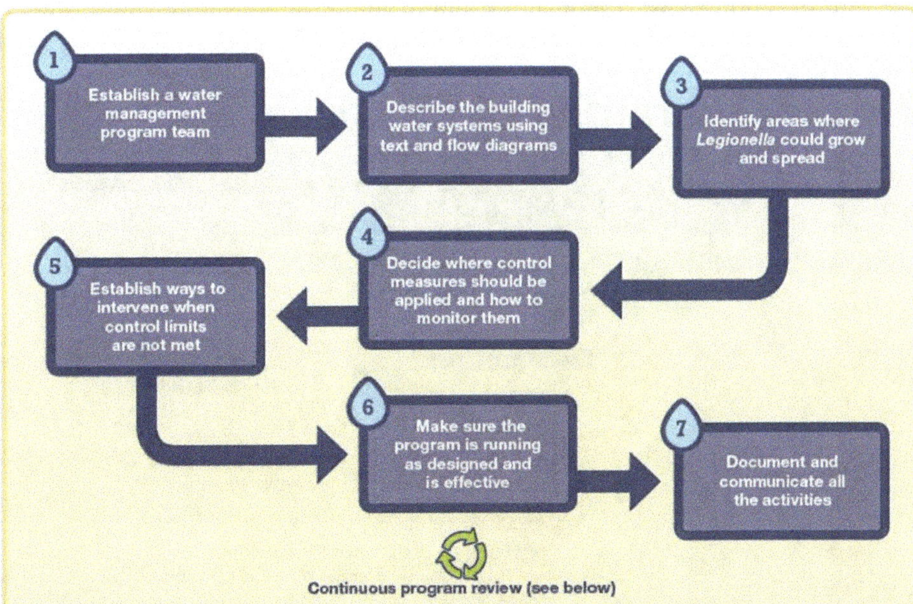


**FIGURE 4.6** Key Steps of Risk Management Plans for *Legionella* spp. That Can Be Adapted to Water and Wastewater Systems (Reprinted with permission from Bill Pearson)

For drinking and recreational waters, the focus of management plans is similar to that described above, but identification of cases of legionellosis is often more challenging unless there are cases of the disease that get recognized by local disease surveillance systems, such as health departments and hospitals.

## 3.0 MANAGING RISK AND PROMOTING PREPAREDNESS

Risk management, especially in occupational settings, requires an integrated approach to (1) prevent workplace injuries and illnesses, (2) improve compliance with laws and regulations, (3) reduce costs, including in workers' compensation premiums, (4) engage workers, (5) enhance worker social responsibility goals, and (6) increase productivity and enhance overall enterprise operations. The core elements of a safety and health program require (1) management leadership, (2) worker participation, (3) hazard identification assessments, (4) hazard prevention and control, (5) education and training, (6) program evaluation and improvement, and (7) communication and coordination involving all workers and stakeholders.

A series of control measures are needed to effectively identify, prevent, and manage risks and hazards, and these can be viewed as an inverted pyramid of control measures, as shown in Figure 4.7.

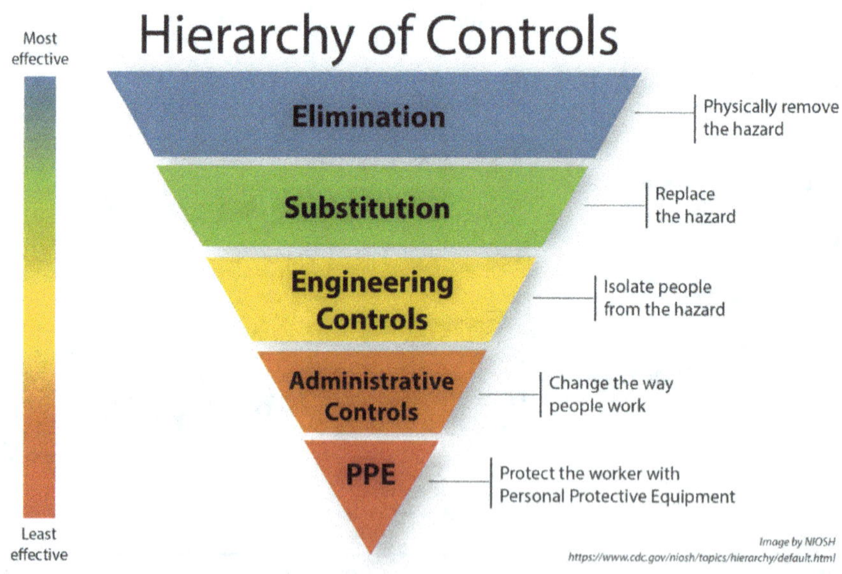


**FIGURE 4.7** Hierarchy of Controls to Identify, Manage, and Reduce Health Risks (NIOSH, 2023)

The health and safety management plans summarized in Figure 4.7 are considered the most effective way to achieve health and safety goals, with a focus on understanding, identifying, assessing, and managing risks from pathogens, as reviewed in this chapter. Of the various control measures available, elimination of all pathogen hazards is impossible because of the multiple and pervasive sources of pathogens in the environment and in people. Likewise, substitution as a control measure is also unrealistic because pathogens are ubiquitous hazards in the environment and in humans, and so substitution is possible. However, engineering controls and PPE are possible and potentially very effective control measures and tools against pathogens and will be addressed later in this chapter.

Administrative controls are also effective measures against pathogens because they involve altering the way work is done and what workers need to do or not do to prevent or minimize pathogen exposures. Administrative control measures include (1) timing and location of work to minimize pathogen exposures; (2) policies and other rules intended to reduce pathogen presence in and risks to workers, such as requirements for vaccinations, isolation of pathogen-infected workers and quarantining of workers who may have been exposed to pathogens (not allowed to work for a specified time period, usually several days to weeks), and regular physical exams and related diagnostic medical tests; and (3) work practices such as standards and operating procedures. Standard operating procedures include training, housekeeping, equipment maintenance, and personal hygiene practices, such as handwashing with soap and bathing. The U.S. Centers for Disease Control and Prevention (CDC) and National Institute for Occupational Safety and Health (NIOSH) have specific guidelines to address risks to wastewater workers potentially exposed to class B biosolids (CDC & NIOSH, 2002). These guidelines can be applied more broadly to all pathogen risks from wastewaters and biosolids.

## 3.1 Identifying, Isolating, and Treating Infections to Prevent and Control Them

Infections caused by pathogens are common in human and animal populations, and all people experience infections on a frequent basis. People can become exposed and infected with pathogens almost anywhere, but especially in their communities and at work.

The period of time or duration of an infection depends on the specific causative pathogen as well as the characteristics of the human host. In some cases, infections may last just a few days or perhaps weeks and then the host recovers from the infection, usually as a result of an immune response, after which the pathogen is eliminated from the host and the host is considered "cured." However, some pathogens cause infections that may continue for months to years or even throughout the rest of the person's life. Infections that continue for a long period of time are considered persistent and those that persist for a very long time or for the rest of the host's life are considered chronic. It is important to know that during infection of a host, the pathogen is in the body, usually increasing in numbers over time, and is being shed from the body in various releases such as fecal matter and urine in the case of intestinal pathogens or saliva, sputum and sometimes eye discharges or tears in the case of respiratory tract pathogens. Some pathogens also get into and even multiply in the bloodstream, resulting in their presence in blood or blood products. Other pathogens cause sexually transmitted infections and diseases and are present in tissues and related bodily discharges, such as semen and menstrual discharges.

Because pathogens are often present in the various bodily discharges and releases from infected hosts, these released microbes can be transmitted to other people, either directly by personal contact (e.g., handshakes, kissing, sexual contact), by exposure to airborne releases of pathogen-containing particles in coughs and sneezes of infected people, or by indirect exposures, such as touching pathogen-contaminated surfaces and objects, called fomites (e.g., door handles or knobs, cell phones, or discarded paper tissues or children's toys) and ingestion of contaminated water or food.

As indicated in Section 1.2 of this chapter, infections may be asymptomatic or inapparent, with no discernable adverse health effects. Other infections may lead to an illness or disease that is experienced by the host as signs and symptoms of adverse health effects in the form of an illness or disease, which can also be observed by others. As a worst case, the illness or disease may result in death. Some illnesses caused by pathogens infecting the intestinal tract (the enteric tract or gut) produce tissue damage and other health effects that result in signs and symptoms common among many different enteric pathogens. These health effects include nausea, abdominal pain and cramps, vomiting, diarrhea (with or without intestinal bleeding), dehydration, and sometimes fever. This enteric illness is often called gastroenteritis, and when it also includes fever as well as blood and mucus in excreted fecal matter, it is called dysentery. Other gastrointestinal diseases caused by specific types of pathogens progress to more severe disease by further pathogen spread and tissue damage to other parts of the body. Examples are spread to the liver, causing infectious hepatitis, or to the central nervous system and related muscles, causing muscle weakness and paralysis; meningitis; severe headache; eye muscle and vision difficulties; swallowing, speaking, or chewing difficulties; and tingling and pricking sensation in the extremities.

Other pathogens, including some that are present in wastewater and biosolids, typically cause respiratory illnesses, such as the common cold—in which the pathogen infects the upper respiratory tract, including the nose, throat, sinuses, and sometimes the eyes—or more severe illnesses such as pneumonia, an illness of the lower respiratory tract, including the lungs and their terminal air sacs (the alveoli). The pneumonia diseases of the lower respiratory tract cause more severe signs and symptoms of illness than those of the upper respiratory tract. Pneumonias can progress to increased tissue damage and even bleeding in the lungs and airways, which is called pulmonary hemorrhage, and hemoptysis (coughing up blood). These severe respiratory disease conditions interfere with breathing, usually require respiratory assistance (mechanical ventilation), and can be life-threatening.

Severe infectious diseases often require medical attention by physicians or other trained health care professionals and may require hospitalization to treat and manage and hopefully cure a patient's disease. Such diseased patients are often kept isolated from other people to prevent the ill person from coming in contact with other people and possibly exposing those other people who may then also become infected. Health care workers of such patients must take precautions to prevent their own exposures to the pathogens being shed as bodily releases from such patients, and the environment of such patients must also be controlled to prevent further direct or indirect spread of the pathogens infecting and being released from these patients. These preventive measures are referred to as *infection prevention and control* (IPC), and they include a range of measures to contain, block and kill the causative pathogens. Health care workers for such patients wear PPE—such as face masks, goggles, gloves, and gowns—and take other measures to minimize their own exposures while also containing and confining and killing pathogens (such as by the use of disinfectants and sanitizers). The patient environment, usually a room or ward, must also be safely configured and managed to prevent pathogen contamination and further release or spread. Such measures include strict control of patient movement, limiting patient access by other people, maintaining effective room ventilation, use of physical barriers to prevent or minimize pathogen release, periodic disinfection and sanitization of the patient environment, and safe disposal or decontamination for reuse of PPE. The pathogen-contaminated wastes from such pathogen-infected patients must also be safely contained, managed, and disposed of by methods that prevent release, escape, and possible spread of the pathogens in such wastes. These measures are further described below.

Some people who become infected with a pathogen may not get treated and cared for in a health care setting but instead get treated and cared for at home. This situation often occurs in low resource settings among people who are unable to get into health care facilities for treatment and care. The burden of home treatment and care for such people may fall on community healthcare workers who can go to the home of ill patients to care for them, or such care may need to be provided by family members or others in the community, or both. Regardless, it is important and necessary that, to the extent possible, the IPC measures used in health care settings be provided for those who are treated and cared for at home. In such situations the physical environment, equipment, and materials available in health care facility (HCFs) need to be provided to the extent possible in the home setting. This may require confining the ill person to a designated room in the home, restricting access to them by others, except for designated care givers, and reconfiguring the home environment to prevent the spread of a pathogen from an ill person to others and minimize the spread of the pathogen from the ill person to the home environment by the same or alternative containment and decontamination methods as used in health care settings, as described above.

To prevent further pathogen spread from an infected and ill person to others, those who may have been exposed to such infected person should be placed in quarantine (be quarantined). Such an exposed person should remain isolated, usually in their home or in another designated location, and kept apart from other people for a sufficient period of time to determine if they themselves become infected with the pathogen and possibly become ill. The duration of such quarantine periods is related to the incubation period, which is the duration of time between host exposure to the pathogen and the development of illness (symptomatic disease), but it is usually 7 to 14 days. However, it may be even longer for pathogens that have a longer incubation period and are slow to produce discernable health effects. Quarantining is a potentially effective preventive measure to control further pathogen spread to others, but it requires timely and efficient identification of those who were exposed (a contact person) to an infected person (a case person). Such contact tracing and resulting quarantining is critical to preventing further spread of pathogens in communities.

The goal of all of these infection prevention and control measures is to identify, diagnose, treat, and cure infected people and prevent the spread of pathogens from them to other people. The steps to achieve this are often referred to as understanding and then breaking (Figure 4.8) the chain of infection. The six links of the chain are (1) the infectious agent, (2) reservoir, (3) portal of exit, (4) mode of transmission, (5) portal of entry, and (6) susceptible host. The way to stop pathogens from spreading is by interrupting this chain at any link, by specific prevention and control measures that can be applied at each link, as shown in Figure 4.8.

## 3.2 Vaccinations and Vaccines Against Known Pathogens

Vaccines are widely used for disease prevention against known pathogens in the Americas and other regions of the world. Some of the pathogens

FIGURE 4.8 Breaking the Chain of Infection (Reprinted with permission from the Association for Professionals in Infection Control and Epidemiology)

causing respiratory and enteric disease, especially those previously causing serious disease and death in infants and young children, are now effectively controlled in the Americas by vaccines that have been developed to provide immunity (immune protection) against infection and disease. Vaccines against pathogens are a preferred prevention and control method.

Many of these vaccines became available in the 20th century, and as new pathogens emerged and were identified and characterized, new vaccines were created against them. The most recent examples are the vaccines against the newly emerged virus, SARS-CoV-2, which caused the COVID-19 pandemic. Listed below in Table 4.6 are the other diseases for which vaccines have been developed as preventive measures. These vaccines have dramatically reduced the disease burdens and deaths caused by these pathogens and in a few cases have eradicated or nearly eliminated the diseases in the Americas.

In addition to the vaccines and pathogens listed in Table 4.6, vaccines for other pathogens are also available but are not widely used in North America. These are vaccines for cholera, a serious enteric disease caused by the bacterium *Vibrio cholerae*, and typhoid fever, a serious enteric disease cause by the bacterium *Salmonella typhi*. There are both live oral vaccines and inactivated vaccines against both of these pathogens that are recommended for travelers who may encounter and be exposed to these pathogens in other parts of the world or may be occupationally to them in North America, such as the bacterium *Salmonella typi*. Another virus that has recently spread globally is monkeypox virus (now called "mpox" as per the WHO), with mpox cases occurring mostly men who have sex with men or their close contacts. Vaccines are available to prevent it. Other diseases for which vaccines are often used in Europe but not North America are smallpox, yellow fever, typhoid fever, Japanese encephalitis, tickborne encephalitis, and malaria (Pavlia & Maltezoub, 2022).

## 3.3 Disinfecting Hospital Waste and Similar Wastes Present in Communities

There are many different kinds of hospital and other health care facility wastes as well as wastes in communities and homes that that may require disinfection. Some of these wastes used widely in HCFs are disposable items, such as (1) PPE, including gloves, face masks, face shields, booties, and gowns; (2) patient wastes such as PPE, dressings, bandages and wrappings, tissues, needles and syringes, intravenous (IV) materials such as bags, tubing, and valves, respiratory therapy materials such as nasal prongs and tubing, nebulizers, PAP (positive airway pressure) devices and accessories, manual resuscitators, and spirometers; (3) bedpans and other patient waste-collection containers; and (4) sponges, wipes, and floor and wall cleaning items. Nearly all of these disposables are solid waste. They are typically collected in biohazard bags and then transported by housekeeping staff for decontamination and then disposal. Similarly, in other community settings and homes, these wastes also must be collected and safely managed by facility workers or homeowners.

Other hospital, community, and household wastes, including some of the items listed above as disposables, are potentially reusable, especially in

**TABLE 4.6** Vaccine-Preventable Diseases, Causative Pathogens, Illnesses, and Exposure Sources

| Disease/ Vaccine | Causative Pathogen | Sites/Kinds of Infection | Exposure/ Sources | Comments |
|---|---|---|---|---|
| Polio/Polio vaccine | Polioviruses (3 types) | Enteric, respiratory, systemic; neurological, and paralytic disease | Fecal-oral and respiratory | Causes the paralytic disease, poliomyelitis; rare but recently found again in the United States |
| Tetanus | *Clostridium tetani*, a bacterium | Local to systemic; produces a poison (toxin) causing painful muscle contractions (e.g., "lockjaw") | Environmental media; via cuts, scrapes, and punctures | Causative bacterium is widely present in the natural environment, including soil, sediment, and metal objects |
| The Flu (Influenza) | Influenza virus types A, B, and C | Respiratory tract/respiratory infections; pneumonia | Airborne respiratory particles, fomites, and direct contact | Viruses change over time and require new or modified vaccines annually. |
| Hepatitis B | Hepatitis B virus | Hepatitis (liver disease); blood-borne; systemic | Contaminated blood and bodily fluids, via sex and contaminated needles | Common in injectable drug users, paid sex workers, male homosexuals, and accidental blood exposures |
| Hepatitis A | Hepatitis A virus | Hepatitis (liver disease); bloodborne; systemic | Ingestion of fecal matter or via contaminated blood | In fecal wastes, wastewater, biosolids, and contaminated foods where sanitation is poor |
| Rubella | Rubella virus (Rubevirus) | Upper respiratory illness followed by skin rash; congenital rubella syndrome (CRS) in pregnancy | Respiratory exposure to secretions and airborne particles; direct contact | CRS in pregnant women can cause miscarriages, stillbirths, and birth defects in offspring. |
| Hib (*Haemophilus influenzae* type B) | *Haemophilus influenzae* type B, a bacterium | Meningitis (inflammation of tissue covering brain and spinal cord); pneumonia; skin infection and arthritis | Respiratory exposure to airborne particles; prolonged direct contact | Serious disease of children |
| Measles | Measles virus, a morbillivirus | Respiratory disease with fever, cough, and skin rash; pneumonia; encephalitis | Respiratory exposure to secretions airborne particles; direct contact | Serious disease of children; highly contagious |

(*continued*)

**TABLE 4.6** Vaccine-Preventable Diseases, Causative Pathogens, Illnesses, and Exposure Sources (*Continued*)

| Disease/ Vaccine | Causative Pathogen | Sites/Kinds of Infection | Exposure/ Sources | Comments |
|---|---|---|---|---|
| Whooping cough (pertussis) | *Bordetella pertussis*, a bacterium | Respiratory disease with distinctive cough (whoop) | Respiratory exposure to secretions, airborne particles, and direct contact | Serious disease of children; very contagious |
| Pneumococcal disease | *Streptococcus pneumoniae*, a bacterium | Respiratory disease that can invade to cause pneumonia, meningitis, and ear and sinus infections | Respiratory exposure to secretions, airborne particles, and direct contact | Serious disease of children and the elderly |
| Rotavirus | Rotavirus types A, B, and C (Reoviridae family); type A is most serious | Enteric disease, especially of children; diarrhea, vomiting, dehydration | Fecal oral-transmission from exposure to human feces; poor sanitation | Serious disease of children |
| Mumps | Mumps virus, a paramyxovirus | Respiratory disease of the salivary glands and ears, with fever | Respiratory exposure to secretions, airborne particles, and direct contact | Serious disease of children that can spread to other sites in the body |
| Chickenpox | varicella-zoster virus (VZV) | Respiratory disease with rash | Respiratory exposure to secretions, airborne particles, and direct contact | Serious disease of children; highly contagious; can reemerge as a rash in the elderly, called shingles |
| Diphtheria | *Corynebacterium diphtheriae*; produces a toxin | Respiratory disease with breathing difficulty due to film formed in back of throat and sinuses | Respiratory exposure to secretions, airborne particles, and direct contact with skin lesions | Serious upper respiratory disease; can lead to paralysis, heart failure, and death |
| Cervical and other genital and throat cancers | Human papilloma virus (HPV) | Cervical, other genital, and throat cancers | Sexual contact and transmission | HPV vaccines are recommended for pre-teens. |

low-resource settings and where materials shortages occur due to supply chain interruption or other access limitations. Such reusable items may be decontaminated by autoclaving if they can tolerate the high temperature and pressure used or they may be cold sterilized or sanitized using physical and/or chemical disinfectants. For additional information on disinfecting hospital waste, see Chapter 3.

A wide range of chemical disinfectants are available to kill microbes in wastes and more generally in communities and our homes. These disinfectants have been identified and reviewed by experts in infection protection and control and have been described in a number of publications. A simple and practical publication that is focused on the virus causing COVID-19 is from the CDC, but its recommendations are broadly applicable (CDC, 2021a). Another publication that is more extensive, comprehensive, and technical focuses on hospitals and healthcare settings, but is a good resource (Rutala et al., 2019). However, only some of the disinfection options described in this resource may be applicable to wastewater reclamation facilities, due to the need for specialized equipment and related facilities that are not usually available and because of the need for specialized training in the use these options. The widely used chemical disinfectants in hospitals and other health care facilities are (1) ethylene oxide gas, (2) hydrogen peroxide plasma (or vaporized hydrogen peroxide), (3) free chlorine/bleach, (4) quaternary ammonium compounds, (5) peracetic acid (and hydrogen peroxide/peracetic acid combinations), (6) iodophors (iodine plus a surfactant), (7) phenolic compounds, and (8) aldehydes such as glutaraldehyde and ortho-phthalaldehyde and alcohols (usually ethanol and sometimes isopropanol). A physical disinfectant besides high temperature such as autoclaving is ultraviolet (UV) radiation, which is used to decontaminate air as well as objects and surfaces). Only some of these disinfectants are widely used in other applications and settings, such as wastewater treatment facilities, public and private institutions, such as schools and colleges, retail commercial establishments and religious buildings such as churches, synagogues, and mosques.

In the United States, disinfectants against various pathogens are regulated by the U.S. EPA, and those that are approved based on independent performance testing are published as a list for each type of pathogen or pathogen group (U.S. EPA, 2021). The U.S. EPA (2021) provides a list of antimicrobial products registered with EPA for claims against common pathogens. It is important to note that because pathogens differ in their susceptibility or resistance to disinfectants, and because chemical disinfectants themselves differ in efficacy against these pathogens, the U.S. EPA has at least 15 such pathogen lists with applicable disinfectants, designated by letters from A through Q:

- List A: EPA's Registered Antimicrobial Products as Sterilizers
- List B: Tuberculocide Products Effective Against *Mycobacterium tuberculosis*
- List C: Products Effective Against Human HIV-1 Virus

- List D: Products Effective Against Human HIV-1 and Hepatitis B Virus
- List E: EPA's Products Effective Against *Mycobacterium tuberculosis* Human HIV-1 and Hepatitis B Virus
- List F: Products Effective Against Hepatitis C Virus
- List G: Products Effective Against Norovirus
- List H: EPA's Products Effective Against Methicillin Resistant *Staphylococcus aureus* (MRSA) and/or Vancomycin Resistant *Enterococcus faecalis* or *faecium* (VRE)
- List J: Products for Medical Waste Treatment
- List K: Products Effective Against *Clostridium difficile* Spores
- List L: Products That Meet the CDC Criteria for Use Against the Ebola Virus
- List M: Products with Label Claims for Avian (Bird) Flu Disinfectants
- List N: Disinfectants for Use Against SARS-CoV-2
- List O: Disinfectants for Use Against Rabbit Hemorrhagic Disease Virus (RHDV2)
- List P: for Claims Against *Candida auris* and Tuberculocidal Products Effective Against *Mycobacterium tuberculosis*
- List Q: Disinfectants for Emerging Pathogens

Because human wastewater can potentially contain almost any or all of the pathogens for which there are EPA disinfectant lists, it is recommended that WRRFs use disinfectants that are effective against the most resistant pathogens, such as *Clostridium difficile* spores, mycotic agents such as *Candida auris*, and tuberculocidal products effective against *Mycobacterium tuberculosis*. Free chlorine is a widely used wastewater disinfectant and is capable of killing all of the EPA pathogens of concern. However, some of the pathogens can be killed only at very high concentrations of free chlorine, specifically 10% HOCl. Because HOCl is highly corrosive, there may be locations and use applications in WRRFs where an EPA-approved alternative disinfectant may have to be used instead.

## 3.4 Disinfecting Surfaces and Tools

Surface disinfection is a critical element of IPC in hospitals and other healthcare settings, as well as in both public and private community settings, including WRRFs. Some of the disinfectants used in hospital and other health care settings are used routinely to disinfect various surfaces, including tables and counters, patient bedrails, doors and door handles (bars) and knobs, floors, walls, shelves, cabinets, sinks, toilets, light switches, lamps, and other objects. The most widely used disinfectants are free chlorine (bleach), alcohol (usually ethanol but also isopropanol), hydrogen peroxide liquid, quaternary ammonium compounds, peracetic acid and peracetic acid/hydrogen peroxide combinations, iodophors, and aldehydes such as glutaraldehyde. Only some of these disinfectants may be suitable for use at wastewater or water reclamation facilities. Ultraviolet radiation is also used to disinfect surfaces, but precautions for its use are required to prevent undue exposures of people. Most surface disinfectants are chemicals that are applied to surfaces by spraying and then wiping. If surfaces have excessive organic load, such as fecal matter, blood, biosolids, or other solids, such surfaces should be cleaned first with a cleaning agent such as a detergent with or without added disinfectant, to first remove or reduce the solids. The cleaned surface is then treated with a disinfectant. In WRRFs, disinfection of surfaces is often done using spray equipment or hoses to deliver the disinfectant. In some applications this may be accompanied by physical wiping or the use of a squeegee to move and then remove the disinfectant or detergent and disinfectant solution, often into a drain.

## 3.5 Hygiene Practices, Such as Handwashing at Critical Times

### *3.5.1 Handwashing at Critical Times*

Hygiene practices are critical for preventing exposures to pathogens. Among the various hygiene practices, handwashing is considered the most important. This is because our hands often acquire our own pathogens, such as by coughing or sneezing into our hands, rubbing our eyes or nose, or by coming in contact with fecal contamination after a bowel movement, especially from anal wiping with toilet paper. Our hands also acquire pathogens from others, such as by shaking hands with another person, by coming in contact with (touching) a surface or other fomite (object) that contains pathogens, or by coming in contact with sewage or biosolids when working at a WRRF. Other critical times for handwashing include the following:

- before and after caring for someone at home who is sick with vomiting or diarrhea or with a respiratory pathogen;
- before and after treating a cut or other wound;
- after changing diapers or cleaning up a child who has used the toilet;
- after touching an animal, animal feed, or animal waste;
- after handling pet food or pet treats; and
- after touching garbage or solid waste of unknown origin and content.

Handwashing with soap needs to be thorough and requires sufficient time to achieve effective removal of pathogens from hands and fingernails. The key steps for handwashing are wet hands with clean, running water (warm or cold), apply soap, and lather hands by rubbing them together with the soap, including the backs of the hands, between fingers, and under fingernails. Scrub hands for at least 20 seconds. To achieve this time, hum the "Happy Birthday" song from beginning to end twice. Then, rinse hands well under clean, running water, and dry them using a clean towel or air dry them.

If water and soap are not available for handwashing, hands can be cleaned to reduce many pathogens by using an alcohol-based hand sanitizer containing at least 60% alcohol. Although hand sanitizers can reduce pathogens on hands, they are less effective because (a) they do not kill all pathogens, such as many viruses and protozoan parasites; (b) their effectiveness is reduced when hands are visibly dirty or greasy; and (c) they may not remove many harmful chemicals like pesticides and heavy metals.

### 3.5.2 Other Hygiene Measures

Other hygiene measures are also important in reducing pathogen exposures, including the following:

- If a person is ill, it is recommended that they stay home to prevent the spread of their pathogen to other people.
- Bathing is an important personal hygiene measure. If possible, bathe daily and after the end of a shift if working with wastewater.
- When one coughs, sneezes, or spits, a tissue should be used to collect what is released; the used tissue should be safely disposed of immediately. Disposable or single tissues are preferred to reusable items like handkerchiefs or towels. If tissues are not available, turn your head to the side and cough, sneeze, or spit directly into your upper arm.
- Do not share cups, glasses, dishes, cutlery, cigarettes, cosmetics/makeup, or cellphones. Child toys may become microbially contaminated and require cleaning and disinfection, especially if handled or used by a child with an infection or illness.
- Food hygiene is also critical to reduce pathogen exposure risks. Reducing risks can be achieved by safe storage of food items, washing foods such as fresh fruits and vegetables, and preparing potentially microbially contaminated foods, such as poultry and meats, in separate

locations and using separate cutting boards, utensils, and other items that come in contact with such foods. After preparing such high-risk foods, all food-preparation items need to be disinfected or sanitized, including sinks. Such foods also need to be thoroughly cooked to reach recommended cooking temperatures. Foods need to be stored at recommended temperatures and used within a short time period to avoid spoilage caused by microbes. Such spoilage due to microbe proliferation or release of metabolic products may also be toxic, due to the production of microbial toxins, as well as potentially infectious due to excessive pathogen growth that can cause infection, disease, and death.

## 3.6 Preventing or Reducing Exposure to Pathogens

In addition to PPE, other protective measures against pathogen exposure include engineering (e.g., covering aeration tanks, having good ventilation in settings where aerosols are generated) and administrative controls.

### *3.6.1 Personal Protective Equipment*

Personal protective equipment is any protective clothing, head coverings, gloves, face shields, goggles, facemasks, and/or respirator or other equipment intended to protect the wearer from exposure to pathogens or other physical and chemical agents of health concern. It also physically reduces the release of pathogens from infected persons. Personal protective equipment provides critical barriers to reduce the release of pathogens by infected people (a pathogen source person) as well as serve as protective barriers to reduce pathogen exposure of others who may get exposed to pathogens released or shed from an infected or ill person (a susceptible recipient).

Personal protective equipment is typically used in WRRFs, hospitals, health care settings and laboratories, as well as in other settings where exposure to pathogens and harmful chemicals and physical agents is a risk and concern. Such locations include occupational settings like WRRFs, food production and processing facilities, and animal handling and care facilities. Personal protective equipment can reduce infection and illness risks and is in widespread use.

When used properly, PPE acts as a physical barrier between pathogens such as viruses, bacteria, and parasites and the exterior parts of the body, including skin, mouth, nose, or eyes (mucous membranes), where pathogen entry and infection begins. These barriers have the potential to block transmission of pathogens in body fluids and other excretions such as feces, as well as respiratory secretions and the skin. Personal protective equipment can protect people who are at risk of being exposed and becoming infected through medical procedures, when visiting patients in health care facilities or who get exposed in communities and public and private places, including at work (occupationally, such as at WRRFs). When used properly with other IPC practices such as handwashing, using alcohol-based hand sanitizers, covering coughs and sneezes, and maintaining physical distance from pathogen-infected sources, PPE minimizes the spread of pathogens from one person (or other source) to another.

Effective use of a mask or respirator as PPE includes proper fit testing to ensure it makes a tight seal against the face, with no gaps at any the edges, training on proper use and maintenance, avoiding touching the respirator or mask surface during use, and properly removing and disposing (don and doff) of contaminated PPE to prevent exposing both the wearer and other people to pathogens that can cause infection. If the respirator or mask is to be reused, it must be properly decontaminated by physical or chemical methods that do not compromise its fit or performance.

Personal protective equipment, and in particular masks and respirators, is addressed by international health agencies such as the WHO; other international organizations such as the ISO; regional agencies such as the European Union; and national government agencies, such as the CDC, the U.S. Food and Drug Administration (FDA), Health Canada, the Public Health Agency of Canada (PHAC), including agencies that address occupational settings such as the U.S. Occupational Safety and Health Administrations (OSHA) and NIOSH. These agencies provide detailed guidance, policies, practices, and in some cases specific regulations on PPE design and especially mask and respirator design and composition, performance properties, and information on correct use and maintenance. In addition, some professional societies and occupational organizations also provide similar guidance, advice and in some cases regulations, including PPE performance testing and certification. Examples of entities that provide mask, respirator and other PPE design, composition, performance requirements, testing protocols, and certification are OSHA, ASTM International (formerly the American Society for Testing and Materials), and the ISO/ANSI.

In the United States, all PPE to be used as a medical device must follow FDA's regulations and meet applicable voluntary consensus standards for protection. These requirements apply to surgical masks, N95 respirators, medical gloves, and gowns. Some PPE is reviewed by the FDA before its legal sale in the United States. Manufacturers have to show that their PPE meets specific performance criteria, labeling, and intended use to demonstrate substantial equivalence. This may be demonstrated by conforming to consensus standards for barrier performance and resistance to tears and snags. Some PPE may need to demonstrate sterility, biocompatibility, fluid resistance, and flammability. Manufacturers must validate the methods used to test conformance to standards and provide appropriate performance test data for the PPE item.

Other critical PPE are respirators that must be used in confined or enclosed spaces where air quality is unsafe or sufficient breathable air is unavailable. Such spaces with unbreathable air include, for example, burning buildings; many sewers, wastewater tanks, and other confined vessels; and some cooling tower interiors. For such conditions, breathable air must be provided by a self-contained breathing apparatus, which is used as PPE. These devices are PPE that provide their own supply of safe air for breathing and are worn by the user. Further details of specific job activities and recommended PPE for them are found in Chapter 6.

### *3.6.2 Engineering Controls of Pathogens*

Engineering controls include process controls for wastewater and biosolids treatment processes, enclosure and isolation of process equipment and use (treatment) measures, and provision of effective ventilation. Process controls in a WRRF include measures to reduce or eliminate pathogen releases from sources, such as covering aeration and other physical mixing tanks or other vessels that harbor pathogens; physically enclosing and isolating processes that contain pathogens such as biosolids separation and enrichment processes using belt screens, strainers, or centrifuges; providing effective ventilation to prevent increased concentrations of airborne pathogens as droplets and aerosols from wastewaters and biosolids; and reduction of pathogen concentrations on surfaces by frequent or continuous cleaning and disinfection where appropriate.

Provision of effective ventilation is necessary for rooms, other enclosed or confined spaces, and cooling towers. Effective ventilation is essential for microbiological and chemical control in buildings and other enclosed spaces, such as aircraft, trains, buses, and other enclosed vehicles and spaces, such as sewers, wastewater tanks, and other enclosed vessels or tanks.

Ventilation measures include the following key elements to reduce airborne pathogens and other contaminants:

1. Supply clean air from an external source directionally to susceptible occupants in a room or other space in which there are pathogens and other contaminant sources, going from clean to dirty zones.
2. Contain/confine contaminated air and/or exhaust it to the outdoors.
3. Dilute the air in a space with cleaner air from outdoors and/or by filtering and/or disinfecting the air.
4. Clean the air within the room or other confined space by filtering and/or disinfecting it (such as with UV radiation) (ASHAE, 2022; CDC, 2021b; OSHA, n.d., WHO, 2021b).

The effectiveness of these ventilation measures is influenced by the rate and direction of clean air delivery, the air change rate per hour in the room or other confined space to reduce the airborne concentrations of pathogens and other contaminants to acceptably low levels, the type and efficacy of air cleaning technology—such as filtration or UV radiation—and achieving and maintaining a specified air flowrate. Where possible, engineered mechanical ventilation and other air-cleaning technologies are used. These approaches include dilution ventilation, laminar and other in-room air-flow regimes, differential room pressurization (positive pressure or negative pressure), personalized ventilation (for sources and/or susceptible), source capture ventilation, filtration (central or unitary), and upper-room UV germicidal irradiation (upper room, in-room, and/or in the airstream). All of these measures are options used in health care facilities, and many of them are now increasingly becoming widely used for other buildings and confined spaces as resources permit. In low-resource settings, natural ventilation is recommended and widely used, especially opening windows and doors to provide cross-ventilation with or without the use of mechanical fans. Further details of ventilation measures, such as air and aerosol monitoring equipment, are found in Chapter 6.

A number of health, safety, engineering, and sanitation resources provide guidance, advice, and (in some cases) requirements for minimum rates of airflow, air change rates, and other ventilation parameters for various rooms and other confined spaces. These metrics differ among settings, such as hospitals, health care facilities, other public and private buildings and their rooms, and other confined spaces. Water resource recovery facilites have their own recommended metrics for these parameters, which are subject to regulation and management by local jurisdictions as well as those at higher administrative levels to which they must answer, such as states, territories, and national government agencies. It is beyond the scope of this chapter to provide these specifications and requirements in detail. However, some of them may be found in Chapter 6.

## 3.7 Awareness, Education, and Communication

Understanding and responding to risks from pathogens that are present in wastewater or could be, as well as those pathogens that are present in or could enter communities is essential, for WRRFs, their management utilities, and the regulatory bodies that set policies and procedures for them. One of the main challenges for the water and wastewater professional community is how pathogen health risks are estimated and communicated. Quantitative risk assessments for pathogens are usually based on the use of dose-response relationships to estimate health risks. As shown graphically in Figure 4.3, the greater the pathogen exposure dose, the greater the health risk, which typically follows a backward *S*-shaped curve.

The efforts to prevent and control pathogen risks in WRRFs requires knowledge, the ability to recognize and respond, preparedness, surveillance, and rapid response. In today's world, a pathogen that arises in human populations far from North America can reach the region and local communities in hours to days via air travel, other modes of transportation, the migration of people and animals, and the transboundary travel of environmental media such as air and water.

The most striking recent example is SARS-CoV-2, the virus that causes COVID-19, which emerged in a province in China in late 2019 and rapidly spread globally via infected people in a matter of weeks to cause the largest pandemic in modern history. Fortunately, this virus is not a significant source of health risk to humans from wastewater or water exposures (Sobsey, 2021). However, the genetic material (RNA) of this virus is readily detectable by molecular methods in the wastewater of any community or specific location (e.g., school or nursing home) in which there are people infected with this virus. The emergence of SARS-CoV-2 has escalated global efforts to improve surveillance for the presence and emergence of pathogens in efforts to detect them rapidly and inform communities and the world of their emergence and presence in locations around the world. The detection of the genetic material of SAR-CoV-2 and other pathogens in wastewater has become increasingly important and more widely used to detect pathogens in communities and their specific settings, such as hospitals, schools and colleges, nursing homes, and private enterprises.

The WHO as well as regional, national, and local governments and their online networks and communication channels attempt to provide timely information about pathogens of health concern, the extent to which they are present or emerging and where, and how we can collectively respond rapidly to the health-risk threats they pose. These responses and actions include public education, awareness, preparedness, rapid response measures, and communication. News media, the internet, social media outlets (e.g., TikTok, Facebook, Twitter, Instagram, LinkedIn), and other communication outlets attempt to keep us informed about these pathogens, their risks, and the various measures we can take as individuals, communities, and enterprises, including wastewater reclamation utilities and facilities, to prevent their entry or reduce the risks from them if they are already present.

Over the decades, similar efforts have provided timely information and mobilized efforts to give guidance, education, and training on recognizing and responding to other pathogens, such as the bacterium *Vibrio cholerae*, the cause of cholera, and viruses such as influenza, the cause of the flu, and poliovirus, the cause of poliomyelitis. It is through such local and global communication resources to detect and respond to these and other pathogens that ongoing efforts are made to recognize, assess, prevent, and reduce the human health risks from all pathogens.

## 4.0 CASE STUDY: USING QMRA TO ASSESS THE RISK OF EBOLA TRANSMISSION TO WASTEWATER WORKERS

An example of the uncertainty and panic and the potential role of risk assessment during infectious disease outbreaks is what happened in 2013 to 2016 when West Africa experienced the biggest outbreak of the Ebola virus ever known (Maal-Bared et al., 2020). Controlling the outbreak in western Africa proved challenging for a variety of social, climactic, and public-health-related factors that were compounded by the lack of access to safe water and sanitation (Kamorudeen et al., 2020). In 2015, a WHO survey of more than 100 public and private health care centers in three Ebola-affected countries in Africa showed that more than half of the surveyed facilities lacked access to clean water or electricity and that more than half of facilities did not have a functioning incinerator to handle medical waste. Waste management was found to be lacking in most centers (Vetter et al., 2016). By October 2015, the WHO reported 28,457 official cases. Almost all were reported in Sierra Leone, Liberia, and Guinea, with about 11,312 deaths, which many experts believe is an underestimate.

When individuals returning from western Africa to the United States and Spain transmitted the infection to three nurses caring for them (two nurses from Dallas and one nurse in Madrid) due to breaches in infection-control practices, many local utilities found themselves at an impasse: Should they accept wastewater from hospitals where suspected cases have been admitted if the hospital does not treat its wastewater? Is this a risk to our wastewater treatment and collection system workers?

The CDC suggested that no additional treatment of Ebola patient waste was necessary before it was introduced into the collection system. The WHO issued two conflicting statements: In August 2014, it suggested patient waste could safely be discharged into the sanitary sewer without additional disinfection. In December 2014, the WHO issued more conservative guidance recommending pretreatment of waste. The CDC provided additional guidance on the management of Ebola-associated waste from hospital facilities a few years later (CDC, 2019a). Throughout the wastewater sector, there were reports of scared workers who did not want to handle wastewater until their safety was guaranteed. There also were reports of tension between utilities and hospitals admitting suspected Ebola cases, with utilities planning to shut down collection systems until hospitals treated their effluents. A Water Environment Federation/Water Environment & Reuse Foundation Ebola webinar on this topic was among the most attended WEF webinars of all time, signaling the thirst of the industry for clear direction and evidence to guide practice (Maal-Bared et al., 2020). Although the CDC, the EPA, and the WHO provide valuable guidance and advisories during outbreaks, reaching consensus among such large organizations can be very time consuming and political, as was seen during the 2013 to 2016 Ebola outbreak. Thus, water professionals play a critical role in providing the water sector with needed guidance during emergencies.

Individuals with Ebola virus disease (EVD) develop nonspecific febrile illnesses, which manifest as fever, hemorrhagic fever, malaise, headaches, and myalgia (Legrand et al., 2007). Ebola virus disease has a high case fatality rate of 65.4% and varies by strain (Lefebvre et al., 2014). The causative agent of EVD is pathogenic strains of Ebola virus including Zaire, Sudan, Taï Forest, Reston, and Bundibugyo. Ebola virus disease was potentially transmitted from animal hosts (e.g., old-world fruit bat species, duikers; Alexander et al., 2015; Leroy et al., 2005). The disease first emerged in 1976, and since then, there have been several outbreaks in humans and animals. The outbreaks in humans were limited to countries in Africa, although there were confirmed traveler cases on other continents as well (Formenty, 2014). In the meantime, limitations in public health resources in certain areas of Africa and the unavailability of vaccines are major drivers of continued circulation of the virus in Africa and transmission risks to the rest of the world. Ebola virus disease person-to-person transmission mainly results from direct contact with body fluids (e.g., blood, saliva, breast milk, urine, semen, stool, and tears) from symptomatic infected individuals. Studies have also demonstrated potential transmission through dermal mechanisms, in which infected body fluids were introduced through cuts on skin (Judson et al., 2015) and through contact with the ocular and conjunctival mucosa (i.e., part of eyes) of nonhuman primates (Jaax et al., 1996; Mire et al., 2016). Evidence to support aerosol or fomite transmission is contradictory (Johnson et al., 1995; Mire et al., 2016; Osterholm et al., 2015).

Considering its high case fatality rate and its potential for disease outbreaks and zoonotic transmission, potential transmission of EVD from wastewater has received significant attention from the public and the academic community (Bibby et al., 2015, 2017; Chattopadhyay & Taft, 2018; Lin & Marr, 2017; Zabinski et al., 2018). The detection of Ebola virus genetic materials in stool or an anal swab of an infected individual was reported with range of $10^4$ to $10^{6.5}$ genomic copies/mL (Schibler et al., 2015), although there were no reports of positive viral infectivity. Bibby et al. (2017) reported that the time needed for 90% inactivation of Ebola virus was 2.1 days. The ZEBOV and REBOV strains were found persistent through the observation time period of 90 minutes in aerosols of tissue culture media, and decay rates were estimated at 4.29% per minute and 27.7 % per minute, respectively (Piercy et al., 2010). In a laboratory-scale aeration basin and sewer models, which were filled with mixed liquor of wastewater spiked with Ebola virus surrogates, 0.3 to 79 plaque-forming units (PFU) $min^{-1}$ (from $10^5$ to $10^7$ aerosols) were generated (Lin & Marr, 2017). Despite the evidence suggesting the persistence of the hazard, to date there has been no evidence of individuals being infected with Ebola virus through exposure to water or wastewater, and current wastewater worker protective practices are expected to be effective against Ebola and other waterborne pathogens.

In 2016, a QMRA evaluated the risk of Ebola virus transmission to wastewater workers (Bibby et al., 2016). As described in this chapter, the QMRA consisted of an exposure assessment, dose-response assessment, risk characterization, and sensitivity analysis. The study by Bibby et al. (2016) was based on the scenario that a wastewater worker operating a wastewater collection system interceptor receiving untreated wastewater effluent from a hospital treating an Ebola virus patient inhaled generated bioaerosols during a shift in the presence and absence of PPE. In the study, the exposure assessment, which was used to determine the number of Ebola virus inhaled by a wastewater worker per exposure, was performed using data inputs from the literature with appropriate data distribution models. The data inputs included patient excreta volume and virus concentration, hospital daily outflow, interceptor daily flow, Ebola virus die-off rate in wastewater, transportation time from shedding point to the worker, aerosolization partitioning, respiration fraction of aerosol, inhalation rate, removal fraction by PPE, and period of exposure to bioaerosols. The authors pointed out that there was a 95% possibility that the worker inhaled fewer than $9.34 \times 10^{-8}$ RNA copies per exposure with PPE and $3.25 \times 10^{-6}$ RNA copies per exposure without PPE. The authors adopted the dose-response fitting from another study (Mitchell et al., 2020), which assumed that inhalation of approximately 9 PFUs resulted in 50% likelihood of EVD. The risk characterization showed the risk median of $10^{-9.34}$ for transmission from infectious virus particle to a worker who did not put on PPE and the reduced risk median to $10^{-11}$ for a worker who put on PPE. The authors highlighted the importance Ebola-associated waste management and use of PPE, which corresponds with Zabinski et al. (2018).

Zabinski et al. (2018) used a Bayesian Belief Network Model (a form of graphical modeling) to predict the risk of EVD transmission risks among wastewater workers. The study included consideration of transmission through ingestion and a scenario, in which the Ebola-associated wastes were treated prior to disposal. The output showed that a worker, who accidentally ingested several drops of wastewater, would bear a median risk of $5.8 \times 10^{-4}$, which could be reduced to $3.8 \times 10^{-7}$ when the waste was properly treated prior to disposal. In Western Africa, the disinfection and inactivation of hospital wastes through autoclaving, on-site incineration, and chlorine pretreatment was recommended and practiced as a precautionary measure (Mallow et al., 2018). So, although there is often a misconception that effective infection control and worker protection requires a vast amount of knowledge of the system involved and the pathogen causing the outbreak, conservative and precautionary procedures could be established for many situations in collaboration with public health agencies.

# 5.0 RESOURCES

## 5.1 Publicly Available

The Canadian Centre for Occupational Health and Safety. (2022). *Hazards and risk*. https://www.ccohs.ca/oshanswers/hsprograms/hazard_risk.html

Disability-adjusted life year (DALYs). (2023, January 29). In *Wikipedia*. https://en.wikipedia.org/wiki/Disability-adjusted_life_year

Food and Agriculture Organization. (2004). *The basics of risk assessment*. https://www.fao.org/3/y4722e/y4722e05.htm

Global Burden of Disease Collaborative Network. (2021). *Global burden of disease study 2019 (GBD 2019) results*. Institute for Health Metrics and Evaluation (IHME). http://ghdx.healthdata.org/gbd-results-tool

Health Canada. (2018). *Guidance on the use of quantitative microbial risk assessment in drinking water*. https://www.canada.ca/en/health-canada/programs/consultations-guidance-quantitative-microbial-risk-assessment-drinking-water/document.html

## 5.2 Require Access and/or Cited in Text

Alexander, K. A., Sanderson, C. E., Marathe, M., Lewis, B. L., Rivers, C. M., Shaman, J., Drake, J. M., Lofgren, E., Dato, V. M., & Eisenberg, M. C. (2015). What factors might have led to the emergence of Ebola in West Africa? *PLOS Neglected Tropical Diseases, 9*(6), e0003652.

ASHRAE. (2018). *Legionellosis: Risk management for building water systems*. ANSI/ASHRAE Standard 188-2018.

ASHRAE. (2022). *Ventilation and acceptable indoor air quality*. ANSI/ASHRAE Standard 62.1-2022.

Bibby, K., Aquino de Carvalho, N., & Wigginton, K. (2017). Research needs for wastewater handling in virus outbreak response. *Environmental Science and Technology, 51.*

Bibby, K., Casson, L., Haas, C. N., & Rycroft, T. (2016). *Risks from Ebola discharge from hospitals to sewer workers.* IWA Publishing.

Bibby, K., Fischer, R. J., Casson, L. W., Stachler, E., Haas, C. N., & Munster, V. J. (2015). Persistence of Ebola virus in sterilized wastewater. *Environmental Science & Technology Letters, 2*(9), 245–249.

Centers for Disease Control and Prevention. (n.d.). *Ebola (Ebola virus disease) 2014-2016 Ebola outbreak in West Africa.*

Centers for Disease Control and Prevention. (2015). *Ebola report: Protecting borders.*

Centers for Disease Control and Prevention. (2019a). *Ebola (Ebola virus disease) waste mangement.*

Centers for Disease Control and Prevention. (2019b). *Managing solid waste contaminated with a category A infectious substance.*

Centers for Disease Control and Prevention. (2021a, April). CDC's Cleaning and Disinfecting Guidance (2021), downloaded from https://www.cdc.gov/coronavirus/2019 facility.html ncov/community/disinfecting on April 27, 2021.building

Centers for Disease Control and Prevention. (2021b, June 2). *Ventilation in buildings.* https://www.cdc.gov/coronavirus/2019-ncov/community/ventilation.html

Centers for Disease Control and Prevention & NIOSH. (2002). *Guidance for controlling potential risks to workers exposed to class B biosolids.* https://www.cdc.gov/niosh/docs/2002-149/pdfs/2002-149.pdf

Chattopadhyay, S., & Taft, S. (2018). *Exposure pathways to high-consequence pathogens in wastewater collection and treatment systems.* U.S. Environmental Protection Agency.

Fischhoff, B., Watson, S. R., & Hope, C. (1984). Defining risk. *Policy Sciences, 17*(2), 123–139.

Formenty, P. (2014). Chapter 9. Ebola virus disease. In: *Emerging infectious diseases* Clinical Case Studies (pp. 121–134). Academic Press.

ISO. (2018). *Non-sewered sanitation systems—Prefabricated integrated treatment units—General safety and performance requirements for design and testing.* ISO 30500:2018. https://www.iso.org/standard/72523.html

Jaax, N. K., Davis, K. J., Geisbert, T. J., Vogel, P., Jaax, G. P., Topper, M., & Jahrling, P. B. (1996). Lethal experimental infection of rhesus monkeys with Ebola-Zaire (Mayinga) virus by the oral and conjunctival route of exposure. *Archives of Pathology & Laboratory Medicine, 120*(2), 140–155.

Jahne, M. A., Schoen, M. E., Garland, J. L., & Ashbolt, N. J. (2017). Simulation of enteric pathogen concentrations in locally-collected greywater and wastewater for microbial risk assessments. *Microbial Risk Analysis*, 5 (2017), 44–52. http://www.sciencedirect.com/science/article/pii/S235235221630038X

Johnson, E., Jaax, N., White, J., & Jahrling, P. (1995). Lethal experimental infections of rhesus monkeys by aerosolized Ebola virus. *International Journal of Experimental Pathology, 76*(4), 227.

Judson, S., Prescott, J., & Munster, V. (2015). Understanding Ebola virus transmission. *Viruses, 7*(2), 511–521.

Kamorudeen, R. T., Adedokun, K. A., & Olarinmoye, A. O. (2020). Ebola outbreak in West Africa, 2014–2016: Epidemic timeline, differential diagnoses, determining factors, and lessons for future response. *Journal of Infection and Public Health, 13*(7), 956–962.

Lefebvre, A., Fiet, C., Belpois-Duchamp, C., Tiv, M., Astruc, K., & Glélé, L. A. (2014). Case fatality rates of Ebola virus diseases: A meta-analysis of World Health Organization data. *Médecine et Maladies Infectieuses, 44*(9), 412–416.

Legrand, J., Grais, R. F., Boelle, P.-Y., Valleron, A.-J., & Flahault, A. (2007). Understanding the dynamics of Ebola epidemics. *Epidemiology & Infection, 135*(4), 610–621.

Leroy, E. M., Kumulungui, B., Pourrut, X., Rouquet, P., Hassanin, A., Yaba, P., Délicat, A., Paweska, J. T., Gonzalez, J.-P., & Swanepoel, R. (2005). Fruit bats as reservoirs of Ebola virus. *Nature, 438*(7068), 575–576.

Lin, K., & Marr, L. C. (2017). Aerosolization of Ebola virus surrogates in wastewater systems. *Environmental Science & Technology, 51*(5), 2669–2675.

Maal-Bared, R., Bastian, B., Bibby, K., Brisolara, K., Gary, L., Reimers, R. S., Schaeffer, S., & Swift, J. (2020). Lessons from the Ebola outbreak: How the water sector can transition from panic to preparedness. *Water Environment & Technology January, 2020*: 38–45.

Mallow, M., Gary, L., Jeng, T., Bongomin Jr., B., Aschkenasy, M. T., Wallis, P., Cranmer, H. H., Debasu, E., & Levine, A. C. (2018). WASH activities at two Ebola treatment units in Sierra Leone. *Plos One, 13*(5), e0198235.

Mire, C. E., Geisbert, J. B., Agans, K. N., Deer, D. J., Fenton, K. A., & Geisbert, T. W. (2016). Oral and conjunctival exposure of nonhuman primates to low doses of Ebola Makona virus. *The Journal of Infectious Diseases, 214*(suppl. 3), S263–S267.

Mitchell, J., Dean, K., & Haas, C. (2020). Ebola virus dose response model for aerosolized exposures: Insights from primate data. *Risk Analysis, 40*(11), 2390–2398.

National Institute for Occupational Safety and Health. (2019). *Preventing occupational exposure to Legionella*. DHHS (NIOSH) Publication No. 2019-131. https://doi.org/10.26616/NIOSHPub2019131

National Institute for Occupational Safety and Health. (2023, January 17). *Hierarchy of Controls*. Centers for Disease Control and Prevention. https://www.cdc.gov/niosh/topics/hierarchy/default.html

Osterholm, M. T., Moore, K. A., Kelley, N. S., Brosseau, L. M., Wong, G., Murphy, F. A., Peters, C. J., LeDuc, J. W., Russell, P. K., & Van Herp, M. (2015). Transmission of Ebola viruses: What we know and what we do not know. *mBio, 6*(2), e00137–00115.

Pavlia, A., & Maltezoub, H. C. (2022) Travel vaccines throughout history. *Travel Medicine and Infectious Disease, 46*, 102278. doi:10.1016/j.tmaid.2022.102278.

Piercy, T., Smither, S., Steward, J., Eastaugh, L., & Lever, M. (2010). The survival of filoviruses in liquids, on solid substrates and in a dynamic aerosol. *Journal of Applied Microbiology, 109*(5), 1531–1539.

Pigott, D. M., Golding, N., Mylne, A., Huang, Z., Henry, A. J., Weiss, D. J., Brady, O. J., Kraemer, M. U., Smith, D. L., & Moyes, C. L. (2014). Mapping the zoonotic niche of Ebola virus disease in Africa. *Elife, 3*, e04395.

Regli, S., Rose, J. B., Haas, C. N., & Gerba, C. P. (1991). Modeling the risk from giardia and viruses in drinking water. *Journal of the American Water Works Association, 83*(1), 76–84. https://doi.org/10.1002/j.1551-8833.1991.tb07252.x

Rutala, W. A., Weber, D. J., & the Healthcare Infection Control Practices Advisory Committee (HICPAC). (2019). Guideline for Disinfection and Sterilization in Healthcare Facilities, 2008. Update: May 2019. Accessible version: https://www.cdc.gov/infectioncontrol/guidelines/disinfection/

Schibler, M., Vetter, P., Cherpillod, P., Petty, T. J., Cordey, S., Vieille, G., Yerly, S., Siegrist, C.-A., Samii, K., & Dayer, J.-A. (2015). Clinical features and viral kinetics in a rapidly cured patient with Ebola virus disease: A case report. *The Lancet Infectious Diseases, 15*(9), 1034–1040.

Schoen, M. E., Ashbolt, N. J., Jahne, M. A., & Garland, J. (2017). Risk-based enteric pathogen reduction targets for non-potable and direct potable use of roof runoff, stormwater, and greywater. *Microbial Risk Analysis, 5*, 32–43. https://www.sciencedirect.com/science/article/abs/pii/S2352352216300408?via%3Dihub

Sobsey, M. D. (2021). Absence of virological and epidemiological evidence that SARS-CoV-2 poses COVID-19 risks from environmental fecal waste, wastewater and water exposures. *Journal of Water Health*, jwh2021182. https://doi.org/10.2166/wh.2021.182.

U.S. Environmental Protection Agency. (2012). *Fact sheet. 2012 recreational water quality criteria*. Office of Water EPA—820-F-12-061 4305T D. https://www.epa.gov/sites/default/files/2015-10/documents/rec-factsheet-2012.pdf

U.S. Environmental Protection Agency. (2021). *US EPA—selected EPA-registered disinfectants*. https://www.epa.gov/pesticide-registration/selected-epa-registered-disinfectants#pathogens

U.S. Occupational Safety and Health Administration. (n.d.). *Ventilation*. U.S. Department of Labor. https://www.osha.gov/ventilation

U.S. Occupational Safety and Health Administrations. (2011). *Indoor air quality in commercial and institutional buildings*. OSHA 3430-04. https://www.osha.gov/sites/default/files/publications/3430indoor-air-quality-sm.pdf

Vetter, P., Dayer, J.-A., Schibler, M., Allegranzi, B., Brown, D., Calmy, A., Christie, D., Eremin, S., Hagon, O., & Henderson, D. (2016). *The 2014–2015 Ebola outbreak in West Africa: Hands on*. Springer.

Whelan, G., Wolfe, K., Parmar, R., Galvin, M., Molina, M., Zepp, R., Kim, K., & Duda, P. (2017). *Quantitative microbial risk assessment tutorial - Primer*. U.S. EPA Office of Research and Development, Washington, DC, EPA/600/B-17/323, 2017. https://cfpub.epa.gov/si/si_public_record_report.cfm?dirEntryId=337764&Lab=NERL

World Health Organization. (2006). *Guidelines for the safe use of wastewater, excreta and greywater. Volume 1: Policy and regulatory aspects*. http://apps.who.int/iris/bitstream/handle/10665/78265/9241546824_eng.pdf?sequence=1

World Health Organization. (2016). *Sanitation safety planning. Manual for safe use and disposal of wastewater, excreta and greywater*. https://apps.who.int/iris/bitstream/handle/10665/171753/9789241549240_eng.pdf

World Health Organization. (2021a). *Guidelines on recreational water quality: Volume 1, coastal and fresh waters*. https://www.who.int/publications/i/item/9789240031302

World Health Organization. (2021b). *Roadmap to improve and ensure good indoor ventilation in the context of COVID-19*. https://www.who.int/publications/i/item/9789240021280

Zabinski, J. W., Pieper, K. J., & Gibson, J. M. (2018). A bayesian belief network model assessing the risk to wastewater workers of contracting ebola virus disease during an outbreak. *Risk Analysis, 38*(2), 376–391.

# Chapter 5

# Utility Management and Operation: Learning From Others

*Aditya Ramamurthy, Nija Ali, Adam Baron, Stephanie Corso, Shea Dunifon, Maureen Holman, Alexandra Lampson, Dusti Lowndes, Scott Perry, & Nelson Sims*

## 1.0 UNDERSTANDING THE MACROECONOMY

Economists use many types of data to measure the performance of an economy at any given time. Three macroeconomic variables are especially important: (1) the gross domestic product measures the total income of everyone in the economy, (2) the inflation rate represents how fast the prices are rising, and (3) unemployment rate is the fraction of the labor force that is out of work. Understanding how these variables are determined, why they change over time, and how they interact and affect each other is core to macroeconomics.

As an unintended consequence of COVID-19, there has been a considerable inflationary effect to the global economy. Furthermore, the economic outlook for 2023 around the world, particularly within the United States, indicates the prospect of a mild recession as higher interest rates slow the demand for goods, services, and slow the workforce to produce/deliver them. Figure 5.1 shows factors that affect business performance and the indirect consequences of COVID-19 on macroeconomic variables. Chapter 1 has covered some of the direct and indirect costs of COVID-19 and other outbreaks on the economy, communities, and water resource recovery facilities (WRRFs).

Although each market sector will be affected differently by these macroeconomic factors, the water sector is seeing considerable increases in prices (of chemicals, construction materials, parts, etc.) due to inflation and high interest rates, as well as a shrunken workforce. However, given that public infrastructure management needs are constant, and the services provided by the water sector critical to public and environmental health, the water sector workforce is relatively resilient to both positive and negative macroeconomic effects.

The federal government passed the Infrastructure Investment and Jobs Act (IIJA) in November 2021, which has allocated approximately $55 billion for investment in rehabilitating and updating the nation's water infrastructure (e.g., drinking water, wastewater, stormwater projects). From a macroeconomic standpoint, as of 2023, the water sector will be faced with high prices (more expensive to buy materials as well as building projects) and a limited workforce recruitment pool with a low unemployment rate, resulting in increased workloads for existing staff. With the availability of IIJA funding, the water sector will have an increased number of rehabilitation projects to complete with a minimal available resource pool. The water sector should therefore carefully explore strategies to plan and prioritize their respective infrastructure spending based on risk as well as resource availability.

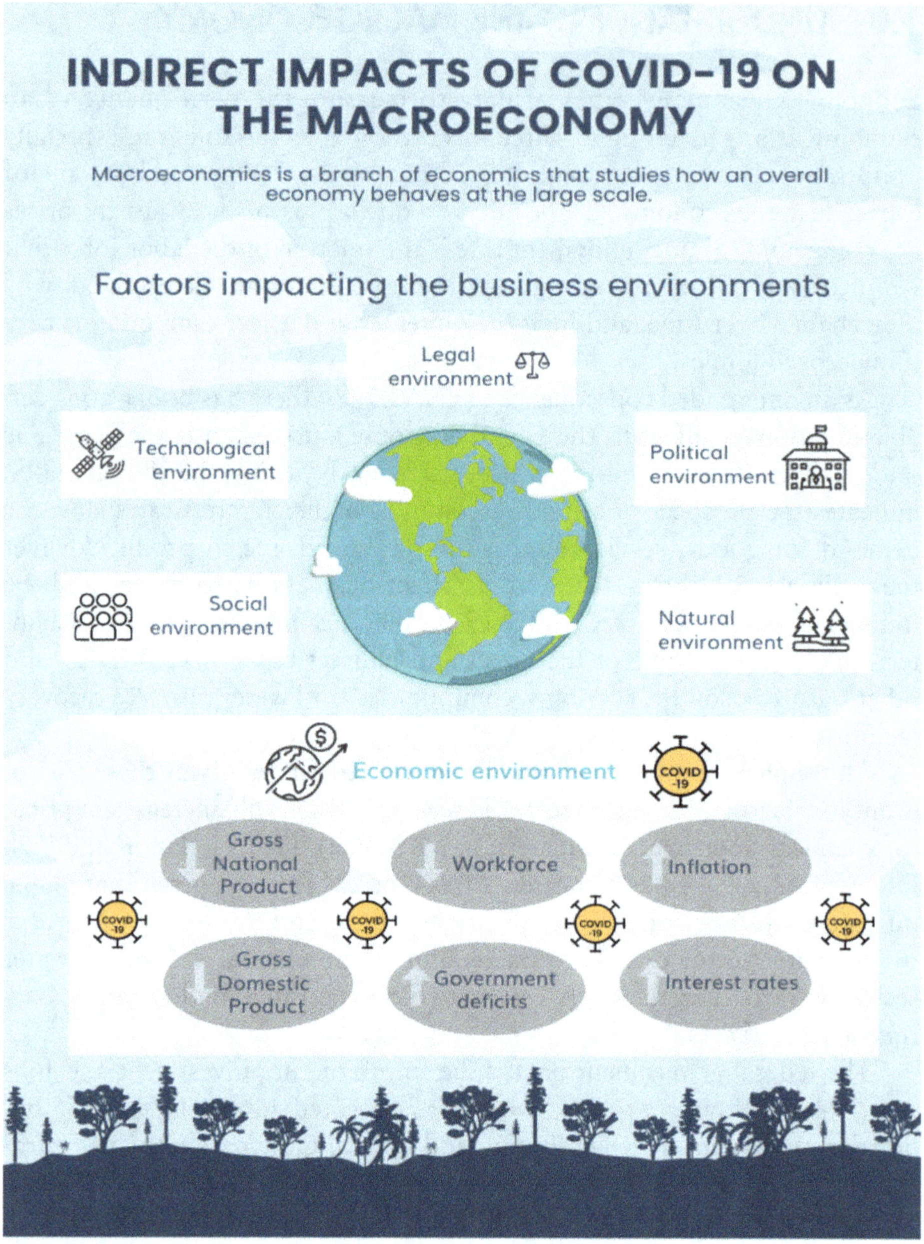


**FIGURE 5.1** Factors That Affect Business Performance and the Indirect Consequences of COVID-19 on Macroeconomic Variables

## 2.0 THE CIRCULAR ECONOMY

Given the current strain on the environment, embracing a systems approach, investing in integration, and adopting circular economy thinking models are necessities rather than choices. The circular economy concept decouples economic growth from the consumption of finite resources. It encourages policy makers and thought leaders to identify ways in which the water, materials, and energy pathways are interconnected. In 2016, the International Water Association (IWA) stated that the water sector was possibly the largest untapped sector in contributing to the circular economy. In the IWA framework *Water Utility Pathways in a Circular Economy*, transitionary pathways for utilities were described that positioned them as engines for sustainable growth. Enabling a circular economy is a key to enabling a sustainable future and will help all of us in the water sector tackle and mitigate issues, such as climate change, sea level rise, loss of biodiversity, energy independence, and waste management in a simple, practical, and impactful manner.

The Water Environment Federation (WEF) is a United Nations partner and supporter of sustainable development goals. In addition, the circular water economy features highly in the WEF 2022 new strategic plan. Water utilities and WRRFs have been leaders in the environmental movement for decades through intrinsic environmental stewardship and extending programs beyond basic compliance, whether through water reclamation, becoming net energy neutral, or producing sustainable and closed-loop fertilizers through reclaimed biosolids. Incorporating the environmental, social, and governance (ESG) mindset to all water sector projects and activities will enable the sector to incorporate a circular economy into the fabric of what we do.

## 3.0 OPERATIONAL EXCELLENCE AND PERFORMANCE

Whether the water sector is exploring asset management options, adopting the ESG mindset, or responding to water sector customers, operational excellence and performance optimization is at the core of everything we do as an industry. Transparency, integrity, safety, respect, initiative, and accountability enable operational excellence and are incorporated into sustainable, effective, and consistent operation and communications to all stakeholders to build trust (see Figure 5.2).

The water sector has established standard key performance indicators and associated benchmarks that each water sector agency can use and adapt to manage, track, and report on progress and performance. With the advent of technology and business intelligence tools, water sector agencies are now able to create on-demand, self-service solutions that can track and report on performance effectively. Several sector specific organizations such as the WEF, the National Association of Clean Water Agencies (NACWA), the Association of Metropolitan Water Agencies, the American Waterworks

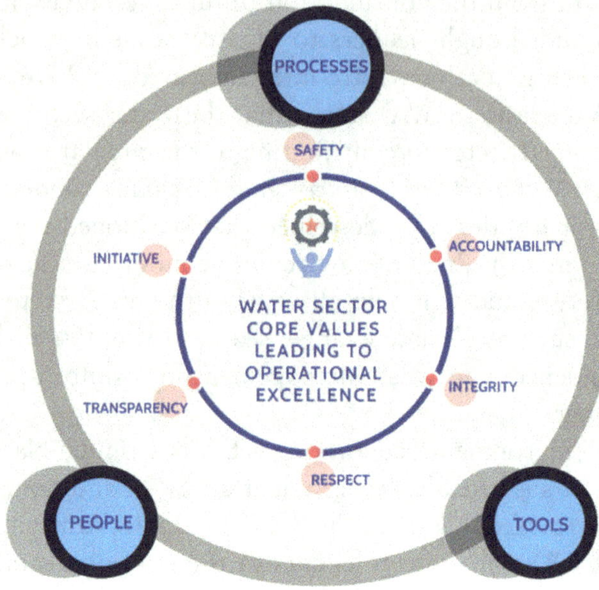


**FIGURE 5.2** Core Values Leading to Operational Excellence and Optimized Performance in the Water Sector

Association (AWWA), and the U.S. Water Alliance are emphasizing the need for performance excellence through Utility of the Future awards and recognitions.

## 4.0 BUSINESS CONTINUITY: THE DC WATER CASE

On March 11, 2020, a State of Emergency was declared in the District of Columbia for COVID-19. Every District agency made changes to their operations to reduce the spread of the virus by modifying services, shifting some or all staff to telework, and making modifications to spaces by adding physical barriers, increasing ventilation, and other changes. District of Columbia's Water and Sewer Authority (DC Water) issued its own emergency declaration on March 18, 2020, which allowed DC Water flexibility in procurement requirements and labor relations while providing initial guidance on telework and the use of administrative leave for staff with COVID-19 illness or exposure. As the District dealt with waves of COVID-19 variants, including the issuance of a new Public Health Emergency Declaration in December 2021, DC Water continued to align with Washington,

DC, officials in terms of our emergency posture and ensuring access to water during the pandemic.

For the protection of customers and employees, DC Water suspended water and sewer service calls that required employees to enter customers' homes, except for emergencies. Nonemergency work was suspended, including scheduled appointments to investigate internal leaks, meter replacements, or new meter sets.

DC Water was slightly below targeted levels of service but, importantly, continued to meet all regulatory requirements, delivering clean drinking water, treating wastewater, and managing the combined sewer overflow system. In some cases, DC Water added services, such as restoring water connections. DC Water finalized a detailed document based on their Continuity of Operations Plan (COOP) and Pandemic Response Plan to align ReOpen DC gating criteria and metrics with DC Water Threat/Service Levels, as can be seen in Figure 5.3.

## 4.1 Financial Measures

DC Water Management adopted a "3 P's" principle of financial guidelines in response to the pandemic: preserve cash, provide essential services to customers, and prioritize employees' safety by providing additional Personal Protective Equipment (PPE), shift and deployment changes, and remote telework access.

Throughout 2020 and 2021, DC Water continued to meet or exceed all critical service levels even though revenue was greatly affected by COVID-19, especially in commercial accounts due to loss of tourism and business events. Immediate actions taken in response to COVID-19 budget impacts included the following:

- delaying noncritical purchases and pausing some hiring;
- deferring proactive inspections, repairs, and maintenance on equipment;
- reducing hours for professional services (e.g., field controllers, calibration and instrumentation, low voltage, automotive maintenance, internal audit, etc.);
- reducing various services based on lower-than-anticipated costs for litigation, claims, biosolids hauling, and software maintenance, with risks being monitored;
- reducing the purchase of office supplies, custodial supplies, promotional items, and specialized equipment;

**RECOVERY CROSSWALK – ReOpen DC and DC Water Threat/Service Levels - June 2020**

| ReOpen DC Stages | Phase/Stage 1/0 | Phase/Stage 2 | Phase/Stage 3 | Phase/Stage 4 |
|---|---|---|---|---|
| ReOpen DC Gating Criteria / Metrics | - Daily Case Rate (>15)<br>- Rate of Transmission (>1.2)<br>- Test Positivity Rate (>10)<br>- New Cases from quarantined contacts (<5)<br>-Percent hospital utilization (90%)<br>Percent COVID-19 patients (10%)<br>Mean test turnaround time (>3 days)<br>- Positive cases with contact attempt (<80%)<br>- Close contacts with contact attempt (<80%)<br>- Positive cases interviewed (<70%) | - Daily Case Rate (5-15)<br>- Rate of Transmission (.8-1.2)<br>- Test Positivity Rate (3%-10%)<br>- New Cases from quarantined contacts (5%-60%)<br>-Percent hospital utilization (80%-90%)<br>Percent COVID-19 patients (5%-10%)<br>Mean test turnaround time (2-3 days)<br>- Positive cases with contact attempt (80%-90%)<br>- Close contacts with contact attempt (80%-90%)<br>- Positive cases interviewed (70%-80%) | - Daily Case Rate (<5)<br>- Rate of Transmission (N/A)<br>- Test Positivity Rate (<3%)<br>- New Cases from quarantined contacts (>60%)<br>-Percent hospital utilization (<80%)<br>Percent COVID-19 patients (<5%)<br>Mean test turnaround time (<2 days)<br>- Positive cases with contact attempt (>90%)<br>- Close contacts with contact attempt (>90%)<br>- Positive cases interviewed (>80%) | District has not released metrics for phase 4 |
| ReOpen DC Universal Safeguards | **Social Distance and Practice Good Hygiene**<br>- Key low-risk activities with strong safeguards<br>- No more than 10 people gathered<br>- 6' of separation whenever possible<br>- Use of face covering required in public space, around groups of people<br>- Exercise precautions when operating near vulnerable populations<br>- Regularly scheduled cleaning of workplaces<br>- Wash hands for at least 20 secs, use hand sanitizer when handwashing isn't possible<br>- Stay home when you are sick | **Social Distance and Practice Good Hygiene**<br>- Additional activities with strong safeguards<br>- No more than 50 people gathered<br>- 6' of separation whenever possible<br>- Use of face covering required in public space, around groups of people<br>- Exercise precautions when operating near vulnerable populations<br>- Regularly scheduled cleaning of workplaces<br>- Wash hands for at least 20 secs, use hand sanitizer when handwashing isn't possible<br>- Stay home when you are sick | **Social Distance and Practice Good Hygiene**<br>- Higher risk activities with safeguards<br>- 6' of separation whenever possible<br>- No more than 250 people gathered<br>- Use of face covering in public spaces<br>- Exercise precautions when operating near vulnerable populations<br>- Regularly scheduled cleaning of workplaces<br>- Wash hands for at least 20 secs, use hand sanitizer when handwashing isn't possible<br>- Stay home when you are sick | **Practice Good Hygiene**<br>- Likely all activities close to normal<br>- No limit on gatherings<br>- Increased telework / remote working as new normal<br>- All travel resumes<br>- Wash hands for at least 20 secs, use hand sanitizer when handwashing isn't possible<br>- Stay home when you are sick |
| DC Water Pandemic Response Level | Level 4 | Level 3 | Level 2 | Level 1 |
| DC Water Operational Safeguards | **FULL IMT - Essential Functions** (EFs) only**<br>- No visitors or non-essential business without prior authorization<br>- Reduced occupancy of common spaces in DC Water Facilities / Leverage outside spaces<br>- Telework maximized<br>- Face coverings required<br>- Enhanced cleaning and disinfection<br>- No travel without approval by CEO<br><br>**Single occupancy** in DC Water vehicles is recommended, if not possible maximize distance and airflow in the vehicle | **PARTIAL IMT**<br>- No visitors or non-essential business without prior authorization<br>- Reduced occupancy of common spaces in DC Water Facilities / Leverage outside spaces<br>- Telework maximized<br>- Face coverings required<br>- Enhanced cleaning and disinfection<br>- No travel without approval by EVP<br><br>**Double occupancy** in DC Water vehicles is recommended, if not possible maximize distance and airflow in the vehicle | **PARTIAL IMT CONSIDERED**<br>- Limited visitors and business with pre-authorization<br>- Reduced occupancy of common spaces in DC Water Facilities / Leverage outside spaces<br>- Telework strongly encouraged pending Facilities adjustments to workstations/HVAC as needed<br>- Enhanced cleaning and disinfection<br>- Travel based on budgetary constraints and necessity<br><br>**Normal occupancy** in DC Water vehicles, while still attempting to maximize distance and airflow in the vehicle | **BACK TO "NORMAL"**<br><br>- No restrictions of gathering or visitors<br>- Telework per policy<br>- Travel based on budgetary constraints and necessity<br><br>**Normal occupancy** in DC Water vehicles |
| Entering inside Customer locations (Residential, Commercial, all other large Customer locations) | -Inside residential work restricted to emergencies for occupied properties, & meter/MTU related work for vacant residential locations.<br>-Entry into hospital locations will require consultation with Safety, along with PPE below.<br>-Commercial & other large customer locations if the property is minimal/unoccupied food service inspection, and financial execution of work has significant impact<br>-Enter the building with the proper PPE (Tyvek suit(optional), face shield, face covering, and non- permeable gloves<br>-Requires appropriate response to health screening questionnaire upon scheduling and entry<br>-Utilize defined disinfection procedures for tools used during the inside location work | -Inside residential work emergencies & meter/MTU related work for vacant and accounts estimated > 1 year, quarterly meter read.<br>-Entry into hospital locations will require consultation with Safety, along with PPE below<br>-All Commercial & other large customer inside locations permitted if work location has less than 50 occupants or the work location is not in a high traffic area (e.g. boiler room).<br>-Enter the building with the proper PPE(Tyvek suit(optional), face shield, face coverings, and nonpermeable gloves.<br>-Requires appropriate response to health screening questionnaire upon scheduling & entry<br>-Utilize defined disinfection procedure for tools used during the inside location work | -Only suspended residential work is internal audits, except for escalated bill complaints.<br>-Utilize face coverings and non-permeable gloves when working in inside locations per District guidance.<br>-Requires appropriate response to health screening questionnaire upon scheduling and entry<br>-Utilize defined disinfection procedures for tools used during the inside location work | -No restrictions to entry to inside locations<br>-Utilize face coverings & non permeable gloves (optional) when working in inside locations per District guidelines & site conditions (e.g. boiler room).<br>-Health screening consistent with current operations for Water Quality residential visits.<br>-Disinfect consistent with normal operations, continued for Sewer Operations. |
| DC Water Potential Level of Service (LOS) | Level of Service 0% - 25% | Level of Service 25% - 50% | Level of Service 50% - 75% | Level of Service 75% - 100% |

**FIGURE 5.3** Recovery Crosswalk—ReOpen DC and DC Water Threat/Service Levels

- suspending DC Water–sponsored events, travels, training, and conferences due to the COVID-19 pandemic; and
- curtailing overtime.

DC Water ended the fiscal year 2020 favorably despite the impact of the COVID-19 pandemic on the Authority's revenues. The year-end estimated revenues were above the revised budget and estimated operating expenditures, and capital disbursements ended below budget due to careful monitoring of expenses and the actions listed above.

In fiscal year (FY) 2021, operating revenues were projected to have a shortfall of about $35 million; the anticipated revenue shortfall was mainly in the commercial and municipal revenue categories due to a decline in consumption in the District. In response to the COVID-19 revenue impacts, DC Water developed a revised near-term Capital Improvement Plan (CIP). The Finance, Accounting, and Budget Department coordinated the development of the new plan with the Operations and Engineering Department, resulting in over $200 million in spending deferred from FY21 and FY22. These near-term changes were then incorporated into the 10-year CIP for the FY21–30 financial plan. DC Water continued to take advantage whenever possible of the municipal bond market and was able to save over $10 million for debt service due to refinancing and lower rates on existing debt.

### 4.1.1 Affordability, Access, and Customer Assistance

The DC Water motto is "Water Is Life," and maintaining their essential services has never been more important to the community than during the COVID-19 pandemic. Recognizing the critical role of water and wastewater services during this health emergency, there were several public initiatives announced within the first 6 to 8 weeks of the DC Water Incident Management Team (IMT) formation to ensure that the people in DC would be healthy, safe, and well.

To ensure every customer in the city had access to clean water, all disconnected residents had their service restored on request. DC Water offered every residential customer flexible payment terms for the duration of the health emergency and halted disconnections for nonpayment. DC Water also suspended all planned water shutoffs for noncritical activities such as plumber or developer connections to the water system and for work associated with capital improvement projects. Crews only performed water shutoffs when necessary to perform emergency repairs.

As the District officially ended the health emergency and DC Water was facing more than $26 million outstanding in unpaid bills, it moved to start disconnecting service for people with unpaid bills in November 2021. DC Water once again reiterated that those who have past-due bills should pay or contact them by December 1 for assistance or to set up a payment plan to avoid any late fees or additional charges. As stated by DC Water CEO David Gadis, "With arrears this significant, it could jeopardize our ability to deliver these critical services. We are resuming disconnections to encourage payment by those who can afford to pay their bills. And to find financial assistance programs for those who cannot afford to pay" (DC Water, 2021).

### 4.1.2 Revenue Streams (e.g., Reduced Usage, Extended Periods of Nonpayment)

Despite an increase in customer delinquencies, and a decrease in permit fees, development contractor fee (water and sewer), and late fees, DC Water remains financially strong, continuing to have more than 250 days of cash in reserve funds, including the Rate Stabilization Fund, and about 25% of overall revenues are fixed, such as the metering fee and Clean Rivers Impervious Area Charge. DC Water benefits from a broad customer base, including the federal government and large institutions within the District as well as suburban wholesale customers.

DC Water announced its plans to resume applying late fees and charging interest on past-due balances beginning September 1, 2021. Any customer with a past-due account may request a flexible payment plan. Customers in good standing on their payment arrangements were exempt from penalties and collection activities. Late fees are waived for our most vulnerable customers—who are protected due to their participation in income-qualified programs such as STAY DC, the Utility Discount Program, which includes DC Water's customer assistance programs; the Low Income Home Energy Assistance Program; the Supplemental Nutrition Assistance Program (SNAP) and Temporary Assistance for Needy Families (TANF); the new Low Income Household Water Assistance Program; and the Medicaid/DC Healthcare Alliance.

### 4.1.3 Customer-Focused Assistance Programs

DC Water enhanced financial assistance programs, created new ones, expanded payment plans to more customers, and waived late fees and liens until the end of the public health emergency. The utility has a long history of lending a helping hand to customers in financial need who are having trouble paying their water bills. Information on traditional DC Water customer assistance programs, which were expanded in response to the COVID-19 pandemic, can be found here: https://www.dcwater.com/customer-assistance.

#### 4.1.3.1 DC Water Cares

DC Water launched DC Water Cares, a new program to assist customers who have fallen behind. DC Water has partnered with a wide range of organizations and agencies to inform the public about this additional assistance. The Authority has presented virtually at dozens of civic groups,

Advisory Neighborhood Commissions, senior centers, homeowner associations, and community groups. DC Water has reached out to the District Office of Aging and other District agencies that provide services and support to vulnerable populations. Local food banks have been inserting DC Water Cares fliers into food bags and boxes. In addition, the Authority uses social media, NextDoor, advertising, and the dcwater.com website to publicize these assistance programs. In addition, they are now offering emergency assistance to people who rent in multifamily buildings.

#### *4.1.3.2 Emergency Residential Relief*

A new program started in early 2021, Emergency Residential Relief is intended to supplement assistance for residential customers who have fallen behind on their bills during the coronavirus (COVID-19) public health emergency. It provides up to $2,000 to help a customer eliminate their past-due balance, provided they are income eligible.

#### *4.1.3.3 Multifamily Assistance Program for Tenants and Renters*

Another new program started in February 2021 to provide assistance for people who rent in multifamily buildings, like apartments or low-income condominium associations, the Multifamily Assistance Program (MAP) targets tenants already income-qualified by other assistance programs. Because their water charges are part of their rent, or HOA dues, they've never been able to receive a discount from DC Water. But under the new program, DC Water provides discounts to owners to pass along to those tenants whose income qualifies them for the help. It is one of a few programs in the nation designed to aid customers who do not pay their water and sewer bills directly. The MAP was expanded in May 2021 so that anyone whose property owner participates can apply for this income-based financial assistance. Most of the credit, 90%, is then passed on to the occupant as a rental or association credit. Occupants may apply for this income-based assistance at any time; however, receipt of any assistance is contingent on property participation. DC Water set aside $7 million earlier this year to fund this initiative, and already more than $2.2 million has been credited. This funding has supported more than 5000 households.

### **4.1.4 Utility Financing Strategies**

DC Water's current financial policies, its credit rating history, financial resilience, and strength all lead to the ability to keep retail rates low through lower borrowing costs while also allowing DC Water to build up a significant operating reserve fund. DC Water's strong finances, flexibility in spending and hiring, and ability to maintain reserves for unexpected situations has helped in weathering the impact of COVID-19, which caused a 29% drop in commercial consumption at the end of FY2020.

While DC Water was making internal decisions to slow down spending wherever possible while maintaining critical service levels and supporting customer assistance programs, it was also working with utility leaders to lobby Congress for outside funding support. DC Water leadership—in coordination with other water utility executives—sent a letter to Congress on March 24, 2020, as the pandemic was just beginning, requesting that Congress consider the impact that COVID-19 has had on communities across the nation. Many water and wastewater systems were experiencing steep revenue losses because of the widespread shelter-in-place orders that essentially froze large parts of the economy. The letter was signed and supported by more than 20 utility leaders from across the country. Water industry organizations coordinated their efforts by sending their own letter signed by different organization leaders, including the NACWA, the AWWA, and the US Water Alliance.

### *4.1.5 Funding and Grants*

Due to COVID-19's ongoing persistence, DC Water's IMT was activated when needed. Upon activation, the finance department set up separate accounting and procurement codes so the organization could best identify and track resources dedicated explicitly to incident and response management. This allowed DC Water to provide granular COVID-19 budgeting information to the Washington, DC Government and DC Council. This tracking also positioned DC Water to easily receive funding based on clearly documented time/materials and activities. DC Water sought reimbursement from the Federal Emergency Management Agency (FEMA) for a declared emergency in conjunction with our District partners, and they will continue to target funding from other federal/state sources to replace affected revenue streams and backfill expenditures they did not foresee when budgeting for FY2020 and FY2021.

## 4.2 Cybersecurity

As a retail utility and a critical infrastructure asset, DC Water is at risk from an array of threat sources, from the casual hacker to a state-sponsored cyber terrorist—a risk that increased as employees shifted to telework for customer service and operational staff. It is the Authority's priority and responsibility to maintain a constantly safe and reliable computing environment.

Preventing unauthorized access to applications, data, and the network is a fundamental security requirement. Users at all levels, from contracting staff to direct reports and executives, are frequently targeted by phishing attacks. The cybersecurity team had to ensure they were able to perform their primary duties of monitoring, analyzing, detecting, and responding to relevant security events during the pandemic.

The cybersecurity framework is now formally known as the NIST Cybersecurity Framework (the "NIST Framework"). The NIST Framework forms the foundation of the Authority's cybersecurity program. Coupled with a layered defense approach, a default-deny strategy, a privilege access control policy, and a comprehensive cyber-awareness program, the Authority strives to ensure the highest level of protection across its computing environment. Additional information about the NIST Framework's five pillars of excellence can be found here: https://www.nist.gov/cyberframework/online-learning/five-functions.

Administrative, call center, and office staff were able to transition to telework. DC Water's Information Technology Department supported up to 655 teleworking staff, distributing hardware and implementing Microsoft Teams to ensure all staff had the ability to communicate securely while teleworking. The transition to Teams was helpful both for ensuring that DC Water could continue to host internal and external meetings and also for collaborating on documents. DC Water has about 1,300 people (staff and contractors) actively using Teams and hosted more than 6,700 Teams meetings in an average month during the pandemic.

Cyber resiliency is also imperative to limiting financial or regulatory impact, as well as damage to business reputation in the event of a data breach (data theft, ransomware, etc.). DC Water continued to evaluate, implement, and exercise technologies and processes that allows the organization to continue to provide critical water and wastewater services during adverse conditions, and that work was more critical during the COVID-19 pandemic when customer data had to be accessed by teleworking employees in multiple states around the region. DC Water's success is achieved by institutionalizing the elements of the NIST Framework, information system, and cyber resilience, so it becomes the way of doing business rather than something else the authority does. The authority also has an established network of resources, which it leverages to proactively assess new and evolving risks including vendors, WaterISAC, the Department of Homeland Security' CyberSecurity & Infrastructure Security Agency, auditors, and peer utilities. These resources contribute to continuously improving the authority's capabilities and proved invaluable during the COVID-19 pandemic response.

## 4.3 Utility Workforce

DC Water developed an operational COVID-19 Recovery Crosswalk (Figure 5.4) to highlight service levels compared to the current phase or stage of COVID-19 measures. Each department was asked to provide the following:

- essential functions deferred or modified: provide a description of how departmental essential functions were either deferred or modified, if applicable;
- noncritical functions deferred: provide a description of nonessential functions that have been deferred due to COVID, if applicable;
- reduced levels of service: detail how levels of services for given functions may have been reduced due to the pandemic, lack of staffing, or inability to perform functions; and
- employees: describe any changes to schedules, posture, or reporting, such as teleworking, staggering shifts, or social distanced in-person reporting.

| DC Water Operational Areas | | | | |
|---|---|---|---|---|
| DC Water Headquarters | | | | |
| DC Water Pandemic Response Level | Level 4 | Level 3 | Level 2 | Level 1 |
| HQO | Facility closed to all but executive office staff, occasional onsite work permitted in Payroll, Facilities, etc. | Facility closed to all but executive office staff, occasional onsite work permitted in Payroll, Facilities, etc. | Facility opening to essential staff with safeguards in place. | Facility fully open to all staff. |
| DC Water Board | Virtual meetings | Virtual meetings | Meetings resume at HQO; public participation virtual | Meetings resume with full public participation |
| General Manager's Office | | | | |
| Finance and Procurement | **EFs Modified/Deferred:** None | **EFs Modified/Deferred:** None | **EFs Modified/Deferred:** None | **EFs Modified/Deferred:** None |
| | **Non-critical functions deferred:** Oracle System Upgrade, Projects deferred due to budget constraints | **Non-critical functions deferred:** Oracle System Upgrade, Projects deferred due to budget constraints | **Non-critical functions deferred:** None; payroll window can be reopened with plexiglass divider or other protection in place | **Non-critical functions deferred:** None |
| | **Reduced LOS:** None | **Reduced LOS:** None | **Reduced LOS:** None | **Reduced LOS:** None |
| | **Employees:** Teleworking, with staff coming into the office on an as-needed basis, awaiting laptops for some staff | **Employees:** Teleworking | **Employees:** Teleworking when able, increased spacing between cubicles | **Employees:** Teleworking |
| Legal Counsel | **EFs Modified/Deferred:** None | **EFs Modified/Deferred:** None | **EFs Modified/Deferred:** None | **EFs Modified/Deferred:** None |
| | **Non-critical functions deferred:** none | **Non-critical functions deferred: none** | **Non-critical functions deferred:** None | **Non-critical functions deferred:** None |
| | **Reduced LOS:** None | **Reduced LOS:** None | **Reduced LOS:** None | **Reduced LOS:** None |
| | **Employees:** Teleworking | **Employees:** Teleworking | **Employees:** Teleworking | **Employees:** Teleworking |
| People and Talent | **EFs Modified/Deferred:** None | **EFs Modified/Deferred:** None | **EFs Modified/Deferred:** None | **EFs Modified/Deferred:** None |
| | **Non-critical functions deferred:** none | **Non-critical functions deferred:** none | **Non-critical functions deferred:** None | **Non-critical functions deferred:** None |
| | **Reduced LOS:** None | **Reduced LOS:** None | **Reduced LOS:** None | **Reduced LOS:** None |
| | **Employees:** Teleworking | **Employees:** Teleworking | **Employees:** Teleworking | **Employees:** Teleworking |
| Communications and Marketing | **EFs Modified/Deferred:** None | **EFs Modified/Deferred:** None | **EFs Modified/Deferred:** None | **EFs Modified/Deferred:** None |
| | **Non-critical functions deferred:** none. | **Non-critical functions deferred: None** | **Non-critical functions deferred:** None | **Non-critical functions deferred:** None |
| | **Reduced LOS:** Community outreach events deferred due to crowd size restrictions, Plant and facility tours suspended, Board meetings held virtually, online water curriculum in lieu of scheduled educational events | **Reduced LOS:** community outreach deferred based on District restrictions, plant and facility tours are suspended, Board of Directors and Town Hall meetings held virtually, Online curriculum in lieu of scheduled educational events | **Reduced LOS:** community outreach deferred based on District restrictions, plant and facility tours are suspended, Board of Directors and Town Hall meetings held virtually, Online curriculum in lieu of scheduled educational events | Reduced LOS: Public outreach, tours, public meetings and educational events dependent on District restrictions and DC Water safety considerations; continue to utilize online meeting tools as much as possible |
| | **Employees:** Teleworking; staff go into the office on an as-needed basis. | **Employees:** Teleworking; staff in office only on as-needed basis; | **Employees:** Primarily Teleworking; leverage shift schedules or other measures to maximize in-person distancing at the office; executive assistant reoccupies the workspace with physical barriers in place | **Employees:** Teleworking according to policy; Rotating office shifts as needed. |

**FIGURE 5.4** Recovery Crosswalk—DC Water Operational Areas

| Administration | | | | |
|---|---|---|---|---|
| DC Water Pandemic Response Level | Level 4 | Level 3 | Level 2 | Level 1 |
| Safety | **EFs Modified/Deferred:** None | **EFs Modified/Deferred:** None | **EFs Modified/Deferred:** None | **EFs Modified/Deferred:** None |
| | **Non-critical functions deferred:** None | **Non-critical functions deferred:** Trainings may be difficult to conduct face-to-face. May need to develop and provide training to staff regarding the "new normal" | **Non-critical functions deferred:** None | **Non-critical functions deferred:** None |
| | **Reduced LOS:** None | **Reduced LOS:** None | **Reduced LOS:** None | **Reduced LOS:** None |
| | **Employees:** Teleworking | **Employees:** Teleworking; looking at administrative techniques to augment staff shift and schedules | **Employees:** Teleworking; looking at administrative techniques to augment staff shift and schedules | **Employees:** Teleworking |
| Security | **EFs Modified/Deferred:** COF Desk, Gate B, city-wide patrol, BP Traffic Control eliminated due to budget, Modified HQO main gate night and weekend officers | **EFs Modified/Deferred:** None | **EFs Modified/Deferred:** None | **EFs Modified/Deferred:** None |
| | **Non-critical functions deferred:** decreased visitor center support hours, decreased badge issuance hours, AB schedule for locksmith | **Non-critical functions deferred:** none | **Non-critical functions deferred:** None | **Non-critical functions deferred:** None |
| | **Reduced LOS:** None | **Reduced LOS: Return to some facilities with Plexiglas dividers in place for in-person interaction** | **Reduced LOS:** None | **Reduced LOS:** None |
| | **Employees:** reduced staffing and hours; some positions cut due to COVID | **Employees:** returning staff to regularly scheduled hours at specific locations (e.g. Blue Plains Visitor Center); continue rotating shifts for locksmiths | **Employees:** returning staff to regularly scheduled hours at specific locations; continue rotating shifts for locksmiths | **Employees:** returning staff to regularly scheduled hours and locations. |
| Fleet | **EFs Modified/Deferred:** cut-off switches added to batteries to slow no starts and cut offs, maintenance, registration, DC inspections were deferred | **EFs Modified/Deferred:** None | **EFs Modified/Deferred:** None | **EFs Modified/Deferred:** None |
| | **Non-critical functions deferred:** preventative maintenance, vehicular maintenance and accident acquisition, shop equipment safety and services, training on CDL, Quality Assurance, and Fleetwave system | **Non-critical functions deferred:** None | **Non-critical functions deferred:** None | **Non-critical functions deferred:** None |
| | **Reduced LOS:** None | **Reduced LOS:** None | **Reduced LOS:** None | **Reduced LOS:** None |
| | **Employees:** alternating shift schedules | **Employees:** alternating shift schedules | **Employees:** alternating shift schedules | **Employees:** alternating shift schedules |
| Emergency Management | **EFs Modified/Deferred:** Virtual incident management, compliance measures for AWIA fast tracked, All staff assigned to response and recovery efforts | **EFs Modified/Deferred:** Virtual incident management, compliance measures for AWIA fast tracked, All staff assigned to response and recovery efforts | **EFs Modified/Deferred:** Virtual incident management, compliance measures for AWIA fast tracked, All staff assigned to response and recovery efforts | **EFs Modified/Deferred:** None |
| | **Non-critical functions deferred:** Training and exercises that are not vital to immediate emergency response, Collaboration and support for stakeholders | **Non-critical functions deferred:** Training and exercises that are not vital to immediate emergency response, Collaboration and support for stakeholders | **Non-critical functions deferred:** Training and exercises that are not vital to immediate emergency response, Collaboration and support for stakeholders | **Non-critical functions deferred:** None |
| | **Reduced LOS:** None | **Reduced LOS:** None | **Reduced LOS:** None | **Reduced LOS:** None |
| | **Employees:** Teleworking | **Employees:** Teleworking | **Employees:** Teleworking and rotating into office | **Employees:** Teleworking |
| Facilities | **EFs Modified/Deferred:** HVAC and plumbing PMs have been modified, janitorial services modified based on site personnel locations and quantities, mail room services | **EFs Modified/Deferred:** HVAC and plumbing PMs modified, janitorial services modified based on site personnel locations and quantities, mail room services modified to reduce personnel. | **EFs Modified/Deferred:** HVAC and plumbing PMs modified, janitorial services modified based on site personnel locations and quantities, mail room services modified to reduce personnel. | **EFs Modified/Deferred:** None |
| | **Non-critical functions deferred:** set up for events, certain groundskeeping activities, construction projects that don't directly impact safety, security, health and hygiene, cafeteria closure, carpentry, painting, copy room services | **Non-critical functions deferred:** set up for events, certain groundskeeping activities, construction projects that don't directly impact safety, security, health and hygiene, cafeteria closure, carpentry, painting, copy room services | **Non-critical functions deferred:** set up for events, certain groundskeeping activities, construction projects that don't directly impact safety, security, health and hygiene, cafeteria closure, carpentry, painting, copy room services | **Non-critical functions deferred:** None |
| | **Reduced LOS:** None | **Reduced LOS:** None | **Reduced LOS:** None | **Reduced LOS:** None |
| | **Employees:** Office staff teleworking; AB shifts implemented | Employees: Office staff teleworking; AB shifts implemented for field crews. | **Employees:** Office staff teleworking, w/ potential for A/B rotation for office staff as needed; field crews A/B shift modification, staggered work hours or alternate reporting locations. | **Employees:** Office staff onsite as needed, teleworking to maintain teleworking technology and skillset |

**FIGURE 5.4** *(Continued)*

### 4.3.1 Recruitment, Retention, and Staff Engagement

As the pandemic began and continued to affect revenues and operations within DC Water, there were several steps taken that had an effect on the recruitment of DC Water employees. There was a temporary hold placed on hiring while DC Water evaluated the impacts of COVID-19, and an internal process was developed to allow critical position requests that would still move forward. Services of temporary staff who were not deemed critical by executive leadership were discontinued, and future hiring of temporary staff required executive approval.

The Department of Occupational Health and Safety (DOSH), in conjunction with the Office of Emergency Management (OEM), lead the efforts

| Customer Experience | | | | |
|---|---|---|---|---|
| **DC Water Pandemic Response Level** | **Level 4** | **Level 3** | **Level 2** | **Level 1** |
| Command & Call Center | **EFs Modified/Deferred:** Reconnection policy & fees, no disconnects, late fees | **EFs Modified/Deferred:** Reconnection policy & fees, no disconnects, late fees | **EFs Modified/Deferred:** None | **EFs Modified/Deferred:** None |
| | **Non-critical functions deferred:** administrative hearings conducted remote, increased estimated bills for all inside locations not transmitting, payment term policy | **Non-critical functions deferred:** administrative hearings conducted remote, increased estimated bills for all inside locations not transmitting, payment term policy | **Non-critical functions deferred:** None | **Non-critical functions deferred:** None |
| | **Reduced LOS:** No access to phone statistics, response time SLA modified; Dispatch work is limited based on work that has been suspended by meter and water services. Phone calls limited to restore & emergency only | **Reduced LOS:** Collection outbound activity pending 30 days after PHE, billing inquiries limited by field work restrictions | **Reduced LOS:** None | **Reduced LOS:** None |
| | **Employees:** Teleworking, except midnight -8am shift | **Employees:** Teleworking, except midnight -8am shift | **Employees:** Teleworking | **Employees:** Teleworking |
| Customer Service (Billing, Collection, Support, COE) | **EFs Modified/Deferred:** No late fees, no pay disconnects, or penalties. Key bill investigation tools not available for dispute resolution( inside work including audits). Payment terms for restore and payment arrangements were | **EFs Modified/Deferred:** None | **EFs Modified/Deferred:** None | **EFs Modified/Deferred:** None |
| | **Non-critical functions deferred:** Internal audits suspended, payment terms and restore policy modified, receivership processing has been suspended | **Non-critical functions deferred:** None | **Non-critical functions deferred:** None | **Non-critical functions deferred:** None |
| | **Reduced LOS:** no collection of late fees, penalties, disconnects, outbound calling except Liens, inability to resolve some bill disputes and process receivership | **Reduced LOS:** None | **Reduced LOS:** None | **Reduced LOS:** None |
| | **Employees:** Teleworking | **Employees:** Teleworking | **Employees:** Teleworking | **Employees:** Teleworking |
| Meter Operations | **EFs Modified/Deferred:** AMI and large meter projects suspended, No internal audits, no disconnects, no inside location meter reading | **EFs Modified/Deferred:** AMI Phase 2 on June 1; Realign install plan from aged vault meters to focus on revenue generating potential | **EFs Modified/Deferred:** None | **EFs Modified/Deferred:** None |
| | **Non-critical functions deferred:** no internal activities or audits, inside work restricted to emergencies for inside occupied locations. Commercial & large meter limited to high impact revenue cases | **Non-critical functions deferred:** limited internal activities and disconnects, suspended audits | **Non-critical functions deferred:** limit audits to escalations, internal meter reads every other month | **Non-critical functions deferred:** None |
| | **Reduced LOS:** Staff prioritizing emergencies, rely on water services to conduct shuts, limited service to inside locations | **Reduced LOS:** limits on some internal work and meter reading inside quarterly versus monthly | **Reduced LOS:** None | **Reduced LOS:** None |
| | Employees: Eliminate AB shift and go to staggered schedule; vehicle occupancy | Employees: Modified work schedule no AB shift; vehicle occupancy guidelines implemented | **Employees: Back to Normal** | **Employees: Back to Normal** |
| IT | **EFs Modified/Deferred:** None | **EFs Modified/Deferred:** None | **EFs Modified/Deferred:** None | **EFs Modified/Deferred:** None |
| | **Non-critical functions deferred:** none | **Non-critical functions deferred:** None | **Non-critical functions deferred:** None | **Non-critical functions deferred:** None |
| | **Reduced LOS:** None | **Reduced LOS:** None | **Reduced LOS:** None | **Reduced LOS:** None |
| | **Employees:** Teleworking | **Employees:** Teleworking | **Employees:** Teleworking | **Employees:** Teleworking |

**FIGURE 5.4** (*Continued*)

within DC Water to protect employees from COVID-19 exposure by conducting weekly operational briefs and presentations throughout the organization with virtual meetings and emergency management plans. Safety bulletins were adapted to include COVID-19-specific information and concerns, such as a hot weather bulletin. Additionally, the OEM built out a training library for the emergency management training and exercise program, which will assist with bringing training capabilities in-house by working with people and talent to implement an incident command structure training matrix for DC Water staff.

Given the impacts of the COVID-19 global pandemic—cases fluctuating throughout the region, DC exceeding substantial community spread, and travel quarantine precautions—DC Water experienced more employees needing to go into administrative quarantine over the course of the pandemic. Quarantine can have a disparate impact on the frontline workforce versus telework-eligible employees, so a recommended list of more than 100 hours of asynchronous training opportunities was developed to allow employees to take online trainings and courses while away from the workplace and not use administrative leave. Not only would the workforce be

| Operations and Engineering | | | | |
|---|---|---|---|---|
| **DC Water Pandemic Response Level** | **Level 4** | **Level 3** | **Level 2** | **Level 1** |
| **Water Services** | **EFs Modified/Deferred:** None | **EFs Modified/Deferred:** None | **EFs Modified/Deferred:** None | **EFs Modified/Deferred:** None |
| | **Non-critical functions deferred:** planned shutdowns and water isolation, taps and connections, lead service line replacements, proactive research within Quality and Technology (e.g. daycare facility testing) | **Non-critical functions deferred:** Proactive research within Quality and Technology (e.g. daycare facilities) | **Non-critical functions deferred:** None | **Non-critical functions deferred:** None |
| | **Reduced LOS:** ~75% (impact due to field staff social distancing and reduced water isolations) | **Reduced LOS:** 50% (impact due to field staff social distancing) | **Reduced LOS:** ~25% (impact to due to field staff social distancing) | **Reduced LOS:** None |
| | **Employees:** Teleworking & AB Shifts | **Employees:** Teleworking, AB Shifts, and modified start times (for some teams) | **Employees:** Teleworking, modified start times and alternate locations (as available) | **Employees:** Teleworking |
| **Sewer Services & Pumping Operations** | **EFs Modified/Deferred:** None | **EFs Modified/Deferred:** None | **EFs Modified/Deferred:** None | **EFs Modified/Deferred:** None |
| | **Non-critical functions deferred:** reduced maintenance to emergency repairs only, AB Shifts reduced routine cleanings, conducted routine cleaning and repairs during decreased emergency call period, non-critical maintenance projects deferred to FY21, Contractual services for cleaning of pumping facilities and wet-wells | **Non-critical functions deferred:** Continued AB staffing, non-critical maintenance projects deferred to FY21, Contractual cleaning of pumping stations and wet-wells | **Non-critical functions deferred:** None | **Non-critical functions deferred:** None |
| | **Reduced LOS:** staffing and work location changes have limited full productivity of pumping ops | **Reduced LOS:** staffing changes and work location changes have limited full productivity | **Reduced LOS:** None | **Reduced LOS:** None |
| | **Employees:** Teleworking | **Employees:** Teleworking | **Employees:** Teleworking | **Employees:** Teleworking |
| **Wastewater Operations** | **EFs Modified/Deferred:** None | **EFs Modified/Deferred:** None | **EFs Modified/Deferred:** None | **EFs Modified/Deferred:** None |
| | **Non-critical functions deferred:** Reduced Operations Inspections (OI), Reduced Preventive (PM) and Predictive Maintenance (PDM). Maintenance Activities limited to High Priority Corrective Maintenance (CM) and High Priority PM. Reduced laboratory sampling and testing. Sampling and testing limited to Permit Compliance. Current AB staffing ensures permit compliance but does not allow implementation of Proactive Inspections and Maintenance to prevent failures and sustain reliability of equipment and treatment processes/systems. | **Non-critical functions deferred:** Reduced laboratory sampling and testing. Sampling and testing limited to Permit Compliance. AB staffing for laboratory ensures permit compliance but does not allow effective monitoring of plant processes to optimize processes and use resources and troubleshoot equipment/system problems. | **Non-critical functions deferred:** None | **Non-critical functions deferred:** None |
| | **Reduced LOS:** None | **Reduced LOS:** None | **Reduced LOS:** None | **Reduced LOS:** None |
| | **Employees:** Modified Shift Rotation for Operations, A/B Rotation for Maintenance and Laboratory, A/B Rotation with Telework for Professional, Technical & Management Staff, Teleworking for Admin staff | **Employees:** Regular Shift Rotation for Operations, Regular Schedule for Maintenance, A/B Rotation for Laboratory, A/B Rotation with Telework for some Professional & Management Staff, Regular Schedule with Situational Telework for some Professional, Technical, and Management Staff, Teleworking for Admin Staff | **Employees:** Regular Shift Rotation for Operations, Regular Schedule for Maintenance, Regular Schedule for Laboratory, Regular Schedule with Situational Telework for most Professional, Technical & Management Staff, Teleworking for Admin Staff | **Employees:** Regular Shift Rotation for Operations, Regular Schedule for Maintenance, Regular Schedule for Laboratory, Regular Schedule with Situational Telework for Professional, Technical & Management Staff, Teleworking for Admin Staff |
| **Engineering and Technical Services** | **EFs Modified/Deferred:** None | **EFs Modified/Deferred:** None | **EFs Modified/Deferred:** None | **EFs Modified/Deferred:** None |
| | **Non-critical functions deferred:** condition assessments of pipework, reduced support for other departments | **Non-critical functions deferred:** None | **Non-critical functions deferred:** None | **Non-critical functions deferred:** None |
| | **Reduced LOS:** None | **Reduced LOS:** None | **Reduced LOS:** None | **Reduced LOS:** None |
| | **Employees:** Teleworking | **Employees:** Teleworking | **Employees:** Teleworking | **Employees:** Teleworking |
| **Clean Rivers** | **EFs Modified/Deferred:** None | **EFs Modified/Deferred:** None | **EFs Modified/Deferred:** None | **EFs Modified/Deferred:** None |
| | **Non-critical functions deferred:** none | **Non-critical functions deferred:** None | **Non-critical functions deferred:** None | **Non-critical functions deferred:** None |
| | **Reduced LOS:** None | **Reduced LOS:** None | **Reduced LOS:** None | **Reduced LOS:** None |
| | **Employees:** Teleworking | **Employees:** Teleworking | **Employees:** Teleworking | **Employees:** Teleworking |
| **Permit Operations** | **EFs Modified/Deferred:** limited ability to conduct in-person drop offs at Blue Plains | **EFs Modified/Deferred:** in-person drop off with social distancing modification measures | **EFs Modified/Deferred:** None | **EFs Modified/Deferred:** None |
| | **Non-critical functions deferred:** none | **Non-critical functions deferred:** none | **Non-critical functions deferred:** None | **Non-critical functions deferred:** None |
| | **Reduced LOS:** permit check processing has caused issues | **Reduced LOS:** permit check processing | **Reduced LOS:** None | **Reduced LOS:** None |
| | **Employees:** Teleworking | **Employees:** Teleworking | **Employees:** Teleworking | **Employees:** Teleworking |

| ReOpen DC Phase/Stages | DC Water Pandemic Response Level |
|---|---|
| Stage 1: Public Health Emergency, Stay-at-Home Order Lifted | Level 4: DC Water Emergency (Low-Level Crisis) |
| Stage 2: Open with Safeguards | Level 3: Localized Emergency (High Level Emergency) |
| Stage 3: New Normal | Level 2: National Emergency (Low Level Emergency) |
| Stage 4: Effective Vaccine Available, Long-term Recovery | Level 1: Near-Normal Operations |

| | |
|---|---|
| ****NOTE on Essential Functions:** | Essential Functions (EFs) are defined in the COOP plan as functions that should not be interrupted or deferred throughout an emergency and should be resumed as soon as possible. These are synonymous with critical functions, mission essential functions, critical operations, or other terms denoting the high priority and criticality of a business's |

**FIGURE 5.4** *(Continued)*

safe and the employee experience more equitable, but also DC Water would have a pool of workers who had been educated and trained to add more value to their critical mission.

One of the DC Water Core Values is teamwork. When COVID-19 impacted the Materials Management Department and temporarily shut down the Bryant Street warehouse—where critical operations materials were stored—the Facilities leadership team went into action. They quickly ordered a sanitizing cleaning of the warehouse and searched for replacement warehouse workers with forklift certification because all the warehouse workers had to self-isolate for 14 days. Two maintenance workers in Facilities volunteered without hesitation despite the safety concerns to ensure water and sewer operations were not interrupted.

### 4.3.2 *Motivation and Morale (Internal and Customer Facing)*

Internally, DC Water's People and Talent Department engaged the staff through a variety of campaigns and virtual experiences. Staff were able to connect around a variety of topics—exercising together, learning new recipes, and receiving much-needed mental health support. Unfortunately, DC Water–sponsored employee recognition and holiday events were suspended for 2020 but were brought back in limited fashion in 2021 in accordance with local COVID-19 regulations.

DC Water has an Authority-focused intranet for communication, important documentation, and staff resources. To assist with information sharing during the pandemic, DC Water created a COVID-19-focused page where staff could find the most up-to-date information, policies, procedures, and forms. This site has proved to be invaluable and will certainly be used for future long-term events.

Externally, DC Water's Office of Marketing and Communications (OMAC) continued to shine while engaging their customers and the community. The OMAC was able to conduct thorough community outreach and social media engagement to keep residents and customers aware of ongoing changes to DC Water operations, capturing everything on a dedicate website: https://www.dcwater.com/covid-19-updates-and-resources.

DC Water traditionally holds biannual town hall meetings to present the water authority's proposed rates and fees for the next 2 years while allowing the public to provide feedback or raise other concerns. The meetings are cohosted by the eight ward council members and are typically held at locations across the city. In 2020, due to the ongoing coronavirus pandemic, the meetings were held remotely for the safety of participants for the first time. DC Water used the Microsoft Teams online platform as well as a tele-townhall system that called customers and invited them to join the meeting in their ward. Most participants who completed a short survey said they preferred the virtual town halls over the traditional in-person meetings. More than 4,100 people participated in the eight meetings held over July 2020, far exceeding the total audience for town halls in past years. This, again, was a COVID-19-generated change to DC Water's business operations that will continue long after the pandemic is over.

DC Water unveiled online resources for parents and teachers that include water and environmental education during the time of online learning. The Resources for Teachers website includes lessons, games, and worksheets specific to DC Water about the watershed, drinking water, and wastewater treatment. They are listed by grade level and are appropriate for pre-K through 12th grade. Additional educational resources for distance learning were pulled together into one reference sheet listing dozens of additional online locations where teachers, parents, and students could access environmental education lessons, videos, and activities about water in many contexts, including health, earth, space, and environmental science.

### *4.3.3 Union Engagement and Policy Flexibility*

DC Water People and Talent was able to coordinate with each of the local unions to modify certain policies and procedures and terms of the collective bargaining agreements quickly and effectively. This allowed for more flexibility in work schedules, teleworking, and administrative leave for those adversely affected by COVID-19.

Noncritical travel was initially suspended, and a robust matrix was developed to evaluate whether travel was necessary and safe under different public health conditions. Leading Blue, DC Water's internal leadership-development program for managers, went all-virtual for the 2020 cohort. The DC Water Summer Internship program, which usually involves dozens of college students, was canceled in 2020 and limited in 2021.

# 5.0 RISK MANAGEMENT

## 5.1 Customer Risks

DC Water engaged in targeted communications to warn customers of potential scams related to the pandemic and reassure the public that the water was safe. There was an initial concern about phone scams demanding payment for continued water service or offers of special water filters to provide protection from COVID-19. DC Water not only took the step to halt disconnections but also restored service to those recently disconnected, so it was necessary to educate customers about their rights. As far as the water filter scam, DC Water reminded customers that the disinfectants used to treat District drinking water are effective against COVID-19 and that tap water is safe from the virus.

## 5.2 Risk Identification and Response Strategies

DC Water hazards and risk identification were implemented by all departments, with the requirement that critical service levels must be maintained. As the scale of the pandemic became clear in February 2020, DC Water's OEM began a thorough review of the DC Water COOP and prepared for IMT activation. This allowed staff and management to continue to adapt and make decisions as new information became available about the transmissibility of the COVID-19 virus.

On a national level, DC Water joined a study on wastewater surveillance for COVID-19 overseen by the U.S. Department of Health and Human Services. DC Water was among a handful of wastewater treatment agencies across the country to participate in a broad, national evaluation of sewage to look for clues to COVID-19 infection hotspots.

# 6.0 UTILITY OPERATIONS LEVERAGING THE INCIDENT MANAGEMENT TEAM

## 6.1 Incident Management for Common Operation Picture and Span of Control

Utilities respond to emergencies every day. Although not every utility will have a dedicated emergency management team, most utilities already use incident management principles to respond to everyday emergencies. Please note that while this worked for DC Water, the shared processes in this chapter may require tailoring to best fit into a utility's needs and objectives.

### *6.1.1 Incident Management and Incident Command*

Managing an abnormal and complex incident is challenging. Although most utilities already use some incident-management and incident-coordination principles when responding to typical emergencies, implementing an incident command structure can be greatly beneficial. This structure builds in additional communication, coordination, resource support, and documentation. Utilizing a standalone structure to establish a span of control and a common operating picture is crucial in both planned and unplanned events that are outside of standard operating procedures. Throughout the emergency management community, incident management and incident command concepts are frequently used interchangeably, and both encourage flexibility, standardization, and unity of effort. According to FEMA (2023a):

- Incident Management is "the broad spectrum of activities and organizations providing effective and efficient operations, coordination, and support applied at all levels of government, utilizing both governmental and nongovernmental resources to plan for, respond to, and recover from an incident, regardless of cause, size, or complexity."

- Incident Command is the Incident Command System's organizational element responsible for overall management of the incident and consisting of the Incident Commander (either single or unified command structure) and any assigned supporting staff.

The Incident Command System (ICS) and the National Incident Management System (FEMA, 2023b) explanations can be found in the National Response Framework of FEMA: https://www.fema.gov/emergency-managers/nims and https://training.fema.gov/emiweb/is/icsresource.

### *6.1.2 Incident Management Team*

Due to the unique location and proximity to federal government, DC Water sees numerous planned and unplanned events over the course of each year. To assist with these events and emergencies, DC Water has implemented a Type 3 IMT. With support from the senior executive team, training efforts, and support from state and federal partners, DC Water has been able to train a team, comprised of staff from across the authority, to support the response and dedicated initiatives.

During DC Water's COVID-19 utility response and recovery, the Authority's Type 3 IMT was activated and used for 12 months for initial pandemic-mitigation efforts to recovery planning and eventually demobilized and transitioned into a "return to work" work group, which continued necessary coordination and collaboration within the utility. The IMT's activation allowed the utility to document and capture efforts and actions that were used to account for critical communications, situational awareness, reporting and span of control structure, incident action plans, staffing time, logged activities, procurement needs, procedural developments, policy changes, union inquires, and financial reimbursement requests.

DC Water has been able to offer the following position-specific trainings and exercises for their team: Incident Commander, Liaison Officer, Public Information Officer, Safety Officer, Legal Officer, Security Officer, Operations Section Chief, Planning Section Chief, Logistics Section Chief, and Finance/Admin Section Chief, branch supervisors, strike team leads, task force leads, and unit leads. The IMT is a multidisciplinary team established by staff from across the Authority. The participation and establishment of the IMT team is encouraged by management as a professional and leadership-development effort and is considered a best practice for incident management.

Due to the lengthy activation time, DC Water's IMT usually has at least two people assigned and identified to each position to support different shifts and for continuity to support alternating operational periods. This ensures that staff don't suffer from burnout or fatigue and can continue to work on their normal duties where appropriate. Although this organizational structure illustrated below was implemented for COVID-19, it can easily be adapted for other events.

### 6.1.3 Continuous Evaluation and Action Plans

Throughout the response phase, the DC Water IMT consistently used the "Planning P" to establish the operational and evaluation rhythm (FEMA, 2023c). The Planning P is a tool developed by FEMA and used in incident management to develop a cycle that helps all IMT participants synchronize their operations and facilitate information exchange. The "leg" of the P starts with the notifications, incident briefings, and initial meetings to develop incident objectives. This is then followed by a more cyclical meeting schedule wherein IMT members discuss strategies and tactics to develop an incident action plan for the response, assess progress, and then reassess incident objectives based on the observed changes or completed activities within the response or recovery. This cycle represents a single operational period for the IMT. As the pandemic progressed from initial response to recovery efforts, the DC Water IMT found itself in a consistently steady cycle of weekly planning meetings and operations briefs.

### 6.1.4 Planning Meetings

At the onset of the pandemic, planning meetings were held biweekly to accommodate both the pace of new information that DC Water was receiving and the pace of implemented changes to operations. The planning meetings were a forum to discuss emerging information, determine their effect on DC Water's operation, and develop strategies to adjust the Authority's operational posture to minimize risk to staff and continue vital services to the community. When dealing with a pandemic or communicable disease, it's important to set a series of meetings each week in which staff can review important information, learn about new requirements, and plan for achieving set objectives for the upcoming operational period.

For the pandemic, planning meetings were hosted via Microsoft Teams. Merely weeks before the onset of the pandemic, DC Water's Information Technology Department installed Teams on all DC Water devices. The installation of Teams proved to be invaluable for the IMT. Although the initial transition to Teams was an adjustment, the platform offered a wide array of functions that were relied on daily. The ability to share files allowed the

IMT to collaborate on the development of IMT-related documentation. The chat function provided a more informal means of communicating between IMT members. Most importantly, the meeting function allowed DC Water to convene and capture attendance of over 100 IMT members in planning and operations calls.

The Planning Section of the IMT developed a template for the planning meetings so that each meeting created a common operating picture through a standardized agenda for report outs and an open discussion structure. The Planning Section Chief of the IMT facilitated the meetings, which started with an overview of any recent documents shared, such as the Incident Action Plan and organizational IMT structure changes that were made to ICS form 207. This was particularly useful in both the early and later stages of the pandemic as new IMT branches were being activated or demobilized. The high-level overview of changes that were made to the Incident Action Plan assisted the IMT members as they grew fatigued of reading documents that were often more than 100 pages, on a weekly basis. The Situation Unit then briefed out on the week's weather, local COVID-19 statistics, local policy changes, and noteworthy local, regional, and national news. Following the situation unit brief, the Incident Commander briefly reviewed any changes or reiterated the operational objectives for a clear line of sight to the intent and purpose of the IMT's operational period.

### 6.1.5 Operations Briefs

Although the planning meetings typically focused on looking ahead and forecasting issues and adjusting objectives, the operations briefing typically focused on providing information to a wider audience and using the forum to discuss operations-related issues that field crews observed at the ground level. Operations briefings were initially held weekly and brought together a much larger group than planning meetings, which were targeted for the IMT and key leadership. In addition to the traditional IMT members, attendees of the operations brief included executives, department heads, senior leaders, supervisors of field crews, and union representatives. The expanded attendance was beneficial for IMT members, as it brought different perspectives to the discussion and allowed "boots on the ground" staff to voice issues and concerns that otherwise may have gone unnoticed.

### 6.1.6 Information Gathering and Documentation

Given the nature of COVID-19 and the general need to be well-informed, most DC Water staff conducted their own information gathering to an extent, whether it was participating in calls, watching the local news, or reading national publications. However, the situation unit of the IMT's

Planning Section was tasked with compiling, vetting, and disseminating information to IMT members to keep them abreast of the situation on local, national, and international levels.

The Situation Unit of the IMT was expected to participate in local, regional, and federal situational calls and gather information from local watch desks and news outlets on such items as daily statistics, operational status changes for schools, transportation, and government facilities from around the region. Before each planning meeting and operations brief, the Situation Unit would compile a list of significant events regarding local, regional, and national COVID-19 news and provide relevant links to the IMT. This information, typically shared in the form of linked articles, helped keep IMT members situationally aware of ongoing advancements in the pandemic.

### 6.1.7 Communications

DC Water's Office of Marketing and Communications acts as the IMT's Public Information Officers and is responsible for all internal and external communications to DC Water staff and stakeholders. The office uses a variety of methods to reach staff through daily briefing emails, specific topic emails, posted information, and DC Water's internal website. Each of these options were relied on to provide information to DC Water staff in a timely and effective manner. The OMAC also participated regularly in regional public information meetings with partner agencies.

Early in the pandemic, the Planning Section coordinated with DC Water's Information Technology Department and OMAC to develop a COVID-19 webpage on the DC Water intranet. The intent of the COVID-19 page was to be a one-stop shop for all COVID-19-related information. The page was accessible to all DC Water staff via computer or mobile devices and was updated routinely by the OEM for staff's situational awareness.

The COVID-19 page had several tabs (latest updates, travel and access, wellness, safety guidance, confirmed and close contact cases, return to work, resources) that covered a variety of facets of COVID-19. The tabs were updated as necessary as newer information became available and as both local and authority-wide policies changed. These informational pages continued to be used and served as a reminder for current in-place policies, even after the IMT demobilized.

### 6.1.8 Transition to Long-Term Recovery and Demobilizing of IMT

After 40 operational periods and being activated for a full calendar year, the DC Water IMT was demobilized. The criteria for IMT deactivation were: a steady state of return to work, long-term recovery status with no immerging threats, and assurance that current activities had established procedures and mitigation measures. The demobilization plan outlined the following items:

- purpose
  - long-term recovery objectives
- background
- continued COVID-19 actions
  - COVID-19 universal safeguards
  - Confirmed and Close Contact Task Force
  - facilities management
  - communications
  - work group transition
- IMT close-out activities
- debriefing
- After-Action Conference

As the IMT demobilized, several structures remained in place to continue to keep staff safe and well. These structures included confirmed and close contact tracing, facilities cleaning requests, communications, and health screening efforts. In addition, DC Water and the long-term working groups continued to adjust to vaccination protocols, masking guidance, and social distancing guidelines based on the current upward or downward trending community spread of COVID-19.

### *6.1.9 After-Action Activities and Review*

After each activation that lasted longer than one operational period, the DC Water IMT conducted an after-action review conference. The conference called together the event's IMT to evaluate and touch on areas that the team felt could have been improved and areas where the team performed well. After-action review conferences are best held within days or weeks after an event ends. Holding these conferences while the thoughts and memories are still fresh within minds is critical for a meaningful recap of the activation. Once the conference is complete, the draft after-action report should be shared with the team for review and edits. The report should outline a summary of the event, contain notes from the review conversations, illustrate successes, and capture improvement needs in an improvement plan matrix. Establishing an improvement plan matrix helps summarize needs, identifies departments responsible for addressing those needs, lists support programs or documents to reference, and helps prioritize immediate improvement needs for action. This continuous improvement effort is important to show that a utility is practicing due diligence and building response capabilities.

## 6.2 Utility Operations and Supply Chain Management

### 6.2.1 Initial Awareness and Response

Initial awareness of a potential issue with COVID-19 came about through the daily due-diligence of the procurement department exercises. The DC Water procurement department routinely monitors supply chain risks through its review of trade and professional organizations as well as various media outlets. Members of the procurement department started to hear discussions about the outbreak in China as early as late November/early December 2019 through their daily monitoring of news outlets. As the media reports continued becoming more ominous regarding the situation in China and its potential global impact, the procurement department in late December 2019 began initial contingency and planning activities around chemical supplies.

### 6.2.2 Chemical Supply Security

Due to the critical nature of the chemicals used in water treatment and delivery processes, DC Water started to enhance its chemical risk-mitigation strategies in December 2019 and January 2020. The Authority undertook a systemic review of its contracts to ensure their coverage and currency. Contract renewals or evaluations were triggered before the pandemic hit.

As part of its overall risk-mitigation strategy, DC Water uses dual-sourcing chemicals. During the December 2019/January 2020 period, it undertook a thorough review of all bulk chemical contracts, suppliers, and distribution networks to identify and address potential vulnerabilities.

Both primary and secondary sources of chemicals were asked to: (a) identify any supply issues that they thought could affect them, including freight issues, supply shortages, storage capacity, and manning shortfalls; and (b) provide their emergency action plans. Emergency action plans were internally reviewed, and DC Water then worked jointly with the suppliers as much as possible to address any concerns. Weekly meetings with those suppliers were instituted and maintained until the late summer of 2020. DC Water also made the decision in January 2020 to maximize the quantity of chemicals on hand by working with the operations teams to top off its bulk storage facilities. It maintained that posture for each chemical until the marketplace reestablished the stability of a given chemical's supply chain. That stabilization generally occurred in the late summer of 2020.

### 6.2.3 Centralize the Procurement of PPE and Cleaning Supplies

As DC Water progressed through January and February of 2020, it became readily apparent that the situation with COVID-19 would require a once-in-a-lifetime response. Pre-pandemic, individual departments were responsible for procuring their needed PPE and cleaning supplies. A review was performed and the strategic decision was made that the Materials Management (MM) department within the Department of Procurement and Finance would be responsible for procuring and distributing all the PPE and cleaning supplies. By centralizing the procurement and distribution of the supplies within MM, granularity of usage trends for individual items was gained. The centralization of procurement and issuance of the supplies also streamlined reporting on the supply chain cost effect of COVID-19 to the organization, thus putting DC Water in a good position to seek reimbursement within the various disaster declarations.

To know what was needed, the Authority needed to understand what was available and in use. The MM developed an initial list of 250 potential COVID-19 response items. The list was shortened by eliminating needless variations, such as the color of hard hats and the color of gloves. In collaboration with DOSH and Operations, MM created a list of 40 items that were to be used as PPE and cleaning supplies for COVID-19. That list has subsequently grown and changed as new needs and products have been identified. Upon list completion, the decision was made to centralize the storage of all those materials in one location. Unused PPE and cleaning supplies in unoccupied workspaces were relocated to the warehouse or were repurposed to frontline staff who needed to continue to work onsite or throughout the District during the pandemic.

### 6.2.4 Creation of an Authority-Wide COVID-19 Work Order

After centralizing cleaning and PPE supplies procurement and storage, a key follow-up action was to centralize the requesting of and issuance of COVID-19 cleaning and PPE supplies on an organizational basis using work orders. The organization-wide work order allowed MM to effectively balance the requirements of all departments with the limited supplies of items. As supplies were obtained, MM would canvass all departments on their immediate requirements for the given item that arrived. After receiving individual department requests, MM would enter the requirements on the organizational work order and then would issue the material to departments based on the criticality of the request to maintain core business functions. For example, nitrile/latex gloves requested by the mail rooms were handled at a lower priority than providing the same material to crews responding for fixing issues in the distribution system.

By using the organizational work order, DC Water gained usage information for each of the identified item numbers. The information was used to manage procurement and sourcing activities across the whole Authority. This information then served as the underpinning for reporting, which put DC Water in a good position to seek reimbursement for COVID-19 expenses.

### 6.2.5 Sourcing Team for COVID-19

As MM became the designated response center for PPE and cleaning supplies, DC Water decided to create a sourcing team within MM and Procurement whose primary job from February to May 2020 was to source COVID-19 items. The team consisted of one buyer from MM, two sourcing senior specialists from Procurement, and the MM materials manager. That team worked to procure the necessary supplies required to allow the Authority to continue functioning. It also focused on developing supply arrangements for anticipated requirements beyond May. After the initial response period in February to May 2020, the two sourcing specialists were released from daily accountability for PPE and cleaning supplies, as the authority had established a level of supply sustainability for its reduced in-person presence.

DC Water wanted to engage the entire workforce in sourcing activities during the response to the pandemic. To that end, a general correspondence was sent out to solicit possible sources of supply. Any employee with an idea on sourcing was directed to send that information to the COVID-19 sourcing and procurement team. That message was continually reinforced during subsequent IMT planning and operations briefings. The COVID-19 sourcing and procurement team thus became the Authority's clearinghouse for supply suggestions. Having the experienced individuals of this team vet the numerous possible sources of supply prevented the Authority from entering supply arrangements of dubious nature.

### 6.2.6 Authority Manning and Policy Decisions

Authority manning and policy decisions directly affected the ability to extend on-hand inventory of PPE and cleaning supplies until supply chain sustainability was confirmed. The first impactful decision that extended the inventory was the decision to maximize the number of people teleworking from their homes. This decision immediately reduced the number of individuals who could draw on our inventory by approximately 40%.

Another key manning decision was to break frontline teams into A/B shifts and to develop additional muster locations for those teams to reduce the potential for person-to-person spread of the virus. By setting up these spaces with the appropriate social distances, the amount of PPE required to be worn by personnel while waiting to be dispatched to work was reduced. This decision to not pursue discretionary work during the initial phases of the pandemic further reduced the draw.

One additional policy decision that was made during the initial pandemic response extended the supply of a key required item. Based on guidelines from the Centers for Disease Control and Prevention (CDC), DC Water made the decision in March of 2020 to not supply hand sanitizer to individuals with ready access to restrooms and directed those individuals to clean their hands with soap and water. This decision immediately reduced the strain on the hand sanitizer supply. It also allowed MM to prioritize the dissemination of hand sanitizer to frontline staff as well as providing MM the authority to deny requests for hand sanitizer for individuals who had ready access to restrooms while working at the Authority.

### *6.2.7 Leveraging the Incident Command System Reporting Structure*

The adaptation of the ICS reporting structure early during DC Water's response to the COVID-19 pandemic eased demands for information on the MM team. Before the official activation of the IMT, the sourcing core team handled multiple requests from departments as to item availability, as well as in-bound delivery status. Later on, the required information on item availability as well as order status was provided in operational and planning briefs. More importantly, the response to direct all users to the inventory information contained in the IMT briefs was endorsed and championed by the Senior Executive Team and IMT leadership. This level of support immediately reduced the spurious requests for information and allowed the COVID-19 sourcing team to focus its efforts outwardly on securing a supply of necessary items.

DC Water developed five reports to support the inventory reporting requirements for the ICS structure:

1. an item list report by internal item numbers used in response to the pandemic, updated as new response items were created;
2. a daily on-hand inventory snapshot of those items run at the end of the dayshift operations, saved daily;
3. a daily open-order report showing all open orders for the items, saved daily;
4. a monthly report showing all inventory transactions for the items; and
5. a burn rate report for the items updated monthly.

The systemic collection of purchase and usage information required to support the IMT reporting requirements also positioned the Authority well to report inventory costs associated with its response to the pandemic.

### 6.2.8 Teamwork and Thinking Out of the Box

The hallmarks of the pandemic response were teamwork and thinking outside the box. In February and March of 2020, it became clear that purchasing enough PPE and cleaning supplies to keep the Authority functioning was going to be a challenge given the global competition. The first two areas addressed involved cleaning and disinfecting chemicals. DC Water encountered immediate shortages of pre-mix disinfecting chemicals and various cleaning wipes. Working jointly with the DOSH, cleaning and disinfecting solutions were sought locally. The COVID-19 sourcing team quickly identified a local small business enterprise (LSBE) within the district who could supply bleach and Dawn dishwashing detergent. The team also identified sources for spray bottles that could be used with bleach solutions.

The next supply challenge the COVID-19 sourcing team tackled was that of hand sanitizer. The preliminary actions of stressing the primacy of handwashing for those with ready access to restrooms, as well as relocating PPE and cleaning supplies from all unoccupied workspaces to the main warehouse, extended the internal supply. Containers with spigots (Igloo coolers as an example) were distributed to field crews so they could clean their hands. Crews were provided with two coolers: one contained the soap cleaning solution, and the other contained water so the crews could wash their hands. By using this alternate approach, the supply of actual hand sanitizers (gel and liquid) was prolonged. Additional actions included purchasing gel hand sanitizer and identifying a local distiller who converted their operations to making liquid hand sanitizer. MM and DOSH worked jointly to get the required documentation and approvals to use the liquid hand sanitizer. As part of that switch to liquid hand sanitizer, the DOSH distributed a safety memo to all staff detailing the proper procedures for using all PPE safely and effectively.

The relationship DC Water created with the local distiller had a multiplicative effect on supply within the Metro Washington Area. During COVID-19 coordination calls with the Metropolitan Washington Council of Governments, DC Water was able to point to this business as a potential supplier for other entities struggling with hand sanitizer supply issues. At least three local entities worked with this business to obtain hand sanitizer. DC Water also helped forge a supply relationship between the LSBE from whom it purchased its bleach and dish detergent and the distiller. The LSBE was unable to obtain a ready supply of hand sanitizer for its local storefronts, so DC Water was able to point them to the distiller. The two entities subsequently formed a business relationship. As the distiller added gel hand sanitizer to its portfolio, DC Water received advance notice of availability. As the business switched back to being a full-time distiller in spring 2021, DC Water was able to purchase its remaining two 275-gallon totes of gel hand sanitizer from them at $150.00 or the cost for the business to rent a truck to transport the totes to the authority's Blue Plains location.

The masking requirement announced at the beginning of April 2020 provided the next opportunity to excel. Available N95 respirators did not fulfill the masking requirements due to fit check and use requirements mandated by Occupational Safety and Health Administration (OSHA). An email sent to the COVID-19 sourcing team from a frontline employee suggested that the team reach out to OMAC to get the names of LSBE used for branded clothing. That insight allowed the identification and partnership with a LSBE to obtain enough cloth gaiters to supply DC Water field staff with three gaiters per individual by the third week of April. Parallel approaches were also developed to obtain branded DC reusable cloth masks, N95s and KN95s, as well as disposable masks with other LSBEs.

Like all organizations, DC Water struggled in procuring latex and nitrile disposable gloves. The sourcing team developed parallel approaches after identifying differing needs for gloves. Gloving requirements generally fell into two categories: industrial gloving for work crews and gloving required in office situations. Multiple sources for industrial gloving were identified. Gloving demand for the office requirements was diverted to nonindustrial sources of supply. All available disposable gloves used for interior house painting from the same LSBE providing bleach and dish detergent were purchased. In addition, food service gloves were purchased from a restaurant supply company for nonindustrial usage requirements. Severely affected businesses, such as local hotels, were contacted to ascertain their interest in selling PPE and cleaning supplies to DC Water.

## 6.3 Facility Readiness

DC Water established a Facilities Task Force that was able to triage and respond to facilities requests. This included requests for cleaning due to possible COVID-19 exposure and facilities-related alterations that needed to be made. The existing janitorial contract was modified to allow for "enhanced" cleanings on a daily basis where high-touch areas (elevators, handrails, time clocks) were cleaned more frequently, and a special deep cleaning service was added specifically for those areas affected by the virus. Wastewater treatment added a deep cleaning service to their process area cleaning contract and leveraged that tool for industrial areas at Blue Plains.

With many departments teleworking, remaining staff were able to leverage the opened space. Crews were able to spread out and occupy areas of the facilities that were not currently in use, allowing for more socially distanced work. For those onsite, Facilities installed and provided asset management numbers on all Centex Hand Sanitizer units to identify the locations of the units, so they could be easily refilled upon staff request. Safety information was pushed across all locations via banners, posters, and other creative signage during DC Water's fall 2020 "Not Yet" campaign to remind staff that safeguards were still necessary.

## 6.4 DC Water's Return-to-Workplace Readiness

Once the IMT was deactivated, DC Water instituted a Return to Workplace Recovery group. This task force determined there were 6 key elements necessary to facilitate a successful return to the workplace for employees:

1. ongoing communication with employees;
2. facilities readiness, space planning, and capacity limitations;
3. enhancing office protocols;
4. collaborating with government agency's reopen teams;
5. PPE; and
6. confirming connectivity and IT logistics.

### 6.4.1 Ongoing Communication With Employees

The core team established a weekly cadence to communicate with department leadership to determine accommodations and approach. Staggered returns, alternating schedules, social distancing, workstation spacing, enhanced cleaning and disinfection, and PPE distribution were discussed regularly to establish the new workflow and support the continuity of operations. A brief questionnaire was instituted via a QR code for employees and visitors to scan and complete before entering a facility to determine their current health condition. Later stickers and signage were provided for ease and accessibility to the questionnaire. Continued communications from the Senior Executive Team encouraged active participation in pre-health screenings and vaccination incentives.

### 6.4.2 Facilities Readiness, Space Planning, and Capacity Limitations

A core Facilities team was mobilized to determine alternative work locations, capacity limits and the possibility for alternating and swing shifts to ensure the continuity of operations in a safe and manageable way. Floor plans were distributed to identify conference rooms, gathering places, and other public spaces to leverage for alternative workspace, including outside spaces at our facilities like patios and roof decks when the weather was nice. The team identified workstations that could be used with the addition of plexiglass. This resulted in installation and outfitting of approximately 3000 square feet of plexiglass throughout workstations, reception areas and guard booths. The team focused on janitorial services for deep cleanings, scheduled disinfections, maintenance and upgrade of air filtration and HVAC systems. Updating and posting new and revised emergency routes and signage that included social distancing guidelines and vaccination reminders and incentives. Signs were posted internally and externally throughout campuses by the Facilities team, which was continually updated and redesigned as needed for the duration of the return-to-workplace effort.

In preparation for users to return to worksites, the IT Department restored connectivity to original spaces, added connectivity to newly identified alternative workspaces, upgraded equipment such as monitors and docking stations, and increased Wi-Fi capacity by upgrading Wi-Fi access points throughout the authority.

Weekly meetings served as the platform for task-force reports that kept the larger group apprised of new developments and addressed all departmental concerns. As policies and protocols were determined, information was disseminated to all staff via email and posted to the central pipeline intranet. Live links to request forms for deep cleaning and floor plans to indicate specific location were included.

### 6.4.3 Collaborating With Government Agency's Reopen Teams

DC Water collaborated with several local and federal agencies for support and guidance to return staff safely to our worksites. The DC Department of Health provided information on vaccine distribution and vaccination clinics. The agency assisted with setting up clinics for employees to get vaccinated on site and offered locations and availability throughout the city for employees to access. Additionally, the Metropolitan Washington Council of Governments (COG) was a forum used to support and inform the readiness and recovery effort. The COG was instrumental in sharing information about COVID-19-related issues and programs in the National Capitol region.

- The OSHA provided guidelines and documented safeguards that the authority used to draft the framework for constructing new policies and procedures for continued operation throughout the pandemic.

- The CDC and the World Health Organization (2023) provided data on the efficacy of the virus on surfaces and transmission, pandemic-travel guidelines, precautions for home and work, and deep cleaning and disinfection.
- The American Industrial Hygiene Association (2023) provided information pertinent to engineering controls such as space planning, ventilation requirements, and appropriate social distancing.

# 7.0 CHANGES IN LABORATORY TESTING

## 7.1 Drinking Water Testing

At the height of the pandemic, drinking water testing locations inside private/commercial properties were not used, and DC Water was able to work with U.S. Environmental Protection Agency (U.S. EPA) Region III to use outside/hydrant locations to meet drinking-water-quality testing requirements. Across water services operations inside work was restricted to emergencies for inside occupied locations, and noncritical water-quality research functions, such as the daycare facility testing program, were deferred.

## 7.2 Wastewater Testing

The DC Water Blue Plains wastewater treatment plant maintained its National Pollutant Discharge Elimination System permit limits, leveraging an alternate staffing model with reduced laboratory sampling and testing activities. The modified staffing ensured the permit compliance but did not allow implementation of proactive inspections and maintenance activities to prevent failures and sustain reliability of equipment and treatment processes/systems.

# 8.0 HEALTH AND SAFETY

Chapters 4 and 6 of this book provide a detailed account of how to assess and manage risks to workers during infectious disease outbreaks.

# 9.0 STRATEGIC COMMUNICATIONS

## 9.1 Introduction

An effective communication strategy is the same regardless of the presence infectious disease outbreak. The difference mainly lies in the targeted audience, and the content of the message used to communicate to that audience.

Regardless of who is responsible for executing the communication strategy, everyone plays a critical role. Scientific and technical staff hold the keys to the data, while communication staff are accountable for translating that data to a broad audience. Do not assume this audience has any working knowledge of water systems or infectious diseases.

This section aims to build an understanding of the communication process for noncommunicators. This process includes the elements of effective messaging and the challenges communicators face when translating complex information into stories that people can understand and take action on.

The four critical components of effective communication are assessment, branding, content, and strategy (ABCS), as can be seen in Table 5.1.

Feedback mechanisms are essential. Providing channels to receive feedback from the audience allows an assessment of the messaging to determine the effectiveness and any necessary adjustments.

Remember to prioritize communication efforts and have essential communication staff at the table from the beginning to reduce the possibility of surprises later down the road that could snowball into more significant issues.

Start communication efforts as soon as possible. Capturing people's attention requires time and repetition. On average, a person is served a piece of content or messaging seven times before it begins to stick. Now is the time to start preparing for the next outbreak. Glean lessons from COVID-19, and integrate those lessons into communication plans. Use situational tabletop exercises annually to practice preparedness.

## 9.2 Identifying the Problem and the Audience

### *9.2.1 The Problem*

Discovery is the most critical phase of the creative process. The goal during this process is to create consensus about the problem, the key stakeholders/audiences, and the organization's role. During this phase, it is likely that the key communication problems identified are different from the key problem

**TABLE 5.1** Critical Components of Effective Communication

| | |
|---|---|
| Assessment | What does success look like? What is our success metric (i.e., reach, engagement, etc.)? |
| Branding | Who are we (as an organization), and what do we stand for? |
| Content | Who are our audiences? What is the message for each audience? |
| Strategy | What are all the different touch points/channels we can use to reach our audiences? What feedback mechanisms do we have in place to measure if our message is effective? |

related to the outbreak itself. For example, for utilities, the initial critical concern with COVID-19 was understanding transmission to keep essential services intact while keeping staff safe. The critical concern for communicators was conveying important factual information within a noisy field focused on drinking-water safety and potential water service disruptions.

The process is easily muddled. A process often used to guide the discussion is called design thinking—a human-centric methodology for creative problem solving that provides a structure for identifying problems. One example is the "What's on Your Radar?" approach, which lays out potential problems at different scales (e.g., primary, secondary, and tertiary importance). Figure 5.5 demonstrates this activity, which begins by asking a statement with the prompt "How might we . . . ?" Examples of communication-related questions during an infectious disease outbreak might be, *How might we communicate in a way that best prepares our community to handle this public health emergency?* or *How might we best communicate what people need to do right now to stay safe?*

Looking at Figure 5.5, the titles around the circle are subcategories. These subcategories can help break down complex issues to identify

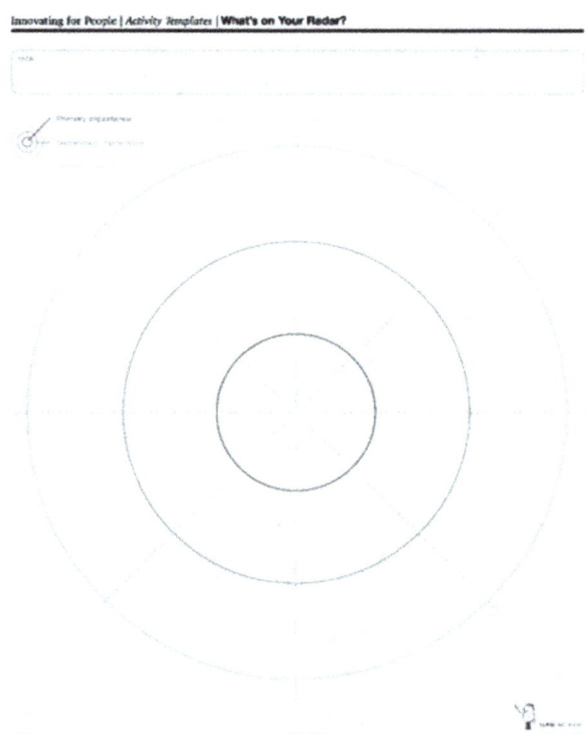


**FIGURE 5.5** "What's on Your Radar?" Template (LUMA Institute, 2021. Reprinted with permission from LUMA Institute.)

communication needs in specific areas. For example, communication subcategories during COVID-19 may have been drinking water, wastewater, or employees. What needs to be explicitly communicated around drinking water? What about wastewater? What information do we need employees to know? What about other internal stakeholders such as elected officials?

The circles within the diagram represent priority. Information in the larger tertiary circle is important but not time-sensitive and could be released later. If the exercise is conducted ahead of an active outbreak, the different circles could represent urgency—the inner circle reflecting priority information during a crisis, the secondary circle reflecting information that's helpful to avoid a crisis, and the tertiary circle reflecting information that is evergreen and serves as a preparedness resource.

Once the problems are laid out and prioritized, the next step is to create an action plan that includes the foundational *ABCS—assessment, branding, content, strategy*. Begin by homing in on key audiences, partners, and stakeholders. Use the data gleaned from the radar exercise as a starting place for the next step: stakeholder mapping.

### 9.2.2 Mapping Stakeholders

Stakeholder mapping ensures that a key audience is included in the communication strategy. Conduct this activity with a small group of representative stakeholders with the goal of further mapping out potential audiences (see Figure 5.6). Be purposeful with the invites to this group.

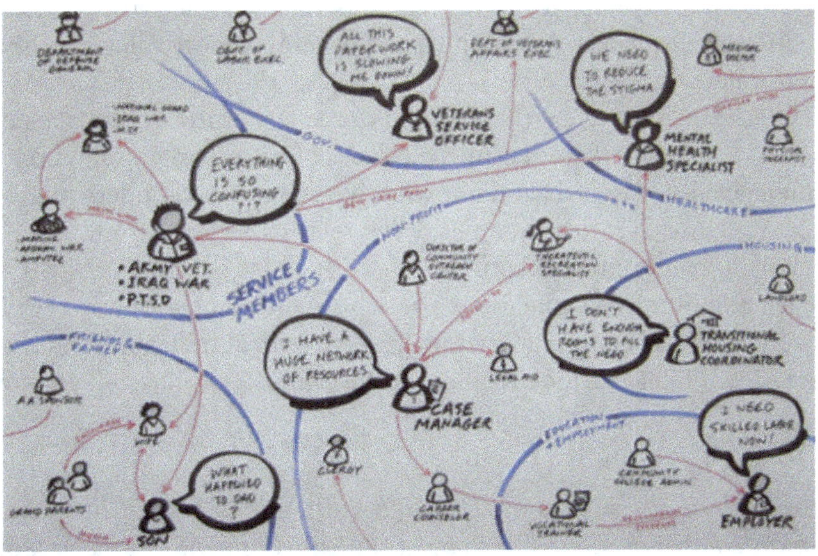


FIGURE 5.6 Stakeholder Mapping Exercise (LUMA Institute, 2021. Reprinted with permission from LUMA Institute.)

This group should be diverse in demographics and organizational roles, especially those with established roles in strategic partnerships and community development. Communication and public education staff are essential. Extend across multiple workgroups and include vital external partners. External partners should be representative of all segments of the community.

Be flexible. This activity can be facilitated both in-person and virtually. Virtual options should include apps with a whiteboard function like Miro or Mural to be more inclusive and collaborative. In-person supplies include a dry erase board, giant sticky notes, sticky notes of varying colors, colored markers, colored dry erase markers, and index cards. Enhance group participation by giving people time to think and write down their responses on index cards individually, followed by a group report out. This method results in more ideas and responses rather than asking people to contribute verbally.

Begin the group stakeholder mapping exercise by brainstorming a list of all potential stakeholders on sticky notes. Once everyone has participated, begin to group any that are similar. The stakeholder groupings should all have a common perspective, motive, or belief. The "general public" is not an acceptable category. The "general public" group needs further breakdown into specific categories such as elderly (with an age range) or immunocompromised.

### 9.2.3 Audience Profiles

Develop tailored audience profiles for each stakeholder group identified in the previous activity. Audience profiles generalize and distill a group of individuals into a snapshot that highlights their characteristics, including behavior, actions, needs, pain points, and general demographic and psychographic data.

These snapshots reveal the problem or scenario through the perspective of the intended audience. It ensures the focus of the messaging stays on what the audience needs to hear and not what we want to say. Understanding the audience at this level keeps empathy for their experience front and center and challenges any preconceptions.

Each community's stakeholders may have distinct traits; however, there are likely universal commonalities. In general, critical stakeholders during an outbreak will likely include regulators, customers, and internal employees.

#### 9.2.3.1 Regulators

The Clean Water Act gives the U.S. EPA authority to implement their own pollution-control programs, with most state governments acting as "authorized states." For these authorized states (excluding Idaho, New Hampshire, Massachusetts, and New Mexico), permitting and enforcement are performed by staff. Depending on the state, regulatory programs for domestic wastewater, industrial wastewater, biosolids, stormwater, and water quality may exist. With both federal and often multiple state regulators involved in wastewater treatment operations' regulations, communication is key.

An example of utilities maintaining strong communications with regulatory agencies is taken from the Florida Water/Wastewater Agency Response Network (FlaWARN). This system offers mutual aid to utilities to provide resources during human-made or natural disasters. The system also acts as a supporting entity to Florida county and state departments of emergency management as well as the Florida Department of Environmental Protection's (FDEP, an authorized state) WATER Tracker database. This enables the FDEP to track events and requests for assistance while connecting utilities to one another. To learn more about the FlaWARN system, visit https://flawarn.pwd.aa.ufl.edu.

### *9.2.3.2 Customers*

As paying individuals or organizations requiring utility services, customers may include diverse groups within a community. Differences such as socioeconomic level or education may present challenges when communicating with customers, especially when communications do not use multiple media streams like television, emails, text alerts, and flyers/notices on doors.

Maintaining communications with customers during unprecedented times can be especially challenging when customers react out of fear or miseducation. Given the media attention of the COVID-19 outbreak, utilities reported increased call volumes from customers asking questions like "Is my tap water still safe to drink?" or "How do you kill COVID-19 in sewage?" Given the publicity the outbreak was receiving and the uncertainty of its effects, there was an understandable sense of concern among customers that their water providers might not be equipped to handle a novel pathogen.

As with regulators, maintaining regular communications with customers is key. Developing a news site, regular TV news interviews, social media posts, emails, and media releases were all communication methods used during the pandemic. Multiple types of media are required, as customers will range in demographics, so understanding their preferred communication methods is useful.

On a precautionary note, any utilities' emergency response plan should include a strategy for communicating with the general public. In addition to identifying a point person (or persons), employees may also benefit from receiving some internal training on communicating with customers, especially fearful or demanding ones. It is essential that employees understand they must maintain a professional yet empathetic nature when dealing with the public or risk having their reputations put on the line.

#### *9.2.3.3 Internal Communications*

During uncertain times, maintaining regular communications with employees is essential. Whether through emails, an internal webpage, text messages, and/or regularly scheduled meetings, keeping employees informed of developments ensures employee confidence. Employees are less likely to trust their employer if they feel uninformed (i.e., they hear information from the media) or misinformed (i.e., as the information is relayed through multiple supervisors from the top down, it changes as it passes from one level/supervisor to the next). Maintaining clear lines of communication between the decision-making employees of an organization and its remaining employees also supports transparency within the organization.

As an example, during the COVID-19 pandemic, the Pinellas County government created a webpage that was updated daily. Not only was this website designed to be a one-stop shop for employees seeking updates, but the website's daily maintenance also helped ensure employees had access to pertinent information first. Information included leave/sick policies, forms if working remote, FAQ's, and links to locations where employees could go to get testing or vaccines.

## 9.3 Develop Campaign Strategies and Content

Strategy development depends on plugging in the core elements of effective communication, the ABCS—assessment, branding, content, and strategy.

### *9.3.1 Assessment*

Assessment begins by identifying the most indicative metric of the problem being solved. This metric should be within the organization's control to influence. For example, a water utility's goal should not be to conquer the wealth of misinformation spread by the media. Instead, the goal should focus on the organization's behavior, such as serving as the trusted expert. Tactics related to this goal might include measuring how often information is shared, outreach efforts to media and partners, the number of touchpoints with internal staff, and the number of touchpoints with external stakeholders. These are all tasks that the organization can directly implement and influence. Goals should be specific, measurable, and time-bound.

For example, be the trusted resource for information pertaining to the water system by releasing hourly updates at the height of the crisis during the first 7 days of the outbreak.

Establishing water industry members as trusted subject-matter experts requires consistency and nurturing. Metrics that measure the inputs and outputs allow for the tracking of these efforts, including open rates on newsletter updates sent to regulators, shares on social media content targeted to at-risk population segments, and the number of responses to an internal survey.

### 9.3.2 Branding

Branding is more than a logo and tagline; it is the organization's permission to operate. It demonstrates the organizational mission, vision, and values; it is foundational to public trust. A strong brand builds clarity around the purpose, expectations, and trustworthiness of an organization. Consistent branding establishes a presence people can recognize and turn to for trusted information.

Branding is a foundational piece of any communication strategy that must be acknowledged across all business functions and levels. Brand guidelines create rules of engagement to ensure brand integrity, and communication staff members are the drivers. A brand strategy and guidelines should be developed before the onset of a crisis.

### 9.3.3 Content

The context and subject matter of the content developed before, during, and after an outbreak will vary. Content should reflect what the audience needs to hear in the wording and language they understand. The content should be consistent, help the organization meet its communication goals, and align with the brand guidelines.

When creating content during a crisis, do not let assumptions rule. The audience will need regular reminders of the organization's identity and role as well as why it can be trusted. Branding plays a role, but strategic outreach and engagement will amplify these efforts.

- Do not assume people know the inner workings or value of the organization.
- Do not assume people know or understand the mission or buy-in.
- Do not assume the organization or the information it disseminates has the audience's trust.

### 9.3.4 Strategy

Communication strategies vary for different phases of a crisis and different intended audiences. However, the following are best practices for any communication effort (see Figure 5.7).

**FIGURE 5.7** Essential Building Blocks of a Communication Strategy

#### *9.3.4.1 Be Omnichannel*

The overall strategy should be omnichannel. *Omnichannel* means using multiple communication channels—TV, radio, digital, print—that work together to reach the audience. This integrated communication approach ensures a strategy that meets people where they are in their understanding of the problem and the expected action. For example, an organization may have different messaging on each channel based on the intended audience's needs (demographic, psychographic), yet all with the same call-to-action driving people to the website.

#### *9.3.4.2 Be Collaborative*

Committees get a bad rap, and some argue they slow things down; however, communication requires collaboration. Hollywood uses writer's rooms to ensure characters and plot lines maintain authenticity. For example, a show about doctors would have a writer's room that included real medical professionals. These professionals would ensure the writing reflected real medical terms and scenarios.

The writer's room could be a trusted circle of stakeholders representing the broader community in this context. This circle of trusted advisors could review potentially sensitive communications in advance to check for confusing messages or jargon. They can also advise on language or information that might not land well with a particular audience. Build a group of stakeholders outside of the industry so their feedback is representative of the audience.

#### *9.3.4.3 Be Fluid*

Use a feedback mechanism to gauge the effectiveness of the communication strategy. The best feedback comes from direct surveys. Another option is to observe social media commentary; however, keep in mind that most social media commentary reflects extremes—people who either love or hate a situation.

Use feedback to assess the strategy and any necessary pivots. Perhaps the feedback demonstrates some unexpected confusion around a specific subject. Use this input as a guide for future content. Remember, people use personal experience to fill in information gaps. Do not ignore these gaps or assume what people know or understand. Be proactive, and work to predict potential gaps and fill them in with accurate information.

## 9.4 Digital Communications: Social Media Engagement and Participation

Digital communication is essential during a crisis, and communication efforts require the same level of preparation as any operational crisis management plan. The majority of people go to social media first for news. Regardless of responsibility in content development, all technical staff play an essential role as conduits of information. That information is central to establishing the organization as the trusted source.

Technical staff can support communication efforts by providing staff with regular updates and information to produce evergreen content in advance of an outbreak. Evergreen content—general and timeless content—ensures no dead space between updates. It should be regularly audited and updated for accuracy. This content is foundational to the organization becoming a trusted source and increases the likelihood that residents and the media will approach the organization first for information.

During a crisis or outbreak, the public is consuming a great deal of information and likely feeling some fear and apprehension. Social media will reflect those feelings and concerns. Social media channels should be monitored constantly. Social media applications such as Sprout Social and Hootsuite alert staff to keywords and comments coming across social media. Tools like Google Alerts should be used to track mentions and comments around keywords to shape messaging and content development.

Engagement is critical during a crisis. Expect there to be questions after updates are posted, and have someone monitoring and responding. Develop a regularly updated "Frequently Asked Questions" page to be shared with each update. This page should include information about how the platform is being monitored for questions and when people can expect a response.

## 9.5 Communicating With the Media

Dealing with the media requires a shift in mindset. The media is not the enemy but rather a partner to help get accurate information disseminated. However, success is more likely if the information sent to them uses language they understand and content aligned with their goal. Their goal is to disseminate information as timely as possible. Extended time between updates and lack of clarity around when updates will occur will cause them to go elsewhere for information.

Share the cadence for updates, and then stick to the schedule, even if an update is that there currently are no updates. Do not be afraid to admit what is unknown, but rather be clear that the information being shared is what is known now and that it will likely evolve, and when it does, the new information will be shared. Be clear in the role of public health representatives and the responsibility not to create unnecessary fear and panic.

### 9.5.1 Before the Crisis

Do not take this time for granted. Use this time to build relationships with members of the media. Introduce the team, the plan, and each individual's role during a crisis so journalists understand the process and their points of contact. Nurture these relationships. Send them content regularly. Even if they do not use it, it does not matter. The goal is to stay on their radar.

### 9.5.2 During the Crisis

Be available, as the technical staff is often the information gatekeepers. Seek to understand the media's perspective. They have the heavy task of translating the information into content that is helpful and informative yet not overwhelming and scary for a nontechnical audience. Work with the media to address the questions of critical concern. The answers to these questions are required to process more comprehensive information. These questions may seem elementary to technical staff. However, if the questions are not addressed, the audience will not process other information offered to them.

### 9.5.3 After the Crisis

Be transparent, and do not debrief in a closed room. Debrief organizationally but also through the lens of the affected community. Be accountable. Show the follow-through actions, and take responsibility.

## 9.6 Deploy and Debrief

Debriefing is the most overlooked step of the communication process but one of the most powerful for future effectiveness. Address any communication successes and failures in any debriefing with internal and external stakeholders. These insights help guide future strategy and preparedness efforts, resulting in a better understanding of what processes and scenarios need regular practice, pinpointing communication gaps, and identifying individual roles that need shifting.

Most importantly, use this activity as a time to celebrate the wins. There will be many corrections and modifications to make, but do not forget to use this time to celebrate what worked but especially to thank the team.

Crises are stressful and can be traumatic. Be sure to support the team during recovery, listen to their feedback, and express gratitude for their work. Leadership should also use this time for their rest, self-reflection, and recovery.

## 10.0 REFERENCES

American Industrial Hygiene Association. (2023, March). *AIHA COVID-19 resources.* https://www.aiha.org/public-resources/consumer-resources/aiha-covid-19-pandemic-efforts

Centers for Disease Control and Prevention. (2023, March). *Isolation and precautions for people with COVID-19.* https://www.cdc.gov/coronavirus/2019-ncov/your-health/isolation.html#:~:text=Avoid%20being%20around%20people%20who,guidance%2C%20see%20CDC's%20Travel%20webpage

DC Water and Sewer Authority. (2021, November 02). *Time is running out! Disconnections for some DC Water customers begin December 1* [Press Release]. https://www.dcwater.com/whats-going-on/news/time-running-out-disconnections-some-dc-water-customers-begin-december-1

DC Water and Sewer Authority. (2023a, March). *COVID-19 Updates and Resources.* https://www.dcwater.com/covid-19-updates-and-resources

DC Water and Sewer Authority. (2023b, March). *Customer Assistance Programs.* https://www.dcwater.com/customer-assistance

Federal Emergency Management Agency. (2023a, March). *ICS Resources.* https://training.fema.gov/emiweb/is/icsresource

Federal Emergency Management Agency. (2023b, March). *NIMS.* https://www.fema.gov/emergency-managers/nims

Federal Emergency Management Agency. (2023c, March). *Operational period planning cycle (the planning P).* https://emilms.fema.gov/is_0200c/groups/80.html

LUMA Institute. (2021, August). https://www.luma-institute.com/

National Institute of Standards and Technnology. (2023, March). *The five functions.* https://www.nist.gov/cyberframework/online-learning/five-functions

Occupational Safety and Health Administration. (2004, August). *Appendix A to § 1910.134 - Fit testing procedures (mandatory).* https://www.osha.gov/laws-regs/regulations/standardnumber/1910/1910.134AppA

World Health Organization. (2023, March). *Coronavirus disease (COVID-19) pandemic.* https://www.who.int/emergencies/diseases/novel-coronavirus-2019

# Chapter 6

# Occupational Health Protection

*Tom Burgess, MS, CSP, CIH; David Gill; Albert Rubin; & Mark Wilson*



## 1.0 INTRODUCTION

Protecting workers from infectious diseases involves understanding potential levels of exposure to infectious agents, the risk associated with the exposures, and the methods to reduce or control exposures, including personal protective equipment (PPE).

Methods supporting this approach have been established and used in occupational safety and health for the anticipation, recognition, evaluation, and control of chemical, physical, and biological agents. These hazard recognition and exposure assessment methods can be applied to understanding and controlling occupational exposures to infectious diseases in the water industry.

During an infectious disease outbreak, especially in the early stages, information may be limited and evolving. This can include important information on the ways the disease can be transmitted, how infectious it is, disinfection methods, severity of the illness, and treatment methods. For the protection of workers in wastewater collection and treatment operations, information regarding the persistence and risk of the pathogen in wastewater may be limited. This may require a very cautious approach to protect workers that can then be adjusted over time as more information becomes available. The Water Environment Association, the Occupational Safety and Health Administration (OSHA), state and provincial safety authorities, the Canadian Centre for Occupational Health & Safety (CCOHS), and other industry and safety and health resources are available sources of information to help assess and control risks to workers. Safety and industrial hygiene professionals are also available to provide assistance.

## 2.0 EXPOSURE ASSESSMENT

### 2.1 Strategy

The strategy for conducting an exposure assessment is to develop an understanding of the exposures and associated health risks within your organization, and then implement controls for tasks or areas with an unacceptable level of risk.

This exposure assessment process includes developing an understanding of what may be a source of exposure—for example, pathogens in wastewater or the air—and determining when workers may be exposed and when the level or duration of those exposures may present an unacceptable health risk. Controls can then be implemented for specific groups, tasks, and risks, and the controls can be tailored to the specific situation and level of risk.

Much of the information-gathering part of the assessment process is qualitative (i.e., based on observations and available information without being able to use a meter or other methods to measure or quantify exposures). Information on pathogens, routes of exposure, and risk of infection may be limited in the early stages of an outbreak or pandemic. Air monitoring or other sampling methods may not be practical, and quantitative exposure limits may not be established.

Occupational epidemiology studies may provide relevant information related to disease transmission associated with specific job categories, but the results of these types of studies are based on scenarios in which exposures, and infections, have already occurred. This retrospective data can be useful for the identification of specific tasks and/or processes that are associated with increased potential for exposure. For new or emerging diseases, we will need to do the best we can under the circumstances to identify health risks and then establish controls to protect workers through primary exposure-prevention strategies.

In addition to assessing risk and developing controls, successful implementation of health and safety programs is supported by management leadership, employee involvement, and other management system elements. Chapter 5 of this book addresses utility management and explores management systems in detail. Management leadership and employee involvement in the exposure assessment process is important for a successful assessment and implementation of controls. The knowledge of collections and treatment operations, maintenance, process engineers, and microbiology will all be valuable in both the assessment of health risks and the implementation of controls.

Occupational risk assessment for biological hazards is particularly challenging because of the high level of variability in exposures, sampling method limitations, differences in worker susceptibility, and a lack of epidemiological data to support developing occupational exposure limits. Although chemical and physical agents are frequently evaluated on a quantitative basis, a qualitative or semi-quantitative approach is typically used for biological agents (Zisook et al., 2020).

Technical support in exposure assessment can be provided by an industrial hygienist. The American Industrial Hygiene Association (AIHA) defines *industrial hygiene* as

> a science and art devoted to the anticipation, recognition, evaluation, control, and confirmation of protection from those environmental factors or stresses arising in or from the workplace which may cause sickness, impaired health and well-being, or significant discomfort among workers or among citizens of the community (AIHA, 2023).

Organizations like AIHA as well as the OSHA or CCOHS can provide helpful information on exposure assessment and controls to protect worker health.

The exposure assessment strategy that will commonly be referenced in this chapter is the AIHA publication *A Strategy for Assessing and Managing Occupational Exposures* (see Table 6.1). The methods in this publication are widely used by industrial hygienists to assess and control exposure to chemical, physical, and biological agents in the workplace (Jahn et al., 2015).

**TABLE 6.1** Exposure Assessment

| Step | Actions/Outcomes |
|---|---|
| Strategy | Engage stakeholders<br>Establish goals |
| Basic Characterization | Hazard information (infection disease)<br>Health effects, infectious dose<br>Sources and pathways of exposure<br>Areas/tasks with worker exposure |
| Exposure Assessment | Worker exposure groups<br>Tasks within exposure groups with exposure<br>Observed/estimated levels of exposure (qualitative)<br>Sampling (quantitative) |
| Exposure Risk | Determine risk<br>Acceptable—no further action<br>Uncertain—conduct additional assessment<br>Unacceptable—implement controls |

## 2.2 Basic Characterization

The characterization step in the exposure assessment process involves identifying potential sources of pathogens and work areas or tasks where workers may be exposed, as well as characterizing the infectious disease risk.

### *2.2.1 Criteria for Acceptable Exposure (Infection Risk)*

Protecting workers from disease means keeping exposures to infectious diseases below exposure levels that can result in disease transmission. To achieve an appropriate level of control, we need to understand what acceptable levels of exposure are. Biological hazards do not have exposure limits like those for chemicals that can be applied to the occupational environment. For new threats, there may be limited information available to assess the level of exposure from inhalation or contact that could transmit the disease, such as occurred in early 2020 during the start of the COVID-19 pandemic.

The amount of virus necessary to make a person sick is called the infectious dose. There may be variability in infectious dose between variants of the same pathogen. Understanding how much and how long a person needs to be exposed to have a risk of infection is important in assessing the risk of our workplace exposure. A lower infectious dose means that it takes less of the virus, bacteria, or other pathogen to make someone sick, increasing the potential for transmission of the disease. Refer to the Quantitative

Assessment section of Chapter 4 Understanding Risk and Knowledge Gaps for additional information on assessing health risks for infectious diseases.

The understanding of levels of infectious dose and contact or inhalation risk may evolve over time. The fate of infectious diseases in the wastewater collections and treatment process will also affect the risk of exposure. How long will the pathogen live in the collection and treatment system? Will it persist on surfaces? Can it multiply, or is it diluted in treatment? With these questions in mind, it becomes clear that developing occupational exposure regulations for biological hazards will be extremely difficult. However, observational data related to specific processes and job tasks can provide insight into occupational scenarios that are likely to facilitate exposure. Examples include processes associated with the production of aerosols or job tasks that cannot provide sufficient physical distance between employees. Chapter 2 of this book provides additional detail on the fate and transport of pathogens. Table 6.2 lists groups of organisms and their potential exposure pathways.

The severity of the disease and treatment availability are also important considerations. If we are responding to an outbreak or pandemic, it is likely because the disease results in a high rate of serious or fatal illness. In the early stages of an outbreak, effective treatments may not be available. As vaccines and treatments are developed, the risk of serious or fatal illness may decrease. This means that level of risk and precautions will change over time. The severity of the disease may also be different based on the level of exposure or group characteristics, such as people with preexisting medical conditions or older/younger age groups.

For pathogens that can be transmitted directly from person to person, it is also important to get information on the routes of transmission, infectious dose, and acceptable exposure. However, to reiterate, there are no existing occupational exposure regulations that define what acceptable versus unacceptable exposures actually are. Individuals may release pathogens, like bacteria or viruses, from blood or bodily fluids, waste, direct person-to-person contact, or from contact or inhalation of exhaled droplets or aerosols. Pathogens may or may not survive outside the human body including in the air

**TABLE 6.2** Organisms of Concern and Potential Exposure Pathways

| Organism | Routes of Exposure |
|---|---|
| Virus | Respiratory, gastro-intestinal, bloodborne |
| Bacteria | Food, water, respiratory |
| Protozoan parasite | Food, water |
| Fungi | Fomite, food, aerosol |
| Helminth | Food, water |

or on surfaces. How the pathogen responds to the environment will affect the viability of transmission from things like contacting doorknobs or other shared surfaces. The potential for disease transmission will be directly related to the amount of pathogen transmitted, its ability to survive in an infective state outside a host, and its ability to get into and infect another person.

Transmission of disease from animals and birds to humans—known as zoonotic transmission—has been an issue for millennia. Zoonotic transmission of disease has followed human populations from human interactions as hunter-gatherers to the present and was discussed in Chapter 1. The organisms of concern include viruses, bacteria, protozoan parasites, fungi, helminths, and others.

### 2.2.2 Workplace Characterization (Exposure Risk)

The next step in the characterization process builds on our understanding of the health risks by assessing and characterizing the workplace to identify the types of tasks or exposures that may have a potential health risk. This involves assessing the processes, work tasks, and their potential for exposure to pathogens and the potential level of exposure. This, along with information on the infectivity and severity of the disease, will help us understand the level of risk.

Information may come from a variety of sources. In process operations like wastewater collection and treatment, process flow diagrams can be helpful. The risks associated with pathogens in wastewater will vary throughout the process, so understanding not just the exposures but at what stage in the process they occur is important.

Input from workers is important. They are an excellent source of information regarding tasks and exposures. Involvement will also be helpful later when we want to implement controls.

When gathering information on exposure, consider splashing, aerosols, individual work practices, and contact with both wastewater and contaminated surfaces. Larger amounts of liquids or solids have the potential to contain larger amounts of pathogen, and a splash could potentially result in liquid getting on the face or hands and entering the body. Aerosols are very small droplets that can also contain pathogens. Although splashes and droplets drop very quickly, the distance will be related to the force of the splash or spray; aerosols are small enough that they can float in the area for some time and move a larger distance. Aeration tanks and areas in the collection and treatment process with turbulent flow or splashing will generate aerosols.

The generation of these aerosol particles may result in contamination of surfaces in and around the wastewater collection, treatment, potential reuse, and biosolids-handling processes. Contaminated surfaces are a potential source of infection, and simple interventions like hand washing and housekeeping are effective tools in mitigating this risk.

Consider nonroutine tasks including equipment maintenance, as well as potential exposures during leaks, spills, or process upsets.

Assessment of person-to-person exposure includes interactions with the public in areas like customer service or working in public areas as well transmission between coworkers. Time is a factor in the amount of pathogen that may be transmitted, so the length of time and frequency of interactions will affect the risk. The workplace characterization should include shared workspaces where physical distancing is difficult including vehicles.

## 2.3 Exposure Assessment

### *2.3.1 Exposure Groups*

One strategy for making sure that exposure for all workers is assessed is to look at groups of workers with similar exposures. For example, we may be able to group collection operations and identify the types of tasks, potential risks, and eventually precautions for collections. Collection exposures and risks may be different than for treatment operations or biosolids depending on the nature of the work, exposures, and risk (see Table 6.3). This approach makes the assessments more specific to the variations in tasks and exposures, and includes all staff, without the time and effort that would be needed to assess each individual.

In addition, as assessment and control may be an ongoing process, it will be easy to manage changes using exposure groups. If data indicate that the infectious disease does not persist into secondary treatment, the assessments and precautions can be updated for the impacted groups and exposures (Water Environment Federation, 2020).

### *2.3.2 Qualitative Assessment*

Qualitative assessment is the process of assessing potential exposures within our exposure groups through observation and other nonquantitative methods. We can gain a lot of information through the qualitative assessment process. For example, exposure to droplets or aerosols will typically be related to some action such as pumping, agitation, or aeration that creates droplets and aerosols. This distance a droplet, or leak from a hose, can travel will be related to the force or pressure. From observing these factors, we can understand a lot about a primary exposure risk.

#### *2.3.2.1 Aerosols*

Wastewater collection and treatment operations may create aerosols that can result in worker exposure through inhalation and skin contact. Aerosols are small droplets, typically less than 100 micrometers (μm) in diameter.

**TABLE 6.3** Categorization of Types of Activities Performed by Wastewater Treatment Workers and Corresponding Wastewater Exposure and Protection in Ascending Order of Exposure Risk (Maal-Bared, 2022)

| Activity Type | Description |
|---|---|
| Type A | Type A activities are primarily located within office spaces or on-site trailers. Typically, no controls are required for these activities beyond basic hygiene. No primary or secondary contact with wastewater is expected at this level. |
| Type B | Type B includes activities that require walking through facility areas and may include inspections. Typical controls for these activities will include basic hygiene, safety glasses, hard hat, and safety boots. No primary contact with wastewater is expected at this level, though some secondary contact is possible through fomites, splashing, or bioaerosols. |
| Type C | Type C activities involve primary—if minor—contact with some wastewater in the form of adjusting valves, working on wiring, isolating equipment, job setup, and other similar work. Typical controls for these activities will include hard hat, safety glasses, steel-toed boots, coveralls, gloves, handwashing, and other standard administrative controls (e.g., safe work plan, hazard assessments). For some tasks, additional PPE, including a face respirator and portable gas monitor, may be deemed necessary. |
| Type D | Type D includes activities undertaken for lab analysis, sampling, pump maintenance, and disconnecting equipment. Moderate levels of primary contact with wastewater can be expected at this level. Typical controls for these activities include hard hat, safety glasses, steel-toed boots, gloves, handwashing, and other standard administrative controls. A portable gas monitor may also be required at times, depending on the task. |
| Type E | Type E activities include working inside various tanks and lift stations, using hoses, doing emergency repairs, adjusting pressurized equipment or valves, and any other work with a high likelihood of splashing. These activities involve the greatest level of primary contact with wastewater. A wide range of controls may be activated for this activity level, depending on the specific tasks, including hard hat, safety glasses, steel-toed rubber boots, face mask, rubber rain suit or coveralls, gloves, and ear protection. |
| Type E1 | A subtype, Type E1, was created to cover the specific activity that could involve accidentally being splashed in the face with a larger quantity of wastewater. |

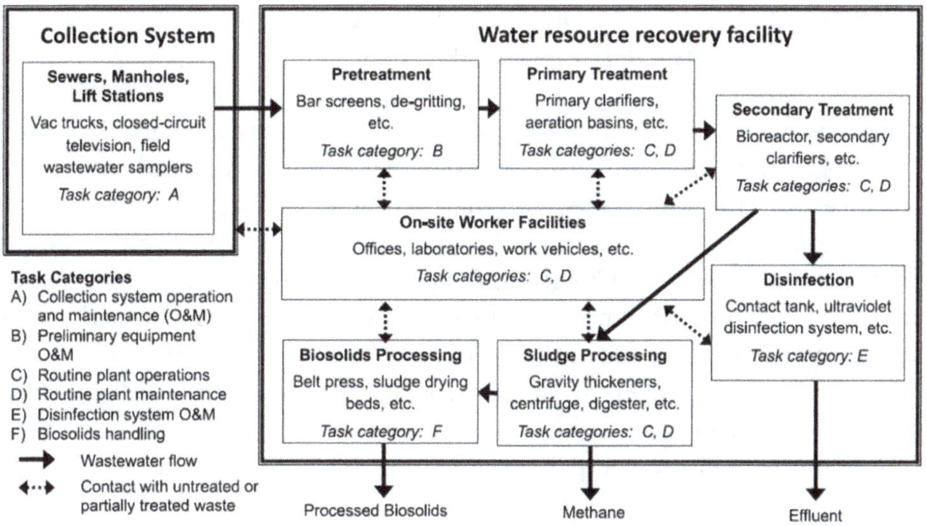


**FIGURE 6.1** Components of the Wastewater Collection System and Water Resource Recovery Facility and Associated Job Task Categories (LeChevallier et al., 2019)

At this small size, the droplets may remain suspended in the air and travel with wind or air currents. Pathogens may be contained within the aerosol. The ability of an aerosol to remain suspended in the air and travel can increase the potential for worker inhalation and skin exposure. Larger droplets almost immediately fall due to gravity, whereas aerosols may be suspended in air and travel based on air movement. The ability of aerosols to remain suspended in the air, potentially exposing workers, will vary based on several factors including size, aerodynamic properties, and weight.

Splashing and agitation typically create larger droplets, not aerosols. The primary mechanism for producing aerosols in wastewater systems is from bubbling. Bubbles breaking at the liquid surface produce very small droplets. This is most pronounced in aeration tanks, but significant turbulence or agitation in other parts of the process can also produce bubbling that can generate aerosols (Baron, 2010).

To create a worker exposure, the aerosol must travel from the surface of the liquid up, against gravity, and be inhaled by the worker or otherwise enter the body.

Although aerosols less than 100 µm have the potential to remain airborne, many factors will affect airborne concentrations and their ability to convey pathogens. These small droplets, and the pathogens within them, are very susceptible to environmental conditions. The water can evaporate, eliminating the environment needed for many pathogens to survive. The aerosol droplets can rapidly change temperature based on the ambient environment. Aerosols also provide little protection from sunlight that can quickly kill UV-sensitive pathogens like viruses. So, although fine aerosols have the aerodynamic potential to stay suspended in air, in reality they will generally evaporate, be diluted, or not provide a suitable environment for a pathogen over an extended period of time. These factors will reduce the distance from the source at which they may present a hazard.

Droplets and aerosols may also be a pathway for person-to-person transmission of disease. Coughs and sneezes can produce droplets greater than 100 μm, but with enough velocity to send them several feet through the air. Normal exhaled air produces aerosols smaller than 10 μm that can stay suspended in air. These small aerosols will normally be diluted with distance, decreasing risk as distance and/or ventilation is increased, but they can accumulate in smaller, poorly ventilated areas.

#### 2.3.2.2 Qualitative Assessment Example

Table 6.4 is an example of the information and risk assessment in a qualitative exposure assessment. Our observations and understanding of the process can help us identify the areas and types of risks.

**TABLE 6.4** Qualitative Assessment Example

| Location | Equipment/Task | Aerosol Risk | Contact Risk |
|---|---|---|---|
| Pretreatment | Screenings conveyor and dumpster | No aerosol generation; screenings have low moisture content and are washed with chlorinated process water. | No increased risk; screenings have low moisture content and are washed with chlorinated process water. |
| Equalization Basin | Operations | Aerosols are generated from mixing. There is visible bubbling at the surface near the center of the basin.<br>The depth of the tank and distance to edge of the tank limit exposure to aerosols adjacent to the basin. | No splashing risk due to level of agitation and depth of surface |
| Equalization Basin | Clean-out | No aerosol generation when basin is drained for entry | Potential for contact with residual solids and liquids |
| In-Facility Pump Station | Submersible pump removal | No aerosol generation in wet well | Potential for contact with solids and liquids on pumps. Pumps can be hosed down with chlorinated water before handling. |

### 2.3.3 Quantitative Assessment

Quantitative methods, like air sampling, may not be feasible. There are significant limitations associated with the ability to detect certain types of pathogens with existing sampling techniques. For instance, the methodology for indoor air sampling for using the National Institute of Occupational Safety and Health (NIOSH) analytical method (NAMM-800) focuses on the detection of culturable microorganisms, such as bacteria and fungi. This method is of limited use if the agent of concern, such as a virus, is not amenable to culture-based assessments or if the agent is novel and we have limited understanding of its growth requirements. Molecular approaches (i.e., polymerase chain reaction or PCR test) and detection of target proteins (i.e., enzyme linked immunosorbent assay or ELISA test) may be used to provide information related to the presence of specific pathogens but do not inform occupational health professionals about the viability of the hazards.

Assessment of the size distribution of an aerosol can give important information related to the likelihood of an inhalation exposure risk. Many direct-reading particle counters can provide size-fractioned assessments in near-real-time. These can be useful tools in exposure evaluations.

As infection requires viable microorganisms, this is a significant limitation of these sampling methods. Recent work focused on overcoming these limitations with the adaptation of existing sampling technologies to detect the influenza virus has shown promise in controlled laboratory settings. However, this approach has not been evaluated or accepted for use in the field (Cao et al., 2011; NIOSH, 2022).

Surface sampling may not be a reliable indicator of worker exposure, as these samples are not from the breathing zone of a worker, and they do not monitor a biological marker of exposure. These are proxies of exposure, and their use should be considered carefully.

Methane, hydrogen sulfide, and other gases may be produced by bacteria and other organisms (i.e., archaea) in wastewater collection and treatment systems. Methane and hydrogen sulfide can be easily measured using commercially available single or multi-gas meters. Although the organisms that produce methane and hydrogen themselves (i.e., methanogens and sulfur reducers) are not pathogenic, the presence of these gases may indicate conditions affecting the viability of infectious disease pathogens within the collection and treatment process.

### 2.3.4 Exposure Risk

The disease "risk" will be a combination of the transmission/exposure potential and the severity of the disease. We must also have special consideration for diseases that may have a low rate of transmission but that cause very severe or disabling illnesses or death.

Our goal in risk assessment and controls is to have an "acceptable" level of risk. What an acceptable level of risk is can be very difficult to assess, particularly when information is limited regarding an emerging disease. A simple example of risk management is a speed limit. This is not a limit of safe or unsafe, but a standard of what is acceptable for the road. A traditional risk matrix is shown in Figures 6.2 and 6.3.

With many chemicals, we have exposure limits that are intended to protect workers from adverse health effects. These exposure limits are defined as the concentration of the substance in air that can be breathed by most people without adverse effect. For some substances like radiation or carcinogens, there is also the practice of keeping exposures "as low as reasonably achievable," or ALARA.

In the earlier stages of an infectious disease outbreak, particularly when potential health effects are severe, we will want to adopt an ALARA approach. This may mean changes in work practices, barriers or other controls, and gloves, respirators, or other PPE.

When information is available regarding the transmission and severity of the disease, the assessment of risk can change to an approach similar to chemical exposure limits. If the disease is associated with severe illness, and treatments are lacking, we may need to stay with an ALARA approach. Intentionally allowing disabling or fatal illnesses is not acceptable.

If the disease is less severe and/or treatable, and we have reasonable information regarding the transmission risk, we can adjust our controls. Deciding what is acceptable can be difficult. Referring to similar risks can be helpful. For example, our goal may be to have disease risk at or below what is experienced during a normal flu season, or from the normal transmission

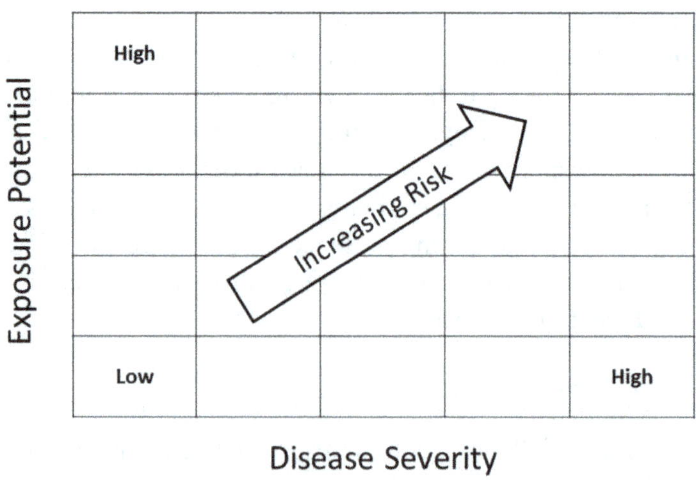


FIGURE 6.2 Risk Matrix

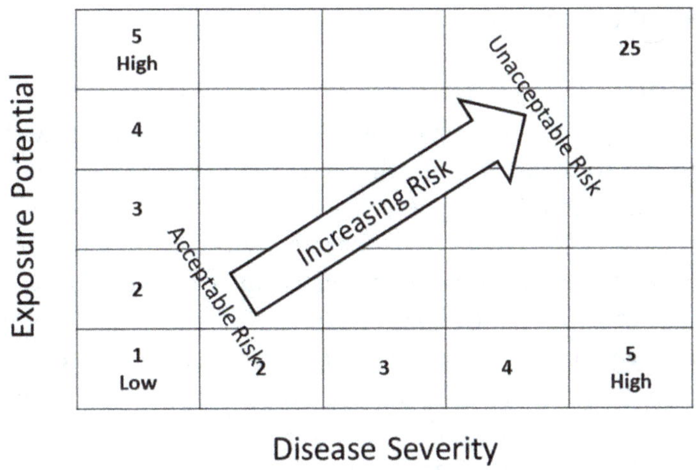


**FIGURE 6.3** Risk Matrix With Risk Ranking

of pathogens in the wastewater. Refer to the Quantitative Assessment section of Chapter 4 Understanding Risk and Knowledge Gaps for additional information on assessing health risks for infectious diseases.

Risk ranking is a method to help organize the information from our assessments and prioritize efforts in areas with the highest risk. A simple risk-ranking process can be to use numeric rating for transmission and severity (see Table 6.5 and Figure 6.3).

## 3.0 CONTROLS

United States' and Canadian safety regulations do not have standards that apply specifically to the protection of workers from pathogens in wastewater. There are general requirements to assess hazards and protect workers even when a hazard is not covered by a specific standard, including the

**TABLE 6.5** Risk Ranking of Tasks

| Task | Exposure | Exposure Rating *Low 1, High 5* | Severity Rating *Low 1, High 5* | Exposure × Severity |
|---|---|---|---|---|
| Opening pump | Contact | 4 | 2 | 8 |
| Opening pump | Ingestion | 3 | 5 | 15 |
| Collecting sample | Contact | 2 | 2 | 4 |
| Enter drywell | n/a | 1 | 1 | 1 |
| Aeration tanks | Aerosols | 3 | 4 | 12 |

OSHA general duty clause and Canadian federal and provincial safety and health codes.

The OSHA Bloodborne Pathogens Standard (29 CFR 1910.1030) provides the requirements for employers having employees with occupational exposure to blood and/or other potentially infectious material. These requirements are intended to prevent worker exposure to infectious bloodborne agents (OSHA, 2019). The operative word in this OSHA standard is *bloodborne*. Contact with raw sewage and wastewater pose a number of worker safety and health hazards, but OSHA does not generally consider contact with raw sewage or wastewater to be an exposure route to bloodborne pathogens. There have been virus infections that are bloodborne (human immunovirus, Ebola) that are transmitted by bodily fluids including blood. Regardless, the employer is responsible for determining which, if any, employees have occupational exposure and are covered under the standard.

Workers handling human waste or sewage in collection, treatment, and residuals management operations may be at increased risk of illness associated with that exposure and contracting illness from waterborne diseases. To reduce this risk and protect against illness, controls and accepted practices associated with wastewater treatment facility operations should be used. These controls and practices can include ventilation, hygiene precautions, specific safe work-site practices, and use of PPE normally required when handling untreated wastewater.

## 3.1 Hierarchy of Controls

The hierarchy of controls shown in Figure 6.4 is an accepted method for determining how to implement feasible and effective controls to protect workers. The idea behind this hierarchy is that the control methods at the

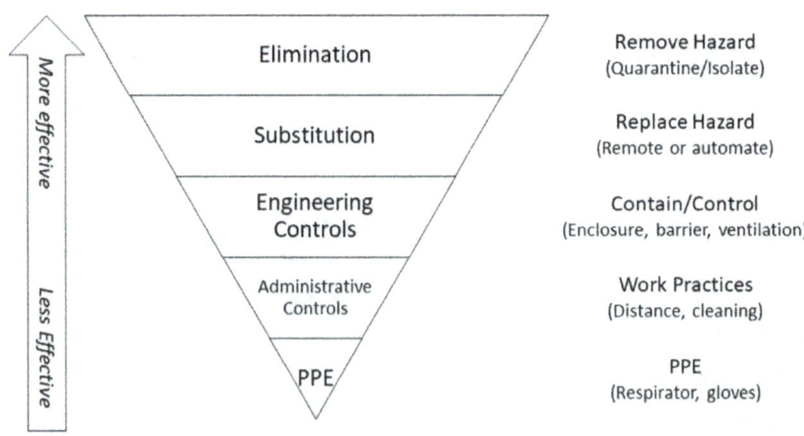


**FIGURE 6.4** Hierarchy of Controls (NIOSH, 2022)

top of graphic below are potentially more effective and protective than those at the bottom. Following this hierarchy normally leads to the implementation of inherently safer systems, where the risk of illness or injury has been substantially reduced (NIOSH, 2022).

The Hazard Prevention Program requirements of the Canadian Occupational Health and Safety Regulations (2022) and provincial or territorial requirements also include the use of a hierarchy of preventive measures to address hazards, including the elimination of the hazard, the reduction of the hazard, PPE, and administrative procedures. Control examples for wastewater operations are shown in Table 6.6.

## 3.2 Elimination/Engineering

Eliminating the source of the disease is a very effective method of protecting workers from illness. However, this may not always be feasible in the wastewater collection and treatment process when the wastewater itself may be a source or pathway for the disease.

In human interaction, there may be opportunities to eliminate the source of disease by eliminating or reducing person-to-person interaction in areas like bill payment and customer service, where electronic or remote systems may be effective.

**TABLE 6.6** Control Examples for Wastewater Operations

| Control Strategy | Example |
|---|---|
| Elimination | • Automated sampling to eliminate manual collection<br>• Eliminate source of splashing in process<br>• One person in vehicle to eliminate person-to-person<br>• Cleaning and disinfection |
| Substitution | • Use of potable water instead of facility water<br>• Change equipment to reduce splashing/aerosols |
| Engineering | • Enclosure or ventilation to contain aerosols<br>• Splash barrier<br>• Barrier between workstations to eliminate person-to-person |
| Administrative | • Physical distancing of employees<br>• Screening and vaccination requirements<br>• Cleaning and disinfection |
| Personal Protective Equipment | • Training on use and supply of appropriate respirators, gloves, and coveralls to employees |

Engineering controls are favored over administrative and PPE for controlling existing worker exposures in the workplace because they are designed to remove the hazard at the source. Control at the source reduces the potential for worker exposure and can be effective without the need for individual compliance, like wearing a respirator. Conceptually engineering controls are similar to a building sprinkler system for fire suppression that will work automatically as designed in the absence of an operator, whereas a fire extinguisher requires a properly trained user to operate. Well-designed engineering controls can be highly effective in protecting workers and will typically be independent of worker interactions to provide this high level of protection. The initial cost of engineering controls can be higher than the cost of administrative controls or PPE, but over the longer term, operating costs are frequently lower and, in some instances, can provide a cost savings in other areas of the process (NIOSH, 2022).

Engineering controls can take time to design and implement, which may not be feasible during a rapidly evolving pandemic. Engineering improvements can be included in plans for future equipment and processes. For example, automated screens to replace manual screens can minimize risk of contact with influent.

Other engineering controls include enclosures/barriers to reduce the potential transmission of disease-causing agents between the source and people.

Vaccines are also a source control, as they reduce the severity and/or transmissibility of a disease. A vaccinated person may present a lower risk of being a source of disease. Vaccines may not be available for a new type or strain of disease. Vaccination recommendations for workers exposed to sewage or human waste should be developed in consultation with local health authorities. Tetanus vaccinations should be up to date, with consideration also given to the need for vaccinations for polio, typhoid fever, hepatitis A and B, COVID-19, and vaccinations developed in the future as knowledge increases regarding pathogens of emerging concern.

## 3.3 Administrative

Administrative controls include management of work practices to minimize worker exposure to sources of disease. Examples include changing work practices, social distancing, cleaning and disinfection, and personal hygiene. Medical management programs may include testing, vaccines, quarantine/isolation, and treatment plans. Workers must also be urged to promptly seek medical attention if displaying any signs or symptoms of the infectious disease of concern or other waterborne diseases such as vomiting, stomach cramps, and watery diarrhea.

Employee education and training on how to conduct their work safely helps to minimize the risk of exposure and is a critical element of any complete workplace health and safety program. Training must cover not only how to do the job safely, but it must also ensure that workers understand the hazards and risks of their job. It is important to note that use of chemicals, UV light, or other disinfecting methods can mitigate issues associated with biohazards but cause new exposures requiring worker protection. Interventions to control hazards must be evaluated in terms of the hazard the intervention may represent as well. Ultimately, training must provide employees with relevant information on how to protect themselves and coworkers from all hazards present in their occupational environments (CCOHS, 2022).

Basic hygiene practices for workers include the following:

- Wash hands with soap and water immediately after working in areas with human waste or sewage or after an exposure to source of potentially infectious disease.
- Wash your hands with soap and water before eating, drinking, or smoking and before and after using the toilet.
- Avoid touching face, mouth, eyes, nose, or open sores and cuts.
- Remove soiled work clothes before entering designated eating areas.
- Keep open sores, cuts, and wounds covered with clean, dry bandages.
- Remove dirty work clothes and shoes before leaving the worksite, and wash contaminated work clothing after use.
- If human waste or sewage comes into contact with your eyes, gently flush them with safe water.

## 3.4 Personal Protective Equipment

Personal protective equipment, commonly referred to as "PPE," is equipment worn to minimize exposure to hazards that cause workplace injuries and illnesses. PPE may include items such as gloves, safety glasses or face shields, earplugs or muffs, hard hats, respirators, boots or safety shoes, and work clothing or protective suits.

In the hierarchy of controls, PPE is rated as the least preferred, as the hazard is still present in the workspace, and worker protection is dependent on the effectiveness and proper use of the PPE. In an emergency, workers are in a potentially hazardous environment, increasing the complexity of emergency response. Controlling hazards at the source or pathway to prevent or minimize the hazards in the workspace is preferred.

Selection and fit of PPE are important for it to work properly and reduce risk. Selection factors include level of protection, comfort and practicality for the work tasks, sizes and fit, and durability. Health and safety professionals or nongovernmental associations can be consulted for assistance with selection and worker training regarding proper fit, use, and inspection of PPE.

The OSHA PPE standard requires documentation of hazard assessments and equipment selection "if hazards are present, or are likely to be present, which necessitate the use of personal protective equipment" and training in use and inspection (OSHA, 2016a). All jurisdictions in Canada have requirements for worker use of PPE.

Basic PPE for wastewater operations include the following:

- Safety glasses, goggles, or face shield to protect the eyes and/or face from aerosols, droplets, or splashes.
- Work clothes or liquid-repellent coveralls to minimize or prevent body contact.
- Work or waterproof gloves to protect hands from cuts or abrasions on contaminated surfaces or direct contact with wastewater or other potential disease sources.
- Work shoes or waterproof boots to provide secure footing and protect feet from contact or immersion in wastewater or other contaminates.
- Respiratory protection for protection from airborne gases, vapors, dusts, or aerosols.

The American National Standards Institute (ANSI) and the Canadian Standards Association (CSA) are nongovernmental associations that have established standards for PPE that are referenced by the applicable regulations. For protective glasses to be considered OSHA-compliant safety glasses, they must meet the requirements of the ANSI Z87 standard. The NIOSH and CSA have requirements for respiratory protection. Look for the applicable ANSI, CSA, or NIOSH reference when selecting PPE.

Respirators are used where hazardous concentrations of gases, vapors, dusts, or aerosols are or may be encountered in the work environment. Aerosols or dust in collection systems, at aeration basins, and on irrigation or land application sites may contain potentially pathogenic organisms and dust particles.

When respirators are required, the OSHA Respiratory Protection Standard and similar Canadian standards require a workplace respiratory protection program including selection, fit testing, and worker training. Fit testing verifies the fit of the respirator; facial hair will interfere is with the seal to the face, reducing the effectiveness of the respirator, and is not permitted. Filtering facepiece respirators, like a N95, can be used on a voluntary basis with fewer requirements. Medical face masks and other face coverings are not considered respirators under the OSHA standard.

In response to the COVID-19 pandemic, the OSHA introduced an emergency temporary standard (1910.504) for respiratory protection for high-risk workers in health care settings. Many aspects of this standard are relevant for other work settings, including wastewater operations. Under the 1910.504 standard, the term *Mini Respiratory Protection Program* (Mini RPP) outlines the requirements. The Mini RPP requires training on respirator use and on user seal checks, with seal checks as an acceptable alternative to fit testing. User seal checks include simple procedures that allow end users to check if a tight-fitting respirator is making a good seal with their face. A tight respirator seal requires the absence of facial hair.

**N95 and Other Respirators**

A N95 filtering facepiece respirator is similar to a dust mask. To be rated "N95" means that the respirator/mask meets certain NIOSH requirements for design and filter efficiency. The term *respirator* is used to differentiate it from dust masks that do not meet NIOSH requirements. NIOSH-approved masks will have a NIOSH certification on the package and the term *N95* on the mask. *N95* means the filter is not oil resistant and has a 95% efficiency for small (0.3 micron) aerosol used for NIOSH certification. The Canadian CAN/CSA-Z94.4 standard is based on the NIOSH standard and uses similar criteria.

There are also N95-rated filters that can be used in rubber or silicone respirators. R95 and P95 also have a 95% filter efficiency and are oil resistant or oil proof. That is important in the workplace if you are working around hydrocarbon solvents that could degrade an N-rated filter. Filters with 99 and 100 ratings have higher efficiency. 100-rated filters may also be called HEPA filters and have 99.7% efficiency.

Having a higher efficiency filter does not necessarily result in a higher level of protection. The OSHA assigned protection factor is the workplace level of respiratory protection that a respirator or class of respirators is expected to provide. Half-face respirators have a protection factor of 10, roughly equivalent to 90%, regardless of if they have a N95 or P100 filter. A full-face or other type of respirator is needed to get more protection.

A surgical mask is not a "respirator," as defined by the OSHA. It is designed to contain aerosols (i.e., saliva) from a cough, sneeze, and so on, by the user. It does not have a tight seal around the face and will provide some but lesser protection than an N95 to the user from inhaling aerosols.

## 3.5 Controlled Risk

The same methodology used to assess and rank risk can be used assess the level of risk with controls implemented. For high or complex risks, a combination of controls may be needed. Table 6.7 highlights some of the tasks with a rating of uncontrolled and controlled and risks.

**TABLE 6.7** Controlled Risk

| Task | Exposure | Exposure Rating *Low 1, High 5* | Severity Rating *Low 1, High 5* | Exposure × Severity |
|---|---|---|---|---|
| Opening pump | Contact | 4 | 2 | 8 |
| *With controls*<br>• Disinfection<br>• Gloves | | *1* | 2 | 2 |
| Opening pump | Ingestion | 3 | 5 | 15 |
| *With controls*<br>• Disinfection<br>• Face shield | | *1* | *3* | *3* |

Controls like training, screenings, vaccinations, stringent testing and isolation policies for known contacts of infected persons, and medical treatment can reduce the potential for disease transmission and/or severity of the illness across all tasks, decreasing risk. The risk matrix with risk reductions through the use of proper controls can be seen in Figure 6.5.

## 4.0 IMPLEMENTATION AND MANAGEMENT

Our outlined exposure assessment process has guided us through the assessment and control of infectious disease risk. A variety of factors, including the effectiveness of the controls and your organization's ability to adapt

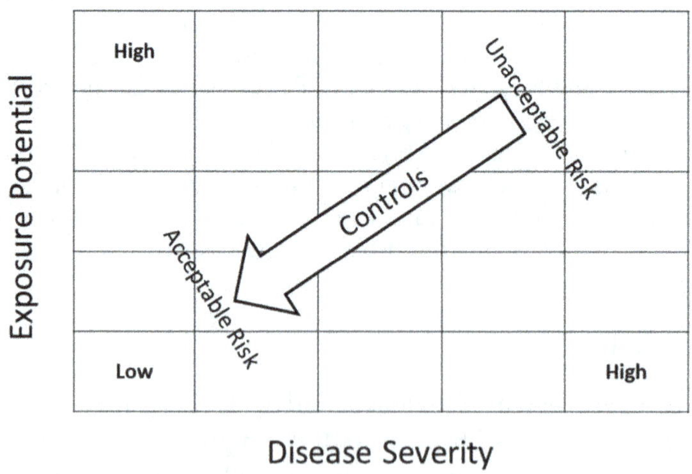


**FIGURE 6.5** Risk Reduction With Controls

and improve, can influence the way health and safety is managed at your organization.

A *safety management systems framework* can be used to assess and improve your organization's health and safety programs and reduce the risk of injury and illness. A systems approach is widely recognized by government and industry experts. In its recommended practices, OSHA (2016b) cites several benefits that organizations have experienced to developing proactive safety processes based on the concept of continuous improvement.

In this section of the chapter, we have outlined elements of the International Organization for Standardization (ISO) 45001 Occupational Health and Safety Management Systems standard. The ISO 45001 standard is a voluntary framework for an occupational health and safety management system developed to help organizations managing occupational health and safety risks and opportunities. ISO 45001 is applicable to all organizations, regardless of size, industry, or nature of business. It is designed to be integrated into an organization's existing management processes and follows the same high-level structure as other ISO management system standards, such as ISO 14001 for environmental management.

The ISO 45001 Occupational Health and Safety managements systems are based on the Plan-Do-Check-Act approach, a cycle of continuous improvement also used in quality management systems (see Figure 6.6).

- Plan—assess risks and opportunities; then develop objectives to achieve the desired results.

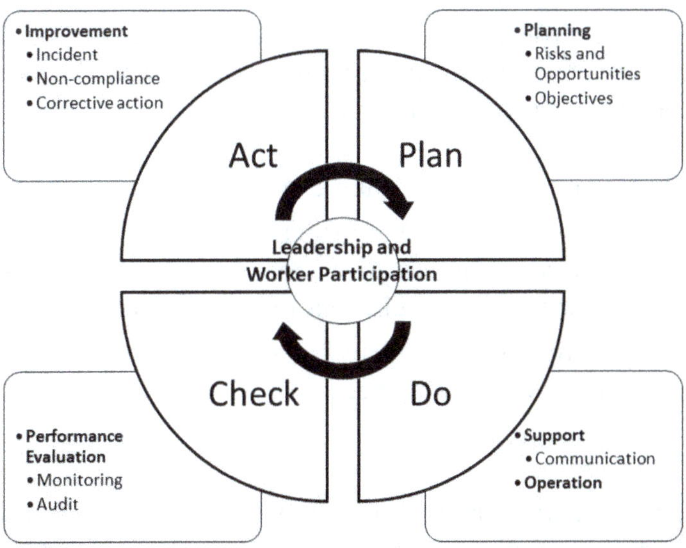


**FIGURE 6.6** ISO 45001 Framework

- Do—implement the process.
- Check—monitor and evaluate activities and processes; compare to the objectives, and report the results.
- Act—take action to continually improve and achieve the intended outcomes.

The effective assessment and control of infectious disease risks is an ongoing process. This is especially true for new or emerging diseases where information on risk and efficacy of control strategies can change regularly. We must continually assess the effectiveness of our controls and make improvements as needed. For example, we could find problems with worker use of PPE, and may need to make changes to equipment, training, or supervision.

## 4.1 Leadership and Employee Participation

> Leadership and commitment from the organization's top management, including awareness, responsiveness, active support and feedback, are critical for the success of the OH&S management system and achievement of its intended outcomes; therefore, top management has specific responsibilities for which they need to be personally involved or which they need to direct. (ISO, 2018)

During times of uncertainty, leadership commitment is crucial in an industry where shutting down is not an option. It is up to leadership to develop a team to provide them timely, relevant information to make decisions on risks that are known and risks that may be developing, trusting the team to make those decisions and achieve the desired objectives. Leadership commitment takes many shapes during outbreaks, which can be seen through the lens of the hierarchy of controls:

- Workplace modifications (engineering and administrative controls): Telework, alternative work schedules, office/work station relocation
- Financial effects (administrative controls): Planning for lost revenue, budgeting for supplies, equipment rental, PPE, and so on.
- Communication: Open, timely, and clear training that is relevant to specific workers and industries
- Lead by example: Wearing a face covering, getting vaccinated

Discuss employee participation in the context of managing infectious disease risk. For example, engaging workers with assessing risk and controls during pump repair or sampling.

## 4.2 Resources and Communication

Resources will be needed from control strategies, from engineering interventions to PPE to time to conduct exposure assessments and implement controls. Budget and staffing changes may be needed to address the time and costs associated with infectious disease prevention.

Communication will be important, as information during infectious disease outbreaks can be very fluid. Dedicated resources may be needed to track relevant and updated information. Providing information from reputable resources is also vital and maintains consistency in reporting; these resources can include the following agencies:

- Centers for Disease Control and Prevention;
- Occupational Health and Safety Administration;
- American Industrial Hygiene Association;
- National Institute for Occupational Safety and Health; and
- World Health Organization.

Multiple methods of communication are needed to and from various levels of the organization from operators to senior management. Management planning meetings that include input on the infectious disease are essential for planning. Regular communication through management and supervision regarding precautions and requirements will be needed, as these may change regularly, particularly at the early stages of an outbreak.

The effectiveness of general communication can be improved by repletion and using a variety of methods, including the following:

- posters/banners;
- electronic postings and email;
- toolbox talks and information sessions;
- risk assessments or job hazard analysis; and
- written procedures.

Information should be communicated in languages that workers can understand and should account for varying levels of education. There should be an opportunity to ask questions and have them answered.

## 4.3 Operational Planning and Control

A disease outbreak may affect everything from availability of supplies to staffing. The planning process should address the resources needed to maintain the necessary controls. Sudden increased demands on supplies like N95 respirators and disinfectants can be problematic. Changes in staffing, vehicles, maintenance, or sampling may need to be addressed to support the implementation of controls.

Planning will be an ongoing process. The ISO 45001 (2018) standard states, "Planning is not a single event, but an ongoing process, anticipating changing circumstances and continually determining risks and opportunities, both for the workers and for the OH&S management system."

Processes and procedures will need to adjust as more information becomes available. Having systems in place that allow the ability for change without much effort in an ever-changing environment is critical. Risk assessments, including job hazard analysis, are good tools for this. Once the potential risk and exposure have been identified, the risk assessment can be updated as the risk potential fluctuates.

## 4.4 Emergency Preparedness and Response

Emergency planning and response should address complexity that may be added by the infectious disease outbreak. For example, a release in the collection system may require additional PPE and decontamination. Also, in an emergency, precautions may be needed to protect workers who need to be in close proximity, including with the public or other responders.

In an infectious disease outbreak emergency, preparedness and response may involve a variety of organizations. In the United States, the National Incident Management System (NIMS) and Incident Command System (ICS) framework has been adopted for emergency response. "The National Incident Management System (NIMS) guides all levels of government, nongovernmental organizations and the private sector to work together to prevent, protect against, mitigate, respond to and recover from incidents" (FEMA, 2022).

Helpful components from those systems include continuity of operations plans, emergency action plans, emergency response plans, and incident action plans. The standardized framework of NIMS and ICS can aid in stakeholder engagement and define clear lines of communication with local hospitals, police and fire departments, Homeland Security, health departments, and the Cybersecurity and Infrastructure Security Agency. The Federal Emergency Management Agency and other organizations offer a variety of guides and training in emergency planning and the NMIS and ICS systems.

## 4.5 Performance Evaluation

Tracking and evaluating performance provides a measurement of controls and effectiveness. Measures can be proactive or reactive. An example of a proactive measure is the number of exposure assessments completed. This is proactive, as it measures a preventative activity that happens before an illness occurs. Other proactive measures can include implementing a planning meeting, monitoring new health standards and recommendations, obtaining supplies and resources, utilizing communication programs, PPE use, or using disinfectants.

Audits and observations are also ways to evaluate performance. What is the level of compliance with PPE and other precautions? What comments or feedback are provided by operators or supervisors?

Reactive measures to incidents or illnesses can use many of the same processes used to track work-related injuries and illnesses, including tracking positive cases (work related, nonwork related, travel, etc.), close contacts (work related, nonwork related), employees with symptoms, employees on quarantine, open versus closed cases, the corrective actions implemented.

It is important to remember that this is a continuous improvement process, and the results of the performance evaluations should address not just the immediate findings but also the system elements that can prevent reoccurrence. If a worker isn't wearing the required PPE, is it a pattern of behavior with the individual and/or other workers that could be addressed through changes to equipment, information, or supervision? If the equipment is problem, is it comfort, interference with the work, or other issues? Asking multiple layers of questions can help get beyond the immediate problem to the systems issues.

## 5.0 RESOURCES

American Industrial Hygiene Association. (2023). *About AIHA*. https://www.aiha.org/about-aiha

Baron, P. (2010). *Generation and behavior of airborne particles (aerosols)*. National Institute for Occupational Safety and Health. https://www.cdc.gov/niosh/topics/aerosols

Burdsall, A. C., Xing, Y., Cooper, C. W., & Harper, W. F. (2020). Bioaerosol emissions from activated sludge basins: Characterization, release, and attenuation. *Science of the Total Environment, 753*.

Canada Occupational Health and Safety Regulations. (2022). Part XIX, hazard prevention program. https://laws-lois.justice.gc.ca/eng/regulations/sor-86-304/page-40.html

Canadian Centre for Occupational Health & Safety. (2022). *Hazard control*. www.ccohs.ca/oshanswers/hsprograms/hazard_control

Cao, G., Noti, J. D., Blachere, F. M., Lindsley, W. G., & Beezhold, D. H. (2011). Development of an improved methodology to detect infectious airborne influenza virus using the NIOSH bioaerosol sampler. *Journal of Environmental Monitoring, 13*, 3321–3328. doi:10.1039/c1em10607d]

Federal Emergency Management Association. (2022, November 29). *National Incident Management System*. https://www.fema.gov/emergency-managers/nims

International Organization for Standardization. (2018). *Occupational health and safety management systems—Requirements with guidance for use*. ISO 45001. https://www.iso.org/obp/ui/#iso:std:iso:45001:ed-1:v1:en

Jahn, S. D., Bullock, W. H., & Ignacio, J. S. (2015). *A strategy for assessing and managing occupational exposures* (4th ed.). American Industrial Hygiene Association.

LeChevallier, M. W., Mansfield, T. J., & Gibson J. M. (2019). *Protecting wastewater workers from disease risks: Personal protective equipment guidelines*. Water Environment Research.

Maal-Bared, R. (2022, December). Protecting wastewater workers by categorizing risks of pathogen exposures by splash and fecal-oral transmission during routine tasks. *Waste, 1*(1), 95–104. Multidisciplinary Digital Publishing Institute.

Morawska, L., & Buonanno, G. (2021). The physics of particle formation and deposition during breathing. *Nature Reviews Physics, 3*, 300–301.

National Institute of Occupational Safety and Health. (2011). *NIOSH manual of analytical methods* (5th ed.).

National Institute of Occupational Safety and Health. (2022). *Hierarchy of controls*. https://www.cdc.gov/niosh/topics/hierarchy

Occupational Safety and Health Administration. (2009). *Assigned protection factors for the revised respiratory protection standard*. OSHA 3352–3302.

Occupational Safety and Health Administration. (2016a). *Personal protective equipment standard*. 29 CFR 1910 Subpart I.

Occupational Safety and Health Administration. (2016b). *Recommended practices for safety and health programs*.

Occupational Safety and Health Administration. (2019). *Bloodborne pathogens standard*. 29 CFR 1910.1030.

Occupational Safety and Health Administration. (2021). *COVID-19 healthcare ETS*. https://www.osha.gov/coronavirus/ets

Water Environment Federation. (2020). *Blue Ribbon Panel on Biological Hazards and Safety Precautions (2020): Protecting wastewater professionals from COVID-19 and other biological hazards*. https://www.accesswater.org/?id=-10027929

Zisook, R. E., Monnot, A., Parker, J., Gaffney, S., Dotson, S., & Unice, K. (2020). Assessing and managing the risks of COVID-19 in the workplace: Applying industrial hygiene (IH)/occupational and environmental health and safety (OEHS) frameworks. *Toxicology and Industrial Health, 36*(9).

# Chapter 7

# Wastewater-Based Surveillance

*Anna Mehrotra; Tiong Gim Aw, PhD; & Krystin Kadonsky*



## 1.0 BACKGROUND

### 1.1 Definitions

Public health surveillance is the "ongoing, systematic collection, analysis, and interpretation of health-related data essential to planning, implementation, and evaluation of public health practice" (Thacker & Berkelman, 1988). Effective surveillance is the backbone of a strong public health system. It allows public health agencies to identify threats to public health (e.g., emerging and reemerging infectious disease outbreaks, bioterrorism events) and respond quickly by deploying resources effectively and creating evidence-based interventions, policies, and programs (Krishna, 2021; Public Health Agency of Canada, 2022).

During the COVID-19 pandemic, one public health surveillance tool that received worldwide attention, setting the stage for future use, is wastewater-based surveillance (WBS). Wastewater-based surveillance is a tool used to measure pathogenic organisms, biochemical markers, pharmaceuticals, and markers of antimicrobial resistance within the sewershed area of a community (Sims & Kasprzyk-Hordern, 2020). It can provide comprehensive unbiased biological and chemical information on the overall health of a community and act as an early warning system for infectious disease outbreaks and environmental exposure (Kirby et al., 2021).

Wastewater-based surveillance focuses on understanding the distribution of health-related states in a population but not the determinants of this distribution. Epidemiology is the study (scientific, systematic, and data-driven) of the distribution (frequency, pattern) and determinants (causes, risk factors) of health-related states and events (not just diseases) in specified populations (neighborhood, school, city, state, country, global). Given this overlap, the terms *WBS* and *wastewater-based epidemiology* (WBE) are often used interchangeably, resulting in much debate about the appropriate use of each term. This chapter provides an overview of the history, methods, limitations, and ethical considerations related to WBS with a strong focus on SARS-CoV-2 surveillance.

## 1.2 Origins

Wastewater-based surveillance relies on the detection of human excretion products in untreated wastewater to inform public health practice and started with the detection of *Salmonella typhi* in Ireland in the 1920s (Wilson, 1928) and poliovirus in the United States in the 1930s (Trask & Paul, 1942). Later WBS theory focused on illicit drugs as the human excretion products (i.e., analytical target) of interest (Daughton, 2001). The first reported WBS implementation for illicit drug surveillance, referred to as "sewage epidemiology," was in 2005 and 2006 in three European cities (Zuccato et al., 2008). In addition to poliovirus and illicit drugs, WBS research has also considered licit drugs, specific pharmaceuticals, counterfeit medicine, biomarkers of population size, industrial chemicals, biomarkers of stress, antimicrobial resistance, and other viruses. Table 7.1, which is adapted from summaries provided in Choi et al. (2018) and Xagoraraki and O'Brien (2020), shows the wide range of substances investigated for WBS to date.

**TABLE 7.1** Examples of Substances Investigated in Wastewater-Based Surveillance Studies (Adapted from Choi et al., 2018; Xagoraraki & O'Brien, 2020)

| Substance Category | Specific Targets | Selected Relevant Reference |
|---|---|---|
| Illicit Drugs | Amphetamines | Lai et al. (2013) |
| | Cannabis | Bijlsma et al. (2009) |
| | Cocaine | Khan and Nicell (2011) |
| | Heroin | Du et al. (2017) |
| Licit Drugs | Alcohol | Rodriguez-Alvarez et al. (2014) |
| | Caffeine | Gracia-Lor et al. (2017) |
| | Nicotine | Senta et al. (2015) |
| Pharmaceuticals | Antibiotics | Subedi et al. (2017) |
| | Antidepressants | Golovko et al. (2014) |
| | Benzodiazepines | Baker et al. (2012) |
| | Pharmaceutical opioids | Jacox et al. (2017) |
| Counterfeit Medications | Sildenafil | Venhuis et al. (2014) |
| Population Biomarkers | Ammonia | Been et al. (2014) |
| | Creatinine | Rico et al. (2017) |
| | 5-hydroxyindoleacetic acid | |
| Industrial Chemicals | Organophosphate ester flame retardants | O'Brien et al. (2015) |
| | Pesticides | Rousis et al. (2017) |
| | Plasticizers | Gonzalez-Marino et al. (2017) |
| Stress Biomarkers | 8-iso-PGF2α | Daughton (2012b) |
| Antimicrobial Resistance | Various genes | Caucci et al. (2016) |
| Viruses | Adenoviruses | Katayama et al. (2008) |
| | Coronaviruses | Medema et al. (2020b); Gerrity et al. (2021) |
| | Non-polio enteroviruses | Prado et al. (2011) |
| | Poliovirus | Hovi et al. (2012) |
| | Hepatitis viruses (A & E) | Villar et al. (2007) |
| | Noroviruses | Hellmér et al. (2014) |

## 1.3 Historical Uses by Public Health Agencies

### 1.3.1 Poliovirus and the Global Polio Eradication Initiative

Led by the World Health Organization (WHO) working in partnership with Rotary International, the U.S. Centers for Disease Control and Prevention (CDC), the Gates Foundation, and UNICEF, the Global Polio Eradication Initiation (GPEI) was created in 1988 to eradicate the disease poliomyelitis (Asghar et al., 2014). The GPEI has relied on immunization campaigns and robust surveillance to reduce the number of polio-endemic countries from 125 in 1988 to 2 in 2022. Surveillance consists of both acute flaccid paralysis (AFP) surveillance and wastewater (or environmental) testing. Environmental surveillance has played an important role in tracking the reintroduction of wild poliovirus to polio-free areas and identifying the presence of circulating vaccine-derived polioviruses (cVDPVs). cVDPVs account for the majority of AFP cases worldwide (Rachlin et al., 2022). As of 2022, environmental surveillance for polio is ongoing in more than 15 countries, including the United States (Global Polio Eradication Initiative, 2022). Wastewater-based surveillance for poliovirus is complicated by international containment guidance designed to prevent the reintroduction of polio to communities after eradication (WHO, 2022).

### 1.3.2 Illicit Drugs and the European Monitoring Centre for Drugs and Drug Addiction

The European Monitoring Centre for Drugs and Drug Addiction (EMCDDA) maintains information on the type, scale, and demographics of illicit drug use in Europe to inform the design of appropriate management strategies by policy makers. The EMCDDA has been facilitating the use of wastewater monitoring to help track illicit drug use throughout Europe since 2007.

The first annual European wastewater monitoring effort was completed in 2011 by the *Sewage Analysis CORe Group—Europe (SCORE)*. This monitoring occurred simultaneously over a single week at 21 water resource recovery facilities in 19 cities and 11 countries covering 15 million people. It involved the analysis of biomarkers of amphetamine, cannabis, cocaine, ecstasy (3,4-methylenedioxymethamphetamine or MDMA), and methamphetamine using harmonized laboratory protocols based on solid phase extraction followed by liquid chromatography and tandem mass spectrometry (Thomas et al., 2012). Comparable week-long studies have been performed annually since the initial effort but have grown in scope: the 2021 monitoring covered 75 cities in 23 countries (EMCDDA, 2022).

### *1.3.3 COVID-19 and the U.S. Centers for Disease Control and Prevention*

The use of WBS exploded during the COVID-19 pandemic, and many severe acute respiratory syndrome coronavirus 2 (SARS-CoV-2) wastewater-testing programs have been organized at national, state/provincial, and local levels. In September 2020, the CDC launched the National Wastewater Surveillance System (NWSS) to understand COVID-19 spread in U.S. communities (CDC, 2022). Wastewater surveillance is one of many types of public health surveillance used by the CDC for COVID-19, along with virologic, outpatient/emergency department, hospitalization, and mortality surveillance. Although the initial NWSS focus was on COVID-19, testing expanded to include the monkeypox virus (also known as mpox) in September 2022 (*PR Newswire*, 2022). Indeed, the NWSS is designed to be a long-term program with enough flexibility to rapidly adapt to changing public health needs related to emerging infections, pandemic preparedness, and bioterrorism. The CDC supports the participation of state, local, territorial, and tribal health departments in the NWSS through funding, information sharing, and data analytics.

# 2.0 METHODOLOGY

A WBS program necessarily involves four steps (Figure 7.1) undertaken via a collaborative effort between three stakeholder groups: (1) sample collection by wastewater utilities, (2) sample processing and analysis by laboratories, (3) data interpretation and sharing by public health officials, and (4) data application and use, also by public health officials. Each of these steps is described below.

## 2.1 Sample Collection

### *2.1.1 Sampling Locations*

A variety of wastewater sampling locations can be used to capture a range of population sizes and types from the individual building to the community

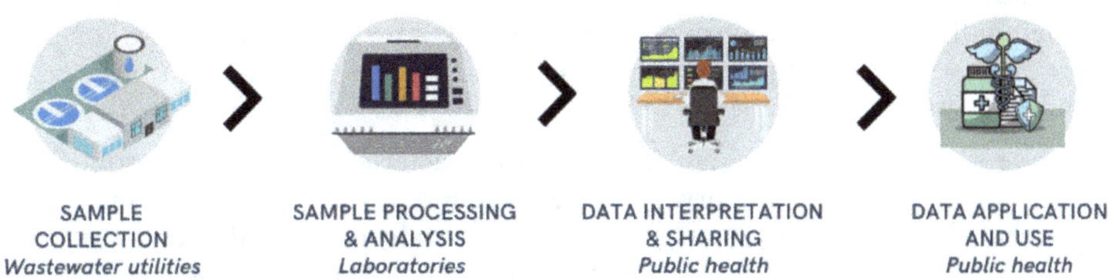


FIGURE 7.1 Wastewater-Based Surveillance Steps and Stakeholder Groups

scale, as shown in Table 7.2. Sampling location selection for a given wastewater surveillance program should be done through collaboration between wastewater utility representatives and public health officials based on (1) the objectives of the wastewater surveillance program, (2) anticipated public health actions implemented in response to wastewater results, (3) access and safety considerations for each sampling site, and (4) staff and funding available for sample collection efforts.

### 2.1.2 Sample Matrices

Most commonly, wastewater analyzed for public health information is collected as untreated raw or whole wastewater, which includes both the

**TABLE 7.2** Types of Wastewater Sampling Locations for Wastewater-Based Surveillance

| Sampling Location Type | Population Captured | Example(s) (selected relevant reference) | Considerations |
|---|---|---|---|
| Water resource recovery facility influent | Community | Hepatitis A virus, norovirus (Hellmér et al., 2014); Poliovirus (Manor et al., 1999); SARS-CoV-2 (Medema et al., 2020b) | Collect wastewater before any equalization, chemical addition, or treatment |
| Pumping station | Neighborhood | SARS-CoV-2 (D'Aoust et al., 2021b) | Collect well-mixed wastewater; avoid dead zones with accumulated settled solids |
| Manhole | Building; facility; neighborhood | Opioids and opioid metabolites (Foppe et al., 2021); SARS-CoV-2 (Reeves et al., 2021) | Collection from very deep (> 25 ft) manholes may be difficult |
| Decentralized system influent | Building; facility; small community | Adenovirus and norovirus (Jahne et al., 2020) | Collect wastewater before any equalization or treatment |
| Latrine | Facility; small community | Variety of bacteria, virus, helminth, and protozoa pathogens (Capone et al., 2021) | Unknown solids residence time makes data interpretation difficult |

liquid portion combined with any suspended solids. This approach has been applied to estimating population exposure to pesticides (Rousis et al., 2016) and phthalate plasticizers (Gonzalez-Marino et al., 2017), use of heroin (Du et al., 2017) and cannabis (Burgard et al., 2019), caffeine consumption (Gracia-Lor et al., 2017), enteric viruses (Bisseux et al., 2018), and SARS-CoV-2 (Gonzalez et al., 2020). One key consideration with this approach is the need to make sure the wastewater sample is well-mixed, both at the point of collection and when dispensing any sample aliquots from the collection container for concentration and analysis.

One alternative to collecting whole wastewater is to filter wastewater at the point of collection, thereby retaining only particles trapped on the filter for analysis (called residue or retentate). The residue is then eluted from the filter surface in the laboratory prior to analysis. This method was described by U.S. Environmental Protection Agency (U.S. EPA, 2001) and has been applied to Cryptosporidium and Giardia (U.S. EPA Method 1623), adenoviruses (Kuo et al., 2010), DNA viruses (O'Brien et al., 2017), and SARS-CoV-2 (Miyani et al., 2020). A second alternative to collecting untreated whole wastewater is to collect settled solids from primary clarifiers (Graham et al., 2021; Wolfe et al., 2021).

### 2.1.3 Sampling Approaches

As summarized in Table 7.3, three approaches are used for collecting wastewater for WBS programs: grab, composite, and passive sampling.

### 2.1.4 Sample Preservation

Sample hold time (the total time that elapses between sample collection and analysis) and preservation are critical aspects of ensuring sample integrity and result accuracy for both chemical and microbial analyses. Maximum hold times vary with the sample matrix, analyte, and analytical methods used to quantify the analyte.

For chemical substances analysis, samples should be filtered using membrane or glass microfiber filters with pore sizes as low as 0.1 mm prior to storage to prevent biotransformation and clogging when performing solid-phase extraction (SPE) (McCall et al., 2016). Samples should be filtered immediately after sampling and stored at –20 °C to prevent any bacterial-induced degradation or biotransformation. Adding preservatives to prevent biodegradation or quench oxidants (if applicable) may be important for some target analytes. Most illicit drugs are stable at neutral pH and –20 °C for at least 3 weeks (McCall et al., 2016). Samples should be processed immediately after thawing, avoiding storage at 4 °C. It is important to add mass-labeled internal standards prior to filtration and storage to evaluate

**TABLE 7.3** Sampling Approaches for Wastewater-Based Surveillance

| | Grab | Composite | Passive |
|---|---|---|---|
| Brief Description | Single portion of wastewater or settled solids taken at one point in time | Series of grab samples taken either automatically or manually over a specified time period to produce an overall averaged sample | Wastewater analyte of interest collected by device deployed in wastewater for a known time period |
| Suitable Matrices | • Whole untreated wastewater<br>• Field-filtered untreated wastewater<br>• Settled primary solids | With automatic samplers:<br>• Whole untreated wastewater<br>With a series of manual grabs:<br>• Settled primary solids | Whole untreated wastewater |
| Required Equipment | • Sampling untreated wastewater requires a bucket dipper or pump.<br>• Taking a field-filtered wastewater sampler requires a pump and a cartridge filter.<br>• Sampling settled primary solids only requires access to a sample tap. | • Sampling untreated wastewater requires either (1) a programmable automatic sampler capable of either time-based sample intervals or flow-paced compositing (for automatic composite sampling) or (2) a bucket dipper or pump (for manual composite sampling).<br>• Sampling settled primary solids only requires access to a sample tap. | Absorbent material (cotton gauze, cotton cheesecloth, cellulose sponges, electronegative filters, or tampons) with housing and/or string |
| Typical Volumes Collected | • 50 to 500 mL for whole wastewater or settled solids<br>• 10 000 to 80 000 mL for field-filtered wastewater | • With automatic samplers: 1000 mL or more total during compositing period, based on sampling 100 mL every 15 minutes<br>• With manual sampling: 50 to 200 mL per grab sample | Not applicable |

(*continued*)

**TABLE 7.3** Sampling Approaches for Wastewater-Based Surveillance (*Continued*)

| | Grab | Composite | Passive |
|---|---|---|---|
| Advantages | Does not require expensive or complex equipment | Generates an "averaged" wastewater sample that is likely more representative of true population loading than grab samples | Least expensive and easiest to deploy |
| Disadvantages | Results are probably not as representative as composite samples. | Higher complexity than grab or passive sampling; more expensive due to higher equipment and/or labor costs | Results are not quantitative (can't give a concentration for the target analyte). |
| Selected examples from the literature for COVID-19 | Randazzo et al. (2020); Scott et al. (2021) | Gonzalez et al. (2020); Ahmed et al. (2021) | Rafiee et al. (2021); Schang et al. (2021) |

for possible losses of the analytes of interest during filtration (Baker & Kasprzyk-Hordern, 2011; Bijlsma et al., 2014).

For microbial analyses, samples should be stored on ice or refrigerated at 1 to 4 °C during transit to the laboratory and analyzed as soon as possible after collection. As per *Standard Methods for the Examination of Water and Wastewater*, hold times change with microbial target but also with tested water matrix intended use (e.g., hold time for bacteria in drinking water for compliance purposes is 30 hours but is 8 hours for bacteria in nonpotable water for compliance purposes). If samples were filtered in the field, hold times can vary by filter manufacturer. For virus analysis, if reliable refrigeration is not possible or long transit times to off-site laboratories are required, the addition of preservatives such as sodium benzoate and calcium propionate or antibiotics to cartridge filters after sample collection could maintain sample integrity (Fagnant et al., 2018). Viruses in wastewater samples can be stored at 4 °C in the dark for up to 10 days (Olson et al., 2004). Freezing and thawing whole wastewater samples may degrade the viral signal (Huge et al., 2022).

## 2.2 Sample Processing

Wastewater is a complex matrix, and the levels of some chemical or biological targets are more diluted in wastewater as compared to clinical specimens. Therefore, preconcentration and extraction steps are often required to successfully recover and subsequently analyze for specific targets of interest.

### 2.2.1 For Chemicals

To date, SPE is one of the most commonly used techniques for chemical compound extraction from wastewater (Fontanals et al., 2019). For SPE, wastewater containing the analyte(s) of interest is exposed to a solid sorbent that selectively adsorbs the target analytes onto the surface. This affinity allows the separation of the target analytes from the interferents. The solid sorbent is subsequently separated from the original sample and analytes are eluted from the sorbent using a solvent. Solid phase extraction is rapid and most of the extraction process can be completed in less than 30 minutes. It can be automated to concentrate target analytes and separate them from unwanted matrix components. Commercially available SPE columns are packed with small quantities of various chromatographic sorbents such as silica, glass beads, activated carbon, cellulose, and polymer (Augusto et al., 2013; Hennion, 1999).

### 2.2.2 For Pathogens

#### 2.2.2.1 Pretreatment

For pathogen detection using molecular methods, heat pasteurization (e.g., at 60 °C for 90 minutes) of wastewater samples can be used to reduce biosafety risk from bioaerosol-generating procedures during sample processing (e.g., vortexing, membrane filtering, centrifuging), although this has been shown to reduce the SARS-CoV-2 RNA signal (Medema et al., 2020a; Palmer et al., 2021).

#### 2.2.2.2 Concentration

For the detection of bacterial pathogens and antibiotic resistance genes, wastewater samples are often filtered using a sterile 0.22- or 0.45-mm pore-size membrane filter to trap and concentrate the microorganisms. The selection of membrane material (e.g., cellulose nitrate, polyethene sulfone, polyvinylidene fluoride, polycarbonate) and porosity are important and can influence bacteria recovery efficiency (Chen et al., 2005; Liang & Keeley, 2013). A pre-filtration step—for example, using nylon filters with a nominal pore size of 20 mm—capable of removing larger particles or solids can help eliminate clogging of membrane filters. Centrifugation can also be used to capture bacteria in a centrifuged pellet, particularly for sewage sludge. Centrifugation methods use centrifugal force to achieve separation of particles from a liquid medium. Centrifuges are generally classified according to their rotors, e.g., swing-bucket rotors, fixed-angle rotors, and vertical rotors. The fixed-angle rotor is commonly used for environmental samples due to its excellent ability in pelleting. Bacteria cells can usually be pelleted with a centrifugation speed of 8000 × *g* (Börjesson et al., 2009). For viruses in wastewater, several concentration methods have been used (Table 7.4).

#### 2.2.2.3 Nucleic Acid Extraction

Nucleic acid extraction is critical to maximize the yield of high-quality DNA/RNA and remove any environmental contaminants (phenolic compounds, humic, or fulvic acids), which may interfere with the downstream analytical methods. Commercial extraction kits are available for the isolation of nucleic acid from concentrated wastewater extracts. Examples of such kits are DNeasy PowerWater or PowerSoil Kits (QIAGEN, Hilden, Germany) and FastDNA Spin Kits (MP Biomedicals, U.S.) for bacteria and QIAamp Viral RNA Kits (QIAGEN, Hilden, Germany) and PureLink Viral RNA/DNA kits (Invitrogen) for viruses. These kits typically use physical bead beating technology (using glass beads, silica beads) followed by chemical lysis of microorganisms, inhibitor removal and DNA/RNA purification. These extraction kits have increased the ease and rapidity of nucleic acid extraction for WBS.

Nucleic acid samples should be stored at freezing temperature (−20 °C, −75 °C, or −80 °C) if immediate analysis is not possible (Hokajärvi et al., 2021). Repeated freezing and thawing the sample could lead to degradation of nucleic acid, and thus, samples should be analyzed immediately after thawing.

## 2.3 Analysis

Rapid development of WBS as a public health tool has increased the importance of appropriate analytical methods for the quantitative analysis of chemical and biological targets in wastewater. Until recently, the analytical methods used to detect these targets are mainly based on expensive equipment that must be housed in a laboratory and operated by highly skilled technicians, for examples, liquid chromatography-tandem mass spectrometry (LC-MS/MS) or gas chromatography-mass spectrometry (GC-MS) for the detection of organic chemicals or their metabolites (Hernández et al., 2018), and polymerase chain reaction (PCR) for the detection of pathogen nucleic acids (Girones et al., 2010).

### 2.3.1 Mass Spectrometry

Many analytical methods developed for WBS use mass spectrometry approaches due to its high sensitivity and selectivity. Liquid chromatography (LC)-MS (both conventional LC and ultra-high-performance LC) is a powerful analytical method and commonly used for separation, identification, and quantification of a wide range of biomarkers, for example, antibiotics,

**TABLE 7.4** Comparison of Different Concentration Methods for Pathogenic Viruses in Wastewater

| Method | Mechanisms | Advantages | Disadvantages | Recovery Efficiency; Reference |
|---|---|---|---|---|
| Adsorption-elution | • Adsorption of viruses to the surface of filter materials such as glass wool, positively or negatively charged membranes<br>• Subsequent recovery of the bound viruses from the filter materials by exposing the adsorbents to an eluant solution | • Inexpensive<br>• No preconditioning of sample is required for glass wool and positively charged membrane | • For negatively charged membrane, pretreatment of the sample is required (e.g., adjustment of the sample pH, addition of salts)<br>• Clogging for turbid water | *Electropositive filter media particle*<br>SARS-CoV: 0%–21.4% (mean 1.02%)<br>Phage f2: 33.6%–260% (mean 127.1%);<br>Wang et al. (2005)<br><br>*Electropositive membranes*<br>Naturally occurring coliphages: 56.5%;<br>Goyal et al. (1980)<br>Poliovirus type 1: 57% (activated sludge effluent); Soto-Beltran et al. (2013)<br><br>*Electronegative membranes*<br>Murine hepatitis virus: 65.7 ± 23.8%;<br>Ahmed et al. (2020) |
| Ultrafiltration | • Retention of viruses in the original sample during a reduction of its volume achieved by particle size separation<br>• Is accomplished by recirculating the sample under pressure through hollow fiber filters | • No preconditioning of sample is required.<br>• Suitable for recovering a wide range of viruses simultaneously | • Costly<br>• Clogging for turbid water and thus, longer processing time<br>• Final retentate volume depends on the apparatus in use. | *Automated InnovaPrep Concentrating Pipette Select*™<br>MS2: 27.7 ± 24.5%<br>Murine hepatitis virus: 7.5 ± 6.1%;<br>Forés et al. (2021)<br><br>*Centricon Plus-70 centrifugal ultrafiltration devices*<br>MS2: 26.3 ± 22.7%<br>Murine hepatitis virus: 24.1 ± 14.5%;<br>Forés et al. (2021)<br><br>*Pseudomonas* phage Phi6: 2.6%–11.6%;<br>Flood et al. (2021) |

(*continued*)

**TABLE 7.4** Comparison of Different Concentration Methods for Pathogenic Viruses in Wastewater (*Continued*)

| Method | Mechanisms | Advantages | Disadvantages | Recovery Efficiency; Reference |
|---|---|---|---|---|
| | | | | Human betacoronavirus OC43: 22 ± 4% (primary treated wastewater); McMinn et al. (2021) |
| Ultra-centrifugation | • Separation of viral particles based on density and size<br>• High sedimental force (30 000–100 000 × *g*) is required. | • A wide range of viruses can be recovered simultaneously.<br>• More rapid for concentrating viruses in sewage<br>• Compared with chemical precipitation | • Limited initial volumes can be processed.<br>• Costly<br>• Lack of portability of equipment | Rotavirus: 34%–60% (mean 47%); Fumian et al. (2010) |
| Physico-chemical flocculation/ precipitation | • Addition of chemicals such as polyethylene glycol (PEG), skimmed milk powder that cause a sample to partition into two parts with one part containing the viruses<br>• Low-speed centrifugation is used to collect precipitates. | • Inexpensive<br>• Well-established concentration method for viruses<br>• No pH manipulation of the sample, suitable for viruses<br>• Sensitive to low pH<br>• Suitable for turbid samples, e.g., raw sewage | • Impractical for large volumes of wastewater samples<br>• Time consuming, overnight incubation for some protocols | *PEG*<br>Poliovirus type 1: 59.5 ± 19.4%; Falman et al. (2019)<br>Feline calicivirus: 62.2%; Barril et al. (2021)<br>Phage Phi6: 29.8%–49.8% and 22.2%–51.5%; Torii et al. (2021)<br><br>*Acidified skim milk powder*<br>Norovirus GII: 10%; Flood et al. (2021)<br>Adenoviruses: 30%–95%<br>JC polyomaviruses: 55%–90%; Hellmér et al. (2014)<br>Poliovirus type 1: 106 ± 24.8%; Falman et al. (2019)<br><br>*Aluminum polychloride*<br>Feline calicivirus: 45%; Barril et al. (2021) |

antidepressants, illicit drugs, and contaminants of emerging concern (Gushgari et al., 2018; Kinyua et al., 2015; Ng et al., 2020; Rossmann et al., 2014). LC-MS encompasses a variety of different LC and MS combinations, including LC-MS, LC-MS/MS (tandem MS) and LC-high resolution MS (Lynch, 2017). LC-MS uses electrospray ionization, which can detect nonvolatile, polar, and thermally labile chemical compounds.

Besides LC-MS, gas chromatography (GC)-MS has been used for qualitative and quantitative drug analysis. Chemical compounds that are nonpolar, volatile, and have a small molecular weight can be analyzed using GC-MS (Bisceglia et al., 2019; Wang et al., 2019). However, for broad-spectrum drug screening and confirmation, lengthy sample preparation procedure including hydrolysis and derivatization is needed prior to GC-MS analysis (Lynch, 2017). Although less commonly used for WBS, GC-MS is more affordable and may be more widely available than LC-MS (Bisceglia et al., 2019).

### 2.3.2 Nucleic Acid Amplification-Based Methods

Molecular methods offer high-throughput, rapid detection of a diverse group of pathogens in wastewater. Polymerase chain rection and reverse transcriptase (RT)-PCR (for the detection of RNA viruses) have emerged as sensitive and specific approaches for the detection and quantification of low levels of pathogen nucleic acids in wastewater samples. Polymerase chain rection is an *in vitro* enzymatic amplification of specific genomic sequences followed by detection of the amplified nucleic acid by gel electrophoresis or fluorescent probe hybridization. This can also allow for developing a targeted search for specific genetic determinants associated with human infections (Le et al., 2016). For all PCR-based methods, quantitative PCR (qPCR) is most widely used for pathogen detection (Girones et al., 2010). Quantitative PCR enables quantification of DNA or complementary DNA (cDNA) targets by monitoring amplified products during PCR cycling as indicated by increased fluorescence. To date, qPCR assays have been developed for the quantitative detection of a wide range of enteric pathogens (Aw & Rose, 2012; Hellmér et al., 2014) as well as enveloped viruses such as Ebola virus and SARS-CoV-2 (Alygizakis et al., 2021; Jääskeläinen et al., 2019). A diverse range of antibiotic resistance genes have been studied in wastewater, typically by using qPCR techniques (Karkman et al., 2016; Pazda et al., 2019).

Despite its wide acceptance, qPCR has several limitations because data interpretation requires: (1) a calibration curve, which is constructed using positive controls that can be difficult to access (e.g., Biosafety Containment Level 3 [BSL 3] organisms like mpox virus or poliovirus) and (2) the calculation of the number of cycles needed to achieve gene amplification in a sample for quantification or to define sample inhibition; the presence of an RNA or DNA target does not imply the presence of viable or intact viruses or bacteria and does not imply risk. Quantitative PCR has been shown to be more susceptible to PCR inhibitors in wastewater, which affect the accuracy of pathogen quantification in wastewater samples (Schrader et al., 2012). Recent advances in digital PCR have enabled an endpoint, sensitive, and accurate quantification of nucleic acid without the need for a standard curve (Hindson et al., 2013; Vogelstein & Kinzler, 1999). These improvements are achieved by partitioning DNA or RNA samples on micro-droplets or reaction wells into individual reactions. The signal in digital PCR is measured after completing amplification, and the absolute number of target nucleic acid molecules in the sample is calculated directly from the ratio of positive to total partitions using binomial Poisson statistics. Due to the partitioning effect, PCR inhibitory substances have been shown to have less of a problem with environmental samples when analyzed using digital PCR (Sedlak et al., 2014). In addition, digital PCR is more sensitive than qPCR for rare targets, which is useful for quantifying low levels of pathogen in wastewater (Jahne et al., 2020). Digital PCR has been increasingly used in the detection of antibiotic resistance genes and various pathogens such as enteric viruses and SARS-CoV-2 in environmental samples (Cavé et al., 2016; Flood et al., 2021; Jahne et al., 2020).

### *2.3.3 Next Generation Sequencing*

Next generation sequencing (NGS) is a high-throughput approach to DNA sequencing using the concept of massively parallel processing, which generates millions to billions of nucleotide short reads in a much shorter time than by the traditional Sanger sequencing method (Pettersson et al., 2009). Although many NGS platforms—such as Illumina, Inc. (San Diego, CA) and Ion Torrent (Thermo Fisher Scientific)—differ in engineering and sequencing chemistry, they share the similar technical approach of massive parallel sequencing via spatially separated and clonally amplified DNA templates in a flow cell to generate sufficient signal for detection during sequencing process. Next generation sequencing has been shown to be useful in genotyping of viruses in wastewaters such as enteroviruses, rotaviruses, noroviruses, and vaccine-like poliovirus strains (Fumian et al., 2019; Shaw et al., 2020) and SARS-CoV-2 variants (Crits-Christoph et al., 2021).

The recently developed nanopore sequencing technology (MinION, Oxford Nanopore Technologies) uses biological or solid-state nanopores to sequence DNA by recording characteristic disruption in ionic current as the strand passes a nanopore (Deamer et al., 2016). This new technology has three main advantages over other sequencing platforms. Firstly, nanopore sequencing can generate long sequence reads, which facilitates the identification of divergent DNA sequences. Nanopore sequencing has been successfully applied in whole genome sequencing of a single virus, for example, a novel species of papillomavirus (Vanmechelen et al., 2017). Secondly, nanopore sequencing exports sequencing reads in real-time, which provides a short turnaround for time-sensitive applications (e.g., rapid response to disease outbreaks or threats). Thirdly, the MinION is a pocket-sized device powered via Universal Serial Bus (USB) portal of a laptop, which is deployable under field conditions (Walter et al., 2017). Several groups have incorporated nanopore sequencing as a crucial part of a mobile workstation during outbreaks, such as those caused by Ebola virus (Hoenen et al., 2016).

Although NGS has made remarkable advances in the past few years, the most critical challenge for its application in WBS is the development of improved user-friendly bioinformatics and visualization platforms to facilitate rapid and robust analysis and interpretation of high-throughput sequencing data (Aw & Rose, 2012).

### *2.3.4 Quality Assurance/Quality Control*

To compare results across varying methods and laboratories, internal standards and reportable metrics are needed. Kantor et al. (2021) suggest these include the following for PCR methods:

- quality controls, such as concentration and extraction method blanks, no-template PCR reaction controls, inhibition controls, and PCR positive controls or standards;
- recovery controls for loss of signal through decay or dilution;
- recovery controls for loss of signal during virus concentration—current protocols typically use a proxy virus of known titer and spike the matrix to assess recovery, and surrogates used include bovine coronavirus (BCoV), bovine respiratory syncytial virus (BRSV), bacteriophage Phi6, human coronavirus OC43, murine hepatitis virus (MHV), F-specific RNA phages, and in some cases, an inactivated form of SARS-CoV-2 itself, among others; and
- recovery efficiency measured by running RNA that is extracted from the proxy virus stock solution and quantified alongside the spiked wastewater samples.

## 2.4 Data Interpretation and Sharing

### *2.4.1 Normalization*

The target signal in wastewater can vary due to changes in the prevalence of a health condition in the contributing population, changes in dilution with nonsewage flows or decay in the collection system, or special events (holidays, music festivals, sporting events) in the sewershed. Normalization of the target signal involves adjusting the wastewater concentration to account for influences other than changes in health condition prevalence so wastewater data can be meaningfully compared over time. Wastewater data can be normalized with a variety of methods, including (1) wastewater flow (Zhang et al., 2019; Zuccato et al., 2008), (2) population (census data), (3) human biomarker concentrations (e.g., creatinine, coprostanol, cholesterol, 5-hydroxyindoleacetic acid, and ammonium; Been et al., 2014; Choi et al., 2018), (4) wastewater quality parameters (e.g., biochemical oxygen demand, chemical oxygen demand, total phosphorus, or total nitrogen; Daughton, 2012a; Zheng et al., 2017; Zuccato et al., 2016), (5) human fecal genetic markers, for example, cross-assembly phage (crAssphage), pepper mild mottle virus (Crank et al., 2020; D'Aoust et al., 2021a), *Bacteroides* HF183, and *Lachnospiraceae* Lachno12 (Feng et al., 2018, 2021), and (6) nonwastewater data approaches (e.g., mobile device-based population activity patterns; Panczak et al., 2020).

### 2.4.2 Complementary Data Sources

Wastewater data are independent of health-care-seeking behavior and, therefore, can potentially provide a more comprehensive assessment of the true community-level extent of a given health condition. However, data collected from WBS programs should be considered in the context of other public health surveillance data. This includes data collected from both active and passive public health surveillance efforts (Nsubuga et al., 2006). Active public health surveillance entails public health officials contacting health care providers, laboratories, schools, nursing homes, and workplaces to obtain information about specific health conditions. *Passive surveillance* refers to automatic reporting by health care providers and laboratories to public health officials. Each country uses its own systems for public health surveillance, and most countries have multiple surveillance systems in place that track cases ("case surveillance") or symptoms ("syndromic surveillance").

### 2.4.3 Data Representation and Display Practices

The presentation of WBS data has evolved over time and is now commonly presented through online dynamic dashboards. Dashboards allow national, state, local, and university surveillance programs to report WBS data in near real-time and show changes in biomarker concentration levels. In addition, they can link data from multiple sources and provide interactive visualizations of information within communities (Keshaviah et al., 2021). Amid the COVID-19 pandemic, state, local, and university surveillance programs have used dashboards as a way to communicate concentration levels of SARS-CoV-2 in wastewater and have published their methodology (along with preliminary results) in peer-reviewed journal articles (Gutierrez et al.,

2021). The CDC (2021) developed guidelines titled "Public Health and Interpretation and Use of Wastewater Surveillance Data" for displaying and sharing WBS data to the public.

### 2.5 Data Application and Use

Information from WBS programs can be used by public health agencies to "assess and characterize the burden and distribution of adverse health events, prioritize public health actions, monitor the impact of control measures, and identify emerging health conditions that may have a significant impact upon population health" (Groseclose & Buckeridge, 2017).

More commonly, wastewater data are used to verify the burden and distribution of adverse health events by corroborating other public health surveillance data. For example, Been et al. (2015) used population survey data, a registry on opioid substitution therapy, and data from a needle and syringe distribution program in combination with wastewater testing results to understand heroin consumption in Lausanne, Switzerland. Local health departments in Wisconsin (United States) used SARS-CoV-2 concentrations in wastewater to confirm that the observed changes in the number of COVID-19 cases reflected true changes in disease incidence and were not artifacts of inadequate testing (McClary-Gutierrez et al., 2021).

Less commonly, specific public health actions are taken directly in response to wastewater data. The nature of these actions depends on the public health condition being monitored via wastewater and the scale at which the wastewater testing is occurring (facility, neighborhood, city, region). Examples of actions taken by state-level health departments in the United States in response to SARS-CoV-2 concentrations in wastewater during the COVID-19 pandemic include (Kirby et al., 2021; McClary-Gutierrez et al., 2021): (1) targeted deployment of mobile testing teams or other testing resources and mask-wearing compliance observers, (2) notification of the public about observed increases in wastewater concentrations, (3) sharing of social media and press resources, and answers to frequently asked questions, with local health departments, and (4) alerting hospitals and other health care providers about possible increases in COVID-19 incidence.

## 3.0 LIMITATIONS

Limitations to WBS implementation include

- Challenges with the wastewater matrix: Wastewater is a complex environment from which to extract chemical or biological targets (Sims & Kasprzyk-Hordern, 2020). Wastewater samples are frequently collected in the early morning when influent volumes to wastewater treatment facilities are at peak levels. The larger influx of flow may provide a higher concentration of biomarker load but results in a more diluted signal (Keshaviah et al., 2021). Combined systems (industrial wastewater or stormwater) may also dilute the signal further.

- Variation in day-to-day biomarker signals: For licit and illicit drug testing, samples collected throughout the week may capture individuals within a population who use drugs habitually, whereas weekend sampling reflects recreational use (Keshaviah et al., 2021).
- Incomplete coverage: On-site treatment (e.g., septic) systems are not connected to a centralized sewer network and, therefore, are excluded from community-level wastewater monitoring (U.S. EPA, 2021).
- Population mobility: Dynamic and fluctuating populations due to tourism and commuters provide unstable signals for pathogen monitoring (CDC, 2021). Population mobility considerations are highest in communities with fewer than 100 000 people, thereby generating higher uncertainty in wastewater surveillance data within these smaller communities. Some authors have recommended tracking with apps or cellular phones (Gudra et al., 2022) to monitor population movement, resulting in some concerns over privacy (Seberger & Pati, 2021).

## 4.0 ETHICAL CONSIDERATIONS

Although WBS has the potential to overcome unintentional biases and inequities in public health surveillance and health care systems, consideration needs to be given to the ethics associated with wastewater monitoring for public health (McClary-Gutierrez et al., 2021). In any public health surveillance activity, public health goals must be achieved without imposing harm on individuals or groups. In 2017, the World Health Organization published public health surveillance guidelines built around four ethical principles:

1. *Common good*: Sufficient oversight from public health agencies is required to ensure that the benefits of surveillance are shared.
2. *Equity*: Public health surveillance can help highlight the health concerns of disadvantaged communities.
3. *Respect for persons:* Information about individuals and groups needs to be protected to minimize harm.

4. *Good governance*: Governance structures need to be accountable and transparent and foster community engagement.

The Canadian Water Network's COVID-19 Wastewater Coalition adapted 14 of the 17 original WHO guidelines to WBS (Canadian Water Network, 2020; Hrudey et al., 2021). Although these guidelines can be useful when implementing WBS programs, care should be taken to manage three specific ethical issues that may arise from WBS programs:

- *Group privacy violation and stigmatization*: Because wastewater data are not collected at the individual level, individual privacy is generally not a concern in WBS programs. Group privacy, however, can be at risk in programs that capture wastewater from small sewersheds, with the ethical concerns increasing as "the number of contributing individuals in a community or institution is smaller" (Canadian Water Network, 2020). Stigmatization and labeling of these groups based on wastewater data is a possibility (Prichard et al., 2016).
- *One-way information transfer*: Early in the COVID-19 pandemic, the data generated from wastewater samples were not always shared in a timely or useful manner with the community, nor were community leaders necessarily involved in the design of WBS programs. Without community input and feedback, WBS programs run the risk of falling short of the goal to equitably capture the health status of populations.
- *Inequitable distribution of the benefits*: As a result of the COVID-19 pandemic, more WBS programs were implemented globally, with the majority of surveillance programs located in high-income countries (Naughton et al., 2023). On average, 80% of the population in high-income countries has access to a sanitation system connected to centralized wastewater treatment, 24% of the population in lower-income countries has access, and only 4% of the population in low-income countries has access (Adhikari & Halden, 2022). The majority of individual households in rural communities, located within all income classifications, rely on onsite wastewater treatment systems (e.g., pit latrines/septic tanks) or sanitation discharges to a local waterbody (e.g., rivers; Shrestha et al., 2021). Lack of centralized wastewater treatment access is only one challenge to establishing a WBS program in these regions. Detection of SARS-CoV-2 in a wastewater sample requires trained personnel, specific reagents, at least biosafety level 2 laboratories, and specialized equipment (e.g., dd-PCR or q-PCR; Aslan et al., 2020; Shrestha et al., 2021).

Adherence to ethical principles during wastewater surveillance efforts can be achieved by engaging social scientists and public health professionals throughout the duration of the program and performing extensive outreach to communities to promote meaningful knowledge exchange (Coffman et al., 2021).

## 5.0 RESOURCES

Adhikari, S., & Halden, R. U. (2022). Opportunities and limits of wastewater-based epidemiology for tracking global health and attainment of UN sustainable development goals. *Environment International,* 107217.

Ahmed, W., Bertsch, P. M., Bivins, A., Bibby, K., Farkas, K., Gathercole, A., Haramoto, E., Gyawali, P., Korajkic, A., & McMinn, B. R. (2020). Comparison of virus concentration methods for the RT-qPCR-based recovery of murine hepatitis virus, a surrogate for SARS-CoV-2 from untreated wastewater. *Science of the Total Environment, 739*, 139960.

Ahmed, W., Bivins, A., Bertsch, P. M., Bibby, K., Gyawali, P., Sherchan, S. P., Simpson, S. L., Thomas, K. V., Verhagen, R., & Kitajima, M. (2021). Intraday variability of indicator and pathogenic viruses in 1-h and 24-h composite wastewater samples: Implications for wastewater-based epidemiology. *Environmental Research, 193*, 110531.

Alygizakis, N., Markou, A. N., Rousis, N. I., Galani, A., Avgeris, M., Adamopoulos, P. G., Scorilas, A., Lianidou, E. S., Paraskevis, D., & Tsiodras, S. (2021). Analytical methodologies for the detection of SARS-CoV-2 in wastewater: Protocols and future perspectives. *TrAC Trends in Analytical Chemistry, 134*, 116125.

Asghar, H., Diop, O. M., Weldegebriel, G., Malik, F., Shetty, S., El Bassioni, L., Akande, A. O., Al Maamoun, E., Zaidi, S., & Adeniji, A. J. (2014). Environmental surveillance for polioviruses in the Global Polio Eradication Initiative. *Journal of Infectious Diseases, 210*(suppl. 1), S294–303.

Aslan, A., Shah, G., Sittaramane, V., & Shankar, P. (2020). Sewage Monitoring in Rural Communities: A Powerful Strategy for COVID-19 Surveillance. *Journal of Environmental Health, 83, 5.*

Augusto, F., Hantao, L. W., Mogollón, N. G., & Braga, S. C. (2013). New materials and trends in sorbents for solid-phase extraction. *TrAC Trends in Analytical Chemistry, 43*, 14–23.

Aw, T. G., & Rose, J. B. (2012). Detection of pathogens in water: From phylochips to qPCR to pyrosequencing. *Current Opinion in Biotechnology, 23*, 422–430.

Baker, D. R., & Kasprzyk-Hordern, B. (2011). Critical evaluation of methodology commonly used in sample collection, storage and preparation for the analysis of pharmaceuticals and illicit drugs in surface water and wastewater by solid phase extraction and liquid chromatography–mass spectrometry. *Journal of Chromatography A, 1218*, 8036–8059.

Baker, D. R., Ocenaskova, V., Kvicalova, M., & Kasprzyk-Hordern, B. (2012). Drugs of abuse in wastewater and suspended particulate matter—further developments in sewage epidemiology. *Environment International, 48*, 28–38.

Barril, P. A., Pianciola, L. A., Mazzeo, M., Ousset, M. J., Jaureguiberry, M. V., Alessandrello, M., Sánchez, G., & Oteiza, J. M. (2021). Evaluation of viral concentration methods for SARS-CoV-2 recovery from wastewaters. *Science of the Total Environment, 756*, 144105.

Been, F., Benaglia, L., Lucia, S., Gervasoni, J. P., Esseiva, P., & Delemont, O. (2015). Data triangulation in the context of opioids monitoring via wastewater analyses. *Drug and Alcohol Dependence, 151*, 203–210.

Been, F., Rossi, L., Ort, C., Rudaz, S., Delemont, O., & Esseiva, P. (2014). Population normalization with ammonium in wastewater-based epidemiology: Application to illicit drug monitoring. *Environmental Science & Technology, 48*, 8162–8169.

Bijlsma, L., Beltrán, E., Boix, C., Sancho, J. V., & Hernández, F. (2014). Improvements in analytical methodology for the determination of frequently consumed illicit drugs in urban wastewater. *Analytical and Bioanalytical Chemistry, 406*, 4261–4272.

Bijlsma, L., Sancho, J. V., Pitarch, E., Ibanez, M., & Hernandez, F. (2009). Simultaneous ultra-high-pressure liquid chromatography-tandem mass spectrometry determination of amphetamine and amphetamine-like stimulants, cocaine and its metabolites, and a cannabis metabolite in surface water and urban wastewater. *Journal of Chromatography A, 1216*, 3078–3089.

Bisceglia, K. J., Kroening, G., & Subedi, B. (2019). GC-MS methods for monitoring illicit drug biomarkers in wastewater: A critical review. In B. Subedi, D. A. Burgard, & B. G. Loganathan (Eds.), *Wastewater-based epidemiology: Estimation of community consumption of drugs and diets* (pp. 51–77). American Chemical Society.

Bisseux, M., Colombet, J., Mirand, A., Roque-Afonso, A. M., Abravanel, F., Izopet, J., Archimbaud, C., Peigue-Lafeuille, H., Debroas, D., Bailly, J.-L., & C. Henquell. (2018). Monitoring human enteric viruses in wastewater and relevance to infections encountered in the clinical setting: A one-year experiment in central France, 2014 to 2015. *Eurosurveillance, 23*.

Börjesson, S., Melin, S., Matussek, A., & Lindgren, P.-E. (2009). A seasonal study of the mecA gene and Staphylococcus aureus including methicillin-resistant S. aureus in a municipal wastewater treatment plant. *Water Research, 43*, 925–932.

Burgard, D. A., Williams, J., Westerman, D., Rushing, R., Carpenter, R., LaRock, A., Sadetsky, J., Clarke, J., Fryhle, H., Pellman, M., & Banta-Green, C. J. (2019). Using wastewater-based analysis to monitor the effects of legalized retail sales on cannabis consumption in Washington State, USA. *Addiction, 114*, 1582–1590.

Canadian Water Network. (2020). *Ethics and communications guidance for wastewater surveillance to inform public health decision-making about COVID-19*. Canadian Coalition on Wastewater-Related COVID-19 Research Ethics Guidance. Canadian Water Network. https://cwn-rce.ca/wp-content/uploads/COVID19-Wastewater-Coalition-Ethics-and-Communications-Guidance-v4-Sept-2020.pdf

Capone, D., Chigwechokha, P., de Los Reyes 3rd, F. L., Holm, R. H., Risk, B. B., Tilley, E., & Brown, J. (2021). Impact of sampling depth on pathogen detection in pit latrines. *PLoS Neglected Tropical Diseases, 15*, e0009176.

Caucci, S., Karkman, A., Cacace, D., Rybicki, M., Timpel, P., Voolaid, V., Gurke, R., Virta, M., & Berendonk, T. U. (2016). Seasonality of antibiotic prescriptions for outpatients and resistance genes in sewers and wastewater treatment plant outflow. *FEMS Microbiology Ecology, 92*, fiw060.

Cavé, L., Brothier, E., Abrouk, D., Bouda, P. S., Hien, E., & Nazaret, S. (2016). Efficiency and sensitivity of the digital droplet PCR for the quantification of antibiotic resistance genes in soils and organic residues. *Applied Microbiology and Biotechnology, 100*, 10597–10608.

Centers for Disease Control and Precention. (2021). *Public health interpretation and use of wastewater surveillance data*. https://www.cdc.gov/healthywater/surveillance/wastewater-surveillance/public-health-interpretation.html

Centers for Disease Control and Precention. (2022). *National wastewater surveillance system*. https://www.cdc.gov/healthywater/surveillance/wastewater-surveillance/wastewater-surveillance.html

Chen, W. T., Hendrickson, R. L., Huang, C. P., Sherman, D., Geng, T., Bhunia, A. K., & Ladisch, M. R. (2005). Mechanistic study of membrane concentration and recovery of Listeria monocytogenes. *Biotechnology and Bioengineering, 89*, 263–273.

Choi, P. M., Tscharke, B., Donner, E., O'Brien, J. W., Grant, S. C., Kaserzon, S. L., Mackie, R., O'Malley, E., Crosbie, N. D., Thomas, K. V., Mueller,

J. F., & Mueller, J. (2018). Wastewater-based epidemiology biomarkers: Past, present and future. *TrAC Trends in Analytical Chemistry, 105*, 453–469.

Coffman, M. M., Guest, J. S., Wolfe, M. K., Naughton, C. C., Boehm, A. B., Vela, J. D., & Carrera, J .S. (2021). Preventing scientific and ethical misuse of wastewater surveillance data. *Environmental Science & Technology, 55*, 11473–11475.

Crank, K., Li, X., North, D., Ferraro, G. B., Iaconelli, M., Mancini, P., La Rosa, G., & Bibby, K. (2020). CrAssphage abundance and correlation with molecular viral markers in Italian wastewater. *Water Research, 184*, 116161.

Crits-Christoph, A., Kantor, R. S., Olm, M. R., Whitney, O. N., Al-Shayeb, B., Lou, Y. C., Flamholz, A., Kennedy, L. C., Greenwald, H., Hinkle, A., Hetzel, J., Spitzer, S., Koble, J., Tan, A., Hyde, F., Schroth, G., Kuersten, S., Banfield, J. F., & Nelson, K. L. (2021). Genome sequencing of sewage detects regionally prevalent SARS-CoV-2 variants. *MBio, 12*, e02703–02720.

D'Aoust, P. M., Mercier, E., Montpetit, D., Jia, J. J., Alexandrov, I., Neault, N., Baig, A. T., Mayne, J., Zhang, X., Alain, T., Langlois, M.-A., Servos, M. R., MacKenzie, M., Figeys, D., MacKenzie, A. E., Graber, T. E., & Delatolla, R. (2021a). Quantitative analysis of SARS-CoV-2 RNA from wastewater solids in communities with low COVID-19 incidence and prevalence. *Water Research, 188*, 116560.

D'Aoust, P. M., Towhid, S. T., Mercier, E., Hegazy, N., Tian, X., Bhatnagar, K., Zhang, Z., Naughton, C. C., MacKenzie, A. E., Graber, T. E., & Delatolla, R. (2021b). COVID-19 wastewater surveillance in rural communities: Comparison of lagoon and pumping station samples. *Science of the Total Environment, 801*, 149618.

Daughton, C. G. (2001). Illicit drugs in municipal sewage: Proposed new nonintrusive tool to heighten public awareness of societal use of illicit-abused drugs and their potential ecological consequences. In C. G. Daughton & T. L. Jones-Lepp (Eds.), *Pharmaceuticals and personal care products in the environment*. American Chemical Society.

Daughton, C. G. (2012a). Real-time estimation of small-area populations with human biomarkers in sewage. *Science of the Total Environment, 414*, 6–21.

Daughton, C. G. (2012b). Using biomarkers in sewage to monitor community-wide human health: isoprostanes as conceptual prototype. *Science of the Total Environment, 424*, 16–38.

Deamer, D., Akeson, M., & Branton, D. (2016). Three decades of nanopore sequencing. *Nature Biotechnology, 34*, 518–524.

Du, P., Zhou, Z., Bai, Y., Xu, Z., Gao, T., Fu, X., & Li, X. (2017). Estimating heroin abuse in major Chinese cities through wastewater-based epidemiology. *Science of the Total Environment, 605–606*, 158–165.

European Monitoring Centre for Drugs and Drug Addiction. (2022). *Wastewater analysis and drugs—A European multi-city study.* https://www.emcdda.europa.eu/publications/html/pods/waste-water-analysis_en

Fagnant, C., Sánchez-Gonzalez, M., Zhou, N., Falman, J., Eisenstein, M., Guelig, D., Ockerman, B., Guan, Y., Kossik, A., & Linden, Y. (2018). Improvement of the bag-mediated filtration system for sampling wastewater and wastewater-impacted waters. *Food and Environmental Virology, 10*, 72–82.

Falman, J. C., Fagnant-Sperati, C. S., Kossik, A. L., Boyle, D. S., & Meschke, J. S. (2019). Evaluation of secondary concentration methods for poliovirus detection in wastewater. *Food and Environmental Virology, 11*(1), 20–31.

Feng, S., Bootsma, M., & McLellan, S. L. (2018). Human-associated Lachnospiraceae genetic markers improve detection of fecal pollution sources in urban waters. *Applied and Environmental Microbiology 84*(14).

Feng, S., Roguet, A., McClary-Gutierrez, J. S., Newton, R. J., Kloczko, N., Meiman, J. G., & McLellan, S. L. (2021). Evaluation of sampling, analysis, and normalization methods for SARS-CoV-2 concentrations in wastewater to assess COVID-19 burdens in Wisconsin communities. *ACS Environmental Science & Technology Water, 1*(8), 1955–1965.

Flood, M. T., D'Souza, N., Rose, J. B., & Aw, T. G. (2021). Methods evaluation for rapid concentration and quantification of SARS-CoV-2 in raw wastewater using droplet digital and quantitative RT-PCR. *Food and Environmental Virology, 13*, 303–315.

Fontanals, N., Pocurull, E., Borrull, F., & Marcé, R. M. (2019). Role of solid-phase extraction in wastewater-based epidemiology. *Current Opinion in Environmental Science & Health, 9*, 26–33.

Foppe, K. S., Kujawinski, E. B., Duvallet, C., Endo, N., Erickson, T. B., Chai, P. R., & Matus, M. (2021). Analysis of 39 drugs and metabolites, including 8 glucuronide conjugates, in an upstream wastewater network via HPLC-MS/MS. *Journal of Chromatography B: Analytical Technologies in the Biomedical and Life Sciences, 1176*, 122747.

Forés, E., Bofill-Mas, S., Itarte, M., Martínez-Puchol, S., Hundesa, A., Calvo, M., Borrego, C., Corominas, L., Girones, R., & Rusiñol, M. (2021). Evaluation of two rapid ultrafiltration-based methods for SARS-CoV-2 concentration from wastewater. *Science of the Total Environment, 768*, 144786.

Fumian, T. M., Fioretti, J. M., Lun, J. H., Dos Santos, I. A., White, P. A., & Miagostovich, M. P. (2019). Detection of norovirus epidemic genotypes in raw sewage using next generation sequencing. *Environment International, 123*, 282–291.

Gerrity, D., Papp, K., Stoker, M., Sims, A., & Frehner, W. (2021). Early-pandemic wastewater surveillance of SARS-CoV-2 in Southern Nevada: Methodology, occurrence, and incidence/prevalence considerations. *Water Research X, 10*, 100086.

Girones, R., Ferrus, M. A., Alonso, J. L., Rodriguez-Manzano, J., Calgua, B., de Abreu Correâ, A., Hundesa, A., Carratala, A., & Bofill-Mas, S. (2010). Molecular detection of pathogens in water—The pros and cons of molecular techniques. *Water Research, 44*, 4325–4339.

Global Polio Eradication Initiative. (2022). *Polio this week*. https://polioeradication.org/polio-today/polio-now/this-week

Golovko, O., Kumar, V., Fedorova, G., Randak, T., & Grabic, R. (2014). Seasonal changes in antibiotics, antidepressants/psychiatric drugs, antihistamines and lipid regulators in a wastewater treatment plant. *Chemosphere, 111*, 418–426.

Gonzalez, R., Curtis, K., Bivins, A., Bibby, K., Weir, M. H., Yetka, K., Thompson, H., Keeling, D., Mitchell, J., & Gonzalez, D. (2020). COVID-19 surveillance in Southeastern Virginia using wastewater-based epidemiology. *Water Research, 186*, 116296.

Gonzalez-Marino, I., Rodil, R., Barrio, I., Cela, R., & Quintana, J. B. (2017). Wastewater-based epidemiology as a new tool for estimating population exposure to phthalate plasticizers. *Environmental Science & Technology, 51*, 3902–3910.

Goyal, S. M., Zerda, K. S., & Gerba, C. P. (1980). Concentration of coliphages from large volumes of water and wastewater. *Applied and Environmental Microbiology, 39*, 85–91.

Gracia-Lor, E., Rousis, N. I., Zuccato, E., Bade, R., Baz-Lomba, J. A., Castrignano, E., Causanilles, A., Hernández, F., Kasprzyk-Hordern, B., Kinyua, J., McCall, A.-K., van Nuijs, A. L. N., Plósz, B. G., Ramin, P., Ryu, Y., Santos, M. M., Thomas, K., de Voogt, P., Yang, Z., & Castiglioni, S. (2017). Estimation of caffeine intake from analysis of caffeine metabolites in wastewater. *Science of the Total Environment, 609*, 1582–1588.

Graham, K. E., Loeb, S. K., Wolfe, M. K., Catoe, D., Sinnott-Armstrong, N., Kim, S., Yamahara, K. M., Sassoubre, L. M., Mendoza Grijalva, L. M., Roldan-Hernandez, L., Langenfeld, K., Wigginton, K. R., & Boehm, A. B.

(2021). SARS-CoV-2 RNA in wastewater settled solids is associated with COVID-19 cases in a large urban sewershed. *Environmental Science & Technology, 55*, 488–498.

Groseclose, S. L., & Buckeridge, D. L. (2017). Public health surveillance systems: Recent advances in their use and evaluation. *Annual Review of Public Health, 38*, 57–79.

Gudra, D., Dejus, S., Bartkevics, V., Roga, A., Kalnina, I., Strods, M., Rayan, A., Kokina, K., Zajakina, A., Dumpis, U., Ikkere, L. E., Arhipova, I., Berzins, G., Erglis, A., Binde, J., Ansonska, E., Berzins, A., Juhna, T., & Fridmanis, D. (2022). Detection of SARS-CoV-2 RNA in wastewater and importance of population size assessment in smaller cities: An exploratory case study from two municipalities in Latvia. *Science of the Total Environment, 823*, 153775.

Gushgari, A. J., Driver, E. M., Steele, J. C., & Halden, R. U. (2018). Tracking narcotics consumption at a Southwestern US university campus by wastewater-based epidemiology. *Journal of Hazardous Materials, 359*, 437–444.

Gutierrez, S., Kazior, K., & Nepal, S. (2021). *A compendium of U.S. wastewater surveillance to support COVID-19 public health response.* U.S. Environmental Protection Agency.

Hellmér, M., Paxéus, N., Magnius, L., Enache, L., Arnholm, B., Johansson, A., Bergström, T., & Norder, H. (2014). Detection of pathogenic viruses in sewage provided early warnings of hepatitis A virus and norovirus outbreaks. *Applied and Environmental Microbiology, 80*, 6771–6781.

Hennion, M.-C. (1999). Solid-phase extraction: Method development, sorbents, and coupling with liquid chromatography. *Journal of Chromatography A, 856*, 3–54.

Hernández, F., Castiglioni, S., Covaci, A., de Voogt, P., Emke, E., Kasprzyk-Hordern, B., Ort, C., Reid, M., Sancho, J. V., Thomas, K. V., van Nuijs, A. L. N., Zuccato, E., & Bijlsma, L. (2018). Mass spectrometric strategies for the investigation of biomarkers of illicit drug use in wastewater. *Mass Spectrometry Reviews, 37*, 258–280.

Hindson, C. M., Chevillet, J. R., Briggs, H. A., Gallichotte, E. N., Ruf, I. K., Hindson, B. J., Vessella, R. L., & Tewari, M. (2013). Absolute quantification by droplet digital PCR versus analog real-time PCR. *Nature Methods, 10*, 1003–1005.

Hoenen, T., Groseth, A., Rosenke, K., Fischer, R. J., Hoenen, A., Judson, S. D., Martellaro, C., Falzarano, D., Marzi, A. & Squires, R. B. (2016). Nanopore sequencing as a rapidly deployable Ebola outbreak tool. *Emerging Infectious Diseases, 22*(2), 331.

Hokajärvi, A.-M., Rytkönen, A., Tiwari, A., Kauppinen, A., Oikarinen, S., Lehto, K.-M., Kankaanpää, A., Gunnar, T., Al-Hello, H., Blomqvist, S., Miettinen, I. T., Savolainen-Kopra, C., & Pitkänen, T. (2021). The detection and stability of the SARS-CoV-2 RNA biomarkers in wastewater influent in Helsinki, Finland. *Science of the Total Environment, 770*, 145274.

Hovi, T., Shulman, L. M., van der Avoort, H., Deshpande, J., Roivainen, M., & De Gourville, E. M. (2012). Role of environmental poliovirus surveillance in global polio eradication and beyond. *Epidemiology and Infection, 140*, 1–13.

Hrudey, S. E., Silva, D. S., Shelley, J., Pons, W., Isaac-Renton, J., Chik, A. H., & Conant, B. (2021). Ethics guidance for environmental scientists engaged in surveillance of wastewater for SARS-CoV-2. *Environmental Science & Technology, 55*, 8484–8491.

Huge, B. J., North, D., Mousseau, C. B., Bibby, K., Dovichi, N. J., & Champion, M. M. (2022). Comparison of RT-dPCR and RT-qPCR and the effects of freeze-thaw cycle and glycine release buffer for wastewater SARS-CoV-2 analysis. *Scientific Reports, 12*, 20641.

Jääskeläinen, A. J., Sironen, T., Diagne, C. T., Diagne, M. M., Faye, M., Faye, O., Faye, O., Hewson, R., Mölsä, M., Weidmann, M. W., Watson, R., Sall, A. A., & Vapalahti, O. (2019). Development, validation and clinical evaluation of a broad-range pan-filovirus RT-qPCR. *Journal of Clinical Virology, 114*, 26–31.

Jacox, A., Wetzel, J., Cheng, S. Y., & Concheiro, M. (2017). Quantitative analysis of opioids and cannabinoids in wastewater samples. *Forensic Sciences Research, 2*, 18–25.

Jahne, M. A., Brinkman, N. E., Keely, S. P., Zimmerman, B. D., Wheaton, E. A., & Garland, J. L. (2020). Droplet digital PCR quantification of norovirus and adenovirus in decentralized wastewater and graywater collections: Implications for onsite reuse. *Water Research, 169*, 115213.

Kantor, R. S., Nelson, K. L., Greenwald, H. W., & Kennedy, L. C. (2021). Challenges in measuring the recovery of SARS-CoV-2 from wastewater. *Environmental Science and Technology, 55*, 6, 3514–3519.

Karkman, A., Johnson, T. A., Lyra, C., Stedtfeld, R. D., Tamminen, M., Tiedje, J. M., & Virta, M. (2016). High-throughput quantification of antibiotic resistance genes from an urban wastewater treatment plant. *FEMS Microbiology Ecology, 92*, fiw014.

Katayama, H., Haramoto, E., Oguma, K., Yamashita, H., Tajima, A., Nakajima, H., & Ohgaki, S. (2008). One-year monthly quantitative survey of noroviruses, enteroviruses, and adenoviruses in wastewater collected from six plants in Japan. *Water Research, 42*, 1441–1448.

Keshaviah, A., Hu, X. C., & Henry, M. (2021). Developing a flexible national wastewater surveillance system for COVID-19 and beyond. *Environmental Health Perspectivse, 129*, 45002.

Khan, U., & Nicell, J. A. (2011). Refined sewer epidemiology mass balances and their application to heroin, cocaine and ecstasy. *Environment International, 37*, 1236–1252.

Kinyua, J., Covaci, A., Maho, W., McCall, A. K., Neels, H., & van Nuijs, A. L. (2015). Sewage-based epidemiology in monitoring the use of new psychoactive substances: Validation and application of an analytical method using LC-MS/MS. *Drug Testing and Analysis, 7*, 812–818.

Kirby, A. E., Walters, M. S., Jennings, W. C., Fugitt, R., LaCross, N., Mattioli, M., Marsh, Z. A., Roberts, V. A., Mercante, J. W., Yoder, J., & Hill, V. R. (2021). Using wastewater surveillance data to support the COVID-19 response—United States, 2020–2021. *Morbidity and Mortality Weekly Report, 70*, 1242–1244.

Krishna, B. (2021). Disease surveillance: The bedrock for control and prevention. *Indian Journal of Critical Care Medicine: Peer-Reviewed, Official Publication of Indian Society of Critical Care Medicine, 25*, 745.

Kuo, D. H., Simmons, F. J., Blair, S., Hart, E., Rose, J. B., & Xagoraraki, I. (2010). Assessment of human adenovirus removal in a full-scale membrane bioreactor treating municipal wastewater. *Water Research, 44*, 1520–1530.

Lai, F. Y., Bruno, R., Leung, H. W., Thai, P. K., Ort, C., Carter, S., Thompson, K., Lam, P. K. S., & Mueller, J. F. (2013). Estimating daily and diurnal variations of illicit drug use in Hong Kong: A pilot study of using wastewater analysis in an Asian metropolitan city. *Forensic Science International, 233*, 126–132.

Le, T.-H., Ng, C., Chen, H., Yi, X. Z., Koh, T. H., Barkham, T. M. S., Zhou, Z., & Gin, K. Y.-H. (2016). Occurrences and characterization of antibiotic-resistant bacteria and genetic determinants of hospital wastewater in a tropical country. *Antimicrobial Agents and Chemotherapy, 60*, 7449–7456.

Liang, Z., & Keeley, A. (2013). Filtration recovery of extracellular DNA from environmental water samples. *Environmental Science & Technology, 47*, 9324–9331.

Lynch, K. (2017). Toxicology: Liquid chromatography mass spectrometry. In *Mass spectrometry for the clinical laboratory*. Elsevier.

Manor, Y., Handsher, R., Halmut, T., Neuman, M., Bobrov, A., Rudich, H., Vonsover, A., Shulman, L., Kew, O., & Mendelson, E. (1999). Detection of poliovirus circulation by environmental surveillance in the absence of clinical cases in Israel and the Palestinian authority. *Journal of Clinical Microbiology, 37*, 1670–1675.

McCall, A. K., Bade, R., Kinyua, J., Lai, F. Y., Thai, P. K., Covaci, A., Bijlsma, L., van Nuijs, A. L. N., & Ort, C. (2016). Critical review on the stability of illicit drugs in sewers and wastewater samples. *Water Research, 88*, 933–947.

McClary-Gutierrez, J. S., Mattioli, M. C., Marcenac, P., Silverman, A. I., Boehm, A. B., Bibby, K., Balliet, M., de Los Reyes 3rd, F. L., Gerrity, D., Griffith, J. F., Holden, P. A., Katehis, D., Kester, G., LaCross, N., Lipp, E. K., Meiman, J., Noble, R. T., Brossard, D., & McLellan, S. L. (2021). SARS-CoV-2 wastewater surveillance for public health action. *Emerging Infectious Diseases, 27*, 1–8.

McMinn B. R., Korajkic A., Kelleher J., Herrmann M. P., Pemberton A. C., Ahmed W., Villegas E. N., & Oshima K. (2021). Development of a large volume concentration method for recovery of coronavirus from wastewater. *Science of the Total Environment, 774*, 145727.

Medema, G., Been, F., Heijnen, L., & Petterson, S. (2020a). Implementation of environmental surveillance for SARS-CoV-2 virus to support public health decisions: Opportunities and challenges. *Current Opinion in Environmental Science & Health, 17*, 49–71.

Medema, G., Heijnen, L., Elsinga, G., Italiaander, R., & Brouwer, A. (2020b). Presence of SARS-Coronavirus-2 RNA in sewage and correlation with reported COVID-19 prevalence in the early stage of the epidemic in the Netherlands. *Environmental Science & Technology Letters, 7*, 511–516.

Miyani, B., Fonoll, X., Norton, J., Mehrotra, A., & Xagoraraki, I. (2020). SARS-CoV-2 in Detroit wastewater. *Journal of Environmental Engineering, 146*.

Naughton, C., Roman, F., Alvarado, A., Tariqi, A., Deeming, M., Kadonsky, K. Bibby, K., Bivins, A., Medema, G., Ahmed, W., Katsivelis, P., Allan, V., Sinclair, R., & Rose, J. (2023). Show us the data: Global COVID-19 wastewater monitoring efforts, equity, and gaps. *FEMS Microbes*, 4.

Ng, K. T., Rapp-Wright, H., Egli, M., Hartmann, A., Steele, J. C., Sosa-Hernández, J. E., Melchor-Martínez, E. M., Jacobs, M., White, B., Regan, F. Parra-Saldivar, R., Couchman, L., Halden, R. U., & Barron, L. P.. (2020). High-throughput multi-residue quantification of contaminants of emerging concern in wastewaters enabled using direct injection liquid chromatography-tandem mass spectrometry. *Journal of Hazardous Materials, 398*, 122933.

Nsubuga, P., White, M. E., Thacker, S. B., Anderson, M. A., Blount, S. B., Broome, C. V., Chiller, T. M., Espitia, V., Imtiaz, R., Sosin, D., Stroup, D.

F., Tauxe, R. V., Vijayaraghavan, M., & Trostle, M. (2006). Public health surveillance: A tool for targeting and monitoring interventions. In D. T. Jamison, World Bank, & Disease Control Priorities Project (Eds.), *Disease control priorities in developing countries*. Oxford University Press; World Bank.

O'Brien, E., Munir, M., Marsh, T., Heran, M., Lesage, G., Tarabara, V. V., & Xagoraraki, I. (2017). Diversity of DNA viruses in effluents of membrane bioreactors in Traverse City, MI (USA) and La Grande Motte (France). *Water Research, 111*, 338–345.

O'Brien, J. W., Thai, P. K., Brandsma, S. H., Leonards, P. E., Ort, C., & Mueller, J. F. (2015). Wastewater analysis of census day samples to investigate per capita input of organophosphorus flame retardants and plasticizers into wastewater. *Chemosphere, 138*, 328–334.

Olson, M., Axler, R., & Hicks, R. (2004). Effects of freezing and storage temperature on MS2 viability. *Journal of Virological Methods, 122*(2), 147–152.

Palmer, E. J., Maestre, J. P., Jarma, D., Lu, A., Willmann, E., Kinney, K. A., & Kirisits, M. J. (2021). Development of a reproducible method for monitoring SARS-CoV–2 in wastewater. *Science of the Total Environment, 799*, 149405.

Panczak, R., Charles-Edwards, E., & Corcoran, J. (2020). Estimating temporary populations: A systematic review of the empirical literature. *Humanities and Social Sciences Communications, 6*, 1–10.

Pazda, M., Kumirska, J., Stepnowski, P., & Mulkiewicz, E. (2019). Antibiotic resistance genes identified in wastewater treatment plant systems—A review. *Science of the Total Environment, 697*, 134023.

Pettersson, E., Lundeberg, J., & Ahmadian, A.. (2009). Generations of sequencing technologies. *Genomics, 93*, 105–111.

*PR Newswire*. (2022). Biobot analytics to expand US Centers for Disease Control and Prevention's National Wastewater Monitoring Program to include monkeypox analysis. https://www.prnewswire.com/news-releases/biobot-analytics-to-expand-us-centers-for-disease-control-and-preventions-national-wastewater-monitoring-program-to-include-monkeypox-analysis–301637010.html

Prado, T., Silva, D. M., Guilayn, W. C., Rose, T. L., Gaspar, A. M., & Miagostovich, M. P. (2011). Quantification and molecular characterization of enteric viruses detected in effluents from two hospital wastewater treatment plants. *Water Research, 45*, 1287–1297.

Prichard, J., Hall, W., Zuccato, E., De Voogt, P., Voulvoulis, N., Kummerer, K., Kasprzyk-Hordern, B., Barbato, A., Parabiaghi, A., Hernandez, F., van Wel, J., Thomas, K., Fent, K., Mardal, M., & Castiglioni, S. (2016). *SCORE Ethical research guidelines for sewage epidemiology*. European Monitoring Centre for Drugs and Drug Addiction.

Public Health Agency of Canada. (2022). *Public health surveillance.* https://www.canada.ca/en/public-health/services/public-health-practice/surveillance.html

Rachlin, A., Patel, J. C., Burns, C. C., Jorba, J., Tallis, G., O'Leary, A., Wassilak, S. G. F., & Vertefeuille, J. F. (2022). Progress toward polio eradication—Worldwide, January 2020–April 2022. *Morbidity & Mortality Weekly Report, 71*, 650–655.

Rafiee, M., Isazadeh, S., Mohseni-Bandpei, A., Mohebbi, S. R., Jahangiri-Rad, M., Eslami, A., Dabiri, H., Roostaei, K., Tanhaei, M., & Amereh, F. (2021). Moore swab performs equal to composite and outperforms grab sampling for SARS-CoV–2 monitoring in wastewater. *Science of the Total Environment, 790*, 148205.

Randazzo, W., Cuevas-Ferrando, E., Sanjuan, R., Domingo-Calap, P., & Sanchez, G. (2020). Metropolitan wastewater analysis for COVID–19 epidemiological surveillance. *International Journal of Hygiene and Environmental Health, 230*, 113621.

Reeves, K., Liebig, J., Feula, A., Saldi, T., Lasda, E., Johnson, W., Lilienfeld, J., Maggi, J., Pulley, K., & Wilkerson, P. J. (2021). High-resolution within-sewer SARS-CoV–2 surveillance facilitates informed intervention. *Water Research, 204*, 117613.

Rico, M., Andres-Costa, M. J., & Pico, Y. (2017). Estimating population size in wastewater-based epidemiology. Valencia metropolitan area as a case study. *Journal of Hazardous Material, 323*, 156–165.

Rodriguez-Alvarez, T., Rodil, R., Cela, R., & Benito Quintana, J. (2014). Ion-pair reversed-phase liquid chromatography–quadrupole-time-of-flight and triple-quadrupole–mass spectrometry determination of ethyl sulfate in wastewater for alcohol consumption tracing. *Journal of Chromatography A, 1328*, 35–42.

Rossmann, J., Schubert, S., Gurke, R., Oertel, R., & Kirch, W. (2014). Simultaneous determination of most prescribed antibiotics in multiple urban wastewater by SPE-LC–MS/MS. *Journal of Chromatography B, 969*, 162–170.

Rousis, N. I., Zuccato, E., & Castiglioni, S. (2016). Monitoring population exposure to pesticides based on liquid chromatography-tandem mass spectrometry measurement of their urinary metabolites in urban wastewater: A novel biomonitoring approach. *Science of the Total Environment, 571*, 1349–1357.

Rousis, N. I., Zuccato, E., & Castiglioni, S. (2017). Wastewater-based epidemiology to assess human exposure to pyrethroid pesticides. *Environment International, 99*, 213–220.

Shrestha, S., Yoshinaga, E., Chapagain, S. K., Mohan, G., Gasparatos, A., & Fukushi, K. (2021). Wastewater-based epidemiology for Cost-Effective Mass Surveillance of COVID-19 in Low-and-Middle-Income Countries: Challenges and Opportunities. *Water, 13*(20), 2897.

Schang, C., Crosbie, N. D., Nolan, M., Poon, R., Wang, M., Jex, A., John, N., Baker, L., & Langeveld, J. G. (2021). Passive sampling of SARS-CoV–2 for wastewater surveillance. *Environmental Science & Technology, 55*, 10432–10441.

Schrader, C., Schielke, A., Ellerbroek, L., & Johne, R. (2012). PCR inhibitors–Occurrence, properties and removal. *Journal of Applied Microbiology, 113*, 1014–1026.

Scott, L. C., Aubee, A., Babahaji, L., Vigil, K., Tims, S., & Aw, T. G. (2021). Targeted wastewater surveillance of SARS-CoV–2 on a university campus for COVID–19 outbreak detection and mitigation. *Environmental Research, 200*, 111374.

Seberger, J. S., & Patil, S. (2021). Post-COVID public health surveillance and privacy expectations in the United States: Scenario-based interview study. *JMIR mHealth and uHealth, 9*, e30871.

Sedlak, R. H., Kuypers, J., & Jerome, K. R. (2014). A multiplexed droplet digital PCR assay performs better than qPCR on inhibition prone samples. *Diagnostic Microbiology and Infectious Disease, 80*, 285–286.

Senta, I., Gracia-Lor, E., Borsotti, A., Zuccato, E., & Castiglioni, S. (2015). Wastewater analysis to monitor use of caffeine and nicotine and evaluation of their metabolites as biomarkers for population size assessment. *Water Research, 74*, 23–33.

Shaw, A. G., Majumdar, M., Troman, C., O'Toole, Á., Benny, B., Abraham, D., Praharaj, I., Kang, G., Sharif, S., & Alam, M. M. (2020). Rapid and sensitive direct detection and identification of poliovirus from stool and environmental surveillance samples by use of nanopore sequencing. *Journal of Clinical Microbiology, 58*, e00920.

Sims, N., & Kasprzyk-Hordern, B. (2020). Future perspectives of wastewater-based epidemiology: Monitoring infectious disease spread and resistance to the community level. *Environment International, 139*, 105689.

Soto-Beltran, M., Ikner, L. A., & Bright, K. R. (2013). Effectiveness of poliovirus concentration and recovery from treated wastewater by two electropositive filter methods. *Food and Environmental Virology, 5*, 91–96.

Subedi, B., Balakrishna, K., Joshua, D. I., & Kannan, K. (2017). Mass loading and removal of pharmaceuticals and personal care products including psychoactives, antihypertensives, and antibiotics in two sewage treatment plants in southern India. *Chemosphere, 167*, 429–437.

Thacker, S. B., & Berkelman, R. L. (1988). Public health surveillance in the United States. *Epidemiologic Reviews, 10*, 164–190.

Thomas, K. V., Bijlsma, L., Castiglioni, S., Covaci, A., Emke, E., Grabic, R., Hernández, F., Karolak, S., Kasprzyk-Hordern, B., & Lindberg, R. H. (2012). Comparing illicit drug use in 19 European cities through sewage analysis. *Science of the Total Environment, 432*, 432–439.

Torii, S., Furumai, H., & Katayama, H. (2021). Applicability of polyethylene glycol precipitation followed by acid guanidinium thiocyanate-phenol-chloroform extraction for the detection of SARS-CoV-2 RNA from municipal wastewater. *Science of the Total Environment, 756,* 143067.

Trask, J. D., & Paul, J. R. (1942). Periodic examination of sewage for the virus of poliomyelitis. *Journal of Experimental Medicine, 75*, 1–6.

Vanmechelen, B., Bertelsen, M. F., Rector, J. A., Van den Oord, J., Laenen, L., Vergote, V., & Maes, P. (2017). Identification of a novel species of papillomavirus in giraffe lesions using nanopore sequencing. *Veterinary Microbiology, 201*, 26–31.

Venhuis, B. J., de Voogt, P., Emke, E., Causanilles, A., & Keizers, P. H. (2014). Success of rogue online pharmacies: Sewage study of sildenafil in the Netherlands. *British Medical Journal, 349*, g4317.

Villar, L. M., de Paula, V. S., Diniz-Mendes, L., Guimaraes, F. R., Ferreira, F. F., Shubo, T. C., Miagostovich, M. P., Lampe, E., & Gaspar, A. M. (2007). Molecular detection of hepatitis A virus in urban sewage in Rio de Janeiro, Brazil. *Letters in Applied Microbiology, 45*, 168–173.

Vogelstein, B., & Kinzler, K. W. (1999). Digital PCR. *Proceedings of the National Academy of Sciences, 96*, 9236–9241.

Wang, X.-W., Li, J.-S., Guo, T.-K., Zhen, B., Kong, Q.-X., Yi, B., Li, Z., Song, N., Jin, M., & Xiao, W.-J. (2005). Concentration and detection of SARS coronavirus in sewage from Xiao Tang Shan Hospital and the 309th Hospital. *Journal of Virological Methods, 128*, 156–161.

Wang, Z., Shao, X.-T., Tan, D.-Q., Yan, J.-H., Xiao, Y., Zheng, Q.-D., Pei, W., Wang, Z., & Wang, D.-G. (2019). Reduction in methamphetamine consumption trends from 2015 to 2018 detected by wastewater-based epidemiology in Dalian, China. *Drug and Alcohol Dependence, 194*, 302–309.

Walter, M. C., Zwirglmaier, K., Vette, P., Holowachuk, S. A., Stoecker, K., Genzel, G. H., & Antwerpen, M. H. (2017). MinION as part of a biomedical rapidly deployable laboratory. *Journal of Biotechnology, 250*, 16–22.

Wilson, W. J. (1928). Isolation of B. typhosus from sewage and shellfish. *British Medical Journal, 1*, 1061–1062.

Wolfe, M. K., Topol, A., Knudson, A., Simpson, A., White, B., Vugia, D. J., Yu, A. T., Li, L., Balliet, M., Stoddard, P., Han, G. S., Wigginton, K. R., & Boehm, A. B. (2021). High-frequency, high-throughput quantification of SARS-CoV-2 RNA in wastewater settled solids at eight publicly owned treatment works in northern California shows strong association with COVID-19 incidence. *mSystems*, 6, e0082921.

World Health Organization. (2017). *WHO guidelines on ethical issues in public health surveillance*. World Health Organization.

World Health Organization. (2022). *WHO global action plan for poliovirus containment*. World Health Organization.

Xagoraraki, I., & O'Brien, E. (2020). Wastewater-based epidemiology for early detection of viral outbreaks. *Women in Water Quality*, 75–97.

Zhang, Y., Duan, L., Wang, B., Du, Y., Cagnetta, G., Huang, J., Blaney, L., & Yu, G. (2019). Wastewater-based epidemiology in Beijing, China: Prevalence of antibiotic use in flu season and association of pharmaceuticals and personal care products with socioeconomic characteristics. *Environment International, 125*, 152–160.

Zheng, Q.-D., Lin, J.-G., Pei, W., Guo, M.-X., Wang, Z., & Wang, D.-G. (2017). Estimating nicotine consumption in eight cities using sewage epidemiology based on ammonia nitrogen equivalent population. *Science of the Total Environment, 590*, 226–232.

Zuccato, E., Castiglioni, S., Senta, I., Borsotti, A., Genetti, B., Andreotti, A., Pieretti, G., & Serpelloni, G. (2016). Population surveys compared with wastewater analysis for monitoring illicit drug consumption in Italy in 2010–2014. *Drug and Alcohol Dependence, 161*, 178–188.

Zuccato, E., Chiabrando, C., Castiglioni, S., Bagnati, R., & Fanelli, R. (2008). Estimating community drug abuse by wastewater analysis. *Environmental Health Perspectives, 116*, 1027–1032.

# Chapter 8

# Looking Forward: Opportunities for the Water Sector and Its Workforce

*Shea Dunifon; Kari Brisolara; Jen Loudon; Bina Nayak, PhD; Leilani Perriatt; & Aditya Ramamurthy*

# 1.0 INTRODUCTION: A WORKFORCE WITH DIVERSE PERSPECTIVES

The challenges facing the water sector during a disease outbreak are as varied as the roles and educational backgrounds of its workforce. From skilled laborers with high school diplomas to executives with terminal college degrees, the water industry employs individuals from all backgrounds. While this diverse workforce brings a variety of perspectives, skillsets, and experiences, it can also present many unique challenges and opportunities during an outbreak.

The objective of this chapter is to highlight the challenges the water sector experienced with a specific focus on critical staff (e.g., field crews, operators, lab staff) and to provide some insight into potential opportunities that come with our lessons learned. It provides a compilation of perspectives brought forth by water professionals from the published literature, surveys, informal interviews, and first-hand experience from utility employees. These perspectives are not comprehensive; they are intended to help readers understand how an outbreak can affect its workforce and how the workforce responds and adapts to the situation. By sharing perspectives, we can create actionable items or opportunities for a more resilient and diverse workforce during future disease outbreaks. The challenges and opportunities that will be discussed are highlighted in Figure 8.1.

# 2.0 WATER SECTOR CHALLENGES DURING COVID-19

## 2.1 Emergency Response Planning

COVID-19 emphasized the importance of proactive, outbreak-specific emergency response plans (ERPs). Although many utilities had ERPs in place to respond to natural disasters (e.g., hurricanes, floods) and a few had response plans for the swine flu and influenza, many were unprepared. The 2020 global pandemic served as a reminder that ERPs must include situations

**FIGURE 8.1** Water Sector Challenges and Opportunities Identified During the COVID-19 Pandemic

like disease outbreaks because, unlike natural disasters, a pandemic is not a single, discrete, and time-bounded event. It is a continuous process, that operates over a time horizon that lasts until there is extinction of the pathogen or introduction of a vaccine (Ammirato et al., 2020). The COVID-19 pandemic allowed utilities with existing ERPs to test and revise those plans to include unique situations like working with skeletal crews over extended periods of time, working with staff to identify projects or assignments that can be completed remotely, failures in supply chains, and improving internal communications among employees.

## 2.2 Knowledge Management and Communications

Knowledge management in pandemics has a higher strategic objective than that in other disasters, aiming to actively save lives, keep workers safe, and maintain the delivery of critical, high-quality services. Thus, pandemic knowledge management has to cross health, socioeconomic, and political boundaries to be effective (Bdeir et al., 2013). This necessitates the use of various communication platforms and tailoring the content and communication frequency to a variety of audiences and stakeholders.

In the case of utilities, the knowledge management and communication strategy should consider how a workforce receives its information—whether via television, radio, newspaper, social media, texts, road signs, word of mouth, or some other source. Workers may be receiving inaccurate or conflicting information, or little to no information at all. This means that during an outbreak, one cannot assume everyone knows the same information or receives information at the same time.

The strategy should also acknowledge the diversity of the workforce's educational background. Although viruses are being described as RNA or DNA, enveloped or naked, and so on, these are foreign terms to most workers and the general public. Depending on the occupation within the industry, these might be terms that are situational and will never be used again. As new information comes to light and scientific discoveries are being made during an outbreak, a large portion of the water workforce may not know or be familiar with many terms used by the medical and scientific communities.

Communication strategies should provide rationales for decisions that affect the workforce, especially frontline staff. Simply making executive decisions behind closed doors and delegating is not enough. For instance, formal and informal field crews and facility operations leaders need to be heard, as they are experts in the boots-on-the-ground operations. Not only do employees need to be made aware of changes to their schedules and their work environments in a timely manner, but they also need to understand why the changes were made.

Effective knowledge management also entails two-way communication and using feedback to improve processes. Managers and supervisors must be empathetic to hearing the concerns of their employees and addressing them. Employee input must be sought. In addition, communications need to be sincere. Attempting to mass produce thank-you cards might backfire as opposed to personalized messages. Moving forward, employee morale should not be overlooked. Employees who feel respected and appreciated will often go above and beyond in their duties.

## 2.3 Pathogen Awareness and Training

According to the Centers for Disease Control and Prevention (CDC), workers handling human waste or sewage are at increased risk of becoming ill from waterborne diseases (CDC, 2002). Many factors—such as work environment, types of wastewater matrices being handled, concentrations and persistence of pathogens in wastewater, engineering controls, and tasks performed—may also increase a worker's risk of exposure (as was described in Chapters 4 and 6). Social distancing may not be possible in some environments where workers may be required to work in close proximity to each other (e.g., trucks, manholes, lift stations). The risk of exposure also increases when workers share equipment, tools, and spaces, thus making breakrooms, restrooms, and maintenance shops areas of possible transmission during an outbreak.

Although the nature and concentration of an infectious agent in human sewage are uncontrollable, training employees to be prepared to work safely during an outbreak is something that can be accomplished at any time. For many workers in utilities or governments, an internal risk and/or safety department may provide on-the-job training, including the identification of occupation-associated hazards such as pathogens (see Chapter 6). If unavailable internally, external trainers can be used, provided they are familiar with the occupation(s) and/or have experience in the industry.

At a minimum, workers of any educational level should receive training that is designed to provide basic, concise information, free of technical jargon but practical enough to apply, including the following:

- an overview of pathogens known to be present in wastewater;
- steps to minimize infection, including personal protective equipment (PPE) and its proper disposal;
- routes of infection and disease transmission;
- signs and symptoms of infection;
- effects of infection on personal (e.g., family) and community health;
- immunizations and treatments available (if applicable);
- testing procedures and available vaccines;
- interpreting test results, false positives/negatives, and the need for retesting (when applicable); and
- a point of contact or a list of resources if they have questions.

This training should be documented by management with acknowledgment from workers that they have received training. Training personnel on the risks involved on the job is essential to developing a safe and proactive mindset, bearing in mind the risks may not all be known at the time of training or that they may change in the future. Communicating these risks as they develop—whether through videos, fact sheets, emails, or other media—and reminding workers to "Practice safety for your loved ones, if not for yourself" both help minimize risks. It is also recommended that any training or communications to employees be updated regularly based on recommendations from the CDC or the local health department to remain consistent and in compliance.

## 2.4 Remote Work

Historically, most agencies and contractors mandated that their staff work in an office setting 5 days a week for a predefined shift (e.g., between 8 AM to 6 PM) with a scheduled lunch break. As a direct consequence of the COVID-19 pandemic and due to staff health and safety requirements, staff had to alter their work schedules regardless of their positions. A study by Adams-Prassl et al. (2022) surveying 24,000 employees across the United Kingdom and the United States highlighted the wide degree of heterogeneity in workers' ability to work from home. Although on average respondents in the United States and United Kingdom report being able to do 42% and 39% of their work tasks from home, respectively, a much smaller but non-negligible share of workers report values of 0% or 100% (Adams-Prassl et al., 2022). Where possible, some positions were asked to work from home completely, and some worked in a hybrid setting (e.g., 2 days in the office and 3 from home), but some were not afforded that flexibility due to their job requirements (e.g., treatment facility operators, maintenance crews).

From a technology standpoint, laptop users were able to transition to work remotely reasonably quickly, whereas those with desktops had more challenges with the transition. For many employees new to remote work, there were shortages in procuring technology. Laptops, webcams, headsets, and microphones were in high demand, but production was limited. Home office furniture, such as ergonomic chairs, in many parts of the United States were hard, if not impossible, to come by. There was also the issue of whether an employee had reliable internet at home creating additional equity issues. For instance, some employees might be sharing their home internet with their children who were attending school virtually or perhaps lived in a rural community where internet was slow or otherwise unreliable.

Some utilities had challenges with supervisory control and data-acquisition (SCADA) software that could not be made available remotely due to security reasons. In contrast, other software (e.g., Teams, Webex, GoToMeeting, Zoom) became increasingly popular for communicating but required employees to learn a new program. Mainstream software presented a unique set of security challenges; these programs became the target of hackers and pranksters "zoombombing" meetings. This forced many government agencies, and utilities, cybersecurity teams to limit the types of software used for external communications, which made communicating a challenge when different agencies adopted different software packages.

One outcome of the remote environment was the need for video conferencing etiquette training. For many individuals working from home, there was a learning curve when it came to being "camera ready," as well as a new sense of vulnerability that came with allowing others to see inside their personal dwellings. With virtual meetings becoming the norm, trainings are encouraged to help remote workers become more polished and professional presenters while minimizing distractions like poor lighting or sound, background objects, dressing for the part, and more. Unlike being in the physical work environment, working remotely can present its own unique set of challenges, including accountability and productivity, that must be addressed by employers to make conditions fair and transparent, especially in the eyes of those who cannot work remotely. Another key consideration for utilities, organizations, and agencies is the imminent need to adopt to the new normal of remote work and the need to develop and institute formalized work-from-home/remote work that can be consistently adopted by the workforce (Lonska et al., 2021). Additionally, contrary to the past, organizations need to consider formal capital investments in making the IT infrastructure resilient, secure, and agile to enable effective remote work.

## 2.5 Reduced Workforce

Utilities have reported some impacts associated with the reduced workforce during the COVID-19 pandemic due to illness, absenteeism, and/or staggered shifts for business continuity. Direct and immediate impacts of COVID-19 include an increase in retirements, attrition, lack of work-life balance, and the blurring of lines in job duties. An unintended post-pandemic consequence in 2022 has been significant cost increases all around due to high inflation. Unfortunately, many utilities have not kept up with inflationary cost adjustments through salary adjustments, benefit enhancements, and so on. This has forced some members of the water sector workforce to either resort to early retirement (if eligible) or switch to alternative job sectors that pay more competitive salaries. This has been further exacerbated by an increase in the number of open positions that have not been filled for extended periods of time. As a result, fewer workers were performing the same job, reducing redundancy and learning opportunities, especially for younger operators. Overall, this has resulted in increased stress, fractured morale, and more tense work environments. It has also highlighted the importance of proactive recruitment and retention, pay adjustment considerations, succession planning, cross-training, and staff redundancy.

## 2.6 Morale and Tension

One of the challenges arising during the COVID-19 outbreak across the water sector—regardless of the work environments (e.g., treatment facilities, laboratories, lift stations, field activities)—was low employee morale. Water resource recovery facility (WRRF) employees often described feeling emotional and mental stress, anger, frustration, exhaustion, and even resentment toward those working remotely.

From an operations standpoint, some facility staff and field crews were working with reduced staffing due to employees out with illness, quarantining due to exposure, and in some instances, crews working in skeletal groups to minimize exposure and maintain business continuity. Utility personnel required to report to a physical location or interact with the public faced unique challenges with regards to heightened exposure. Adopting positive messaging and perspectives became more important than ever. One positive example came from a small field crew that adapted a "We're family" mentality. Members maintained transparency when it came to their status (e.g., an exposure or positive member in their immediate household) and took turns cleaning shared equipment between shifts. Some utilities encouraged the public to show support for field staff using social media and the #wavefortheFrontline campaigns. DC Water also provided several examples in Chapter 5.

Although many reasons for low employee morale exist, a few compiled from speaking with frontline workers in nonexecutive roles included the following:

- longer hours;
- increased workloads;
- lack of recognition from management;
- lack of direction from management;
- lack of communication from management;
- inability (and/or flexibility) to work remote;
- concerns that coworkers were not taking enough safety precautions to protect one another from disease transmission (e.g., not wearing masks);
- lack of transparency (e.g., employees not disclosing they had a direct exposure to someone with COVID-19);
- employees using politics, religion, and/or other personal beliefs to avoid wearing masks, getting vaccinated, maintaining social distancing, and so on;

- limited or lack of PPE;
- lack of training and/or resources on the virus and its associated disease;
- emotional and financial strain due to illness or loss of a family member or decreased family income due to COVID-19 or school closures; and
- lack of in-person work socialization.

Although the COVID-19 outbreak was an unpredictable event that forced many high-level decisions to be made in short periods of time and with incomplete information, not all of these decisions were well-received by employees. This presented a lot of lessons learned about utility resilience planning and employee morale. One of these lessons related to maintaining employee actual and perceived accountability when working from home. Predictably, studies have reported that productivity while working from home varies by worker (Bao et al., 2022; Bloom, 2020). Thus, accountability and adaptation of leadership and management style are important to maintain equity and fairness in the workplace (Stoker et al., 2022). However, some authors have also pointed out that the accountability issue may be more related to perception. A study by Stanford University (Bloom, 2020) investigating the productivity of 16,000 workers over 9 months found that working from home increase productivity by 13% (e.g., fewer disruptions, fewer breaks, quieter work environment, fewer sick days). Tackling perception issues is much more complicated but requires open and transparent dialogue as well.

One other cause of low morale has been work-life balance for those employees working from home (Lonska et al., 2021). *Work-life balance* is defined as the successful combination of work, family responsibilities, and personal life, which has a direct effect on employee mental well-being, stress level, morale, and productivity (Bao et al., 2022). At the beginning of the COVID-19 pandemic, many assumed that working from home could be a positive factor that would promote work-life balance. However, over time, negative effects were also revealed, including that employees were only one call or message away from the employer, that an uncertainty of leisure time with family often created more stress, and that the gender gap was frequently reported to be worsening (Lonska et al., 2021).

Lastly, COVID-19 fatigue is affecting utilities at the individual, managerial, and organizational levels (e.g., sample collectors and laboratory analysts; Water Environment Federation & CDC, 2022). COVID-19 fatigue has frequently been reported in other essential service sectors (e.g., health care; Ahmadipour et al., 2022) but is rarely discussed in the utility context.

COVID-19 fatigue also had direct effect on morale, mental health, and compliance with health restrictions (Ahmadipour et al., 2022). Several utilities in Canada rolled out the Mental Health Commission of Canada's Working Mind Training to help frontline crews, team leads, and managers identify when they or coworkers are under stress and require support. The training is based on a mental health continuum model, which describes the range of mental health—healthy, reacting, injured, and ill—and lists the behaviors associated with each part of this range (see Figure 8.2).

## 3.0 THE INFLUENCE OF WATER SECTOR EXPERIENCE AND ROLE IN PERCEPTIONS OF CHALLENGES AND OPPORTUNITIES IN THE WATER SECTOR

A trend currently observed in the workforce across several industries is the change in job motivators. These are the factors that drive an individual's decision to join an industry, organization, or company and stay put for an extended period of time. In the last 40 to 60 years, job security, benefits, and loyalty to the organization were key motivators. Today, quality of life has taken precedence. As a result, the workforce is constantly on the lookout for an opportunity that will strike an optimal work-life balance and direct effect on the overall quality of life. Therefore, water sector agencies and organizations should explore opportunities and ideas to recruit and retain workers with an emphasis on quality of life or work-life balance rather than just base salary, benefits, and so on.

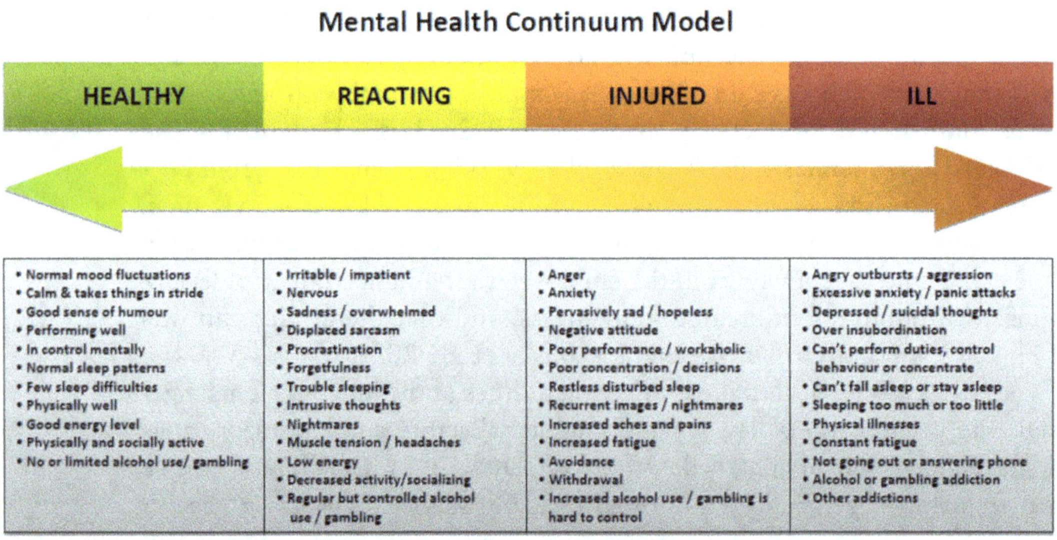


**FIGURE 8.2** Health Canada's Mental Health Continuum Model (Government of Canada, 2011)

To best understand how the recent pandemic affected water professionals, we conducted a series of informal interviews. Although these interviews were originally intended to be administered as surveys, we found the information we gathered to be quite candid at times and, in many situations, unique to the individual's life and work environments. We found the responses were influenced by the region where the interviewee worked or was seeking employment. Regardless of the interviewee's years of experience in the water sector, the perceptions were not all unique; for instance, 7 of the 10 recent graduates (college, trade, or high school) and job seekers expressed finding desirable employment challenging. As one high school graduate stated, "There will always be jobs. Like McDonald's or Dunkin' Donuts. They just aren't the jobs you want to make a career out of." Another recent college graduate stated that she couldn't find a desirable position in a place she wanted to live and said she would "likely end up liking the position or the location but not both."

To present the results of the interviews in a useful but insightful manner, we chose to highlight some of the responses provided by the 42 individuals we interviewed based on (1) relevance, (2) commonality (e.g., same response from multiple respondents), and (3) insightfulness. These interviewees included utility personnel, consultants, recent college graduates, two high school graduates seeking opportunities in the sector, and two new employees to the sector.

## 3.1 Recent College Graduates and Job Seekers (Less Than 1 Year of Experience)

Interviewees stated that finding desirable employment was a perceived challenge. For instance, some interviewees stated that local municipalities had enacted hiring freezes for nonessential personnel, limiting their options to other industries. One interviewee, who was new to the water sector but had 25+ years of work experience, stated that he feared a recession might lead to job cuts in the near future. On the contrary, most recent graduates (Gen Z) saw the opportunity to work from home remotely as an advantage for their tech-savvy generation. Although open to conducting business on online meeting platforms, the group was divided by whether they preferred in-person networking and training versus online, self-paced training and virtual happy hours.

## 3.2 Young Professionals (1–10 Years)

Interviewees seemed more optimistic than their job seeking counterparts. A few stated that they were concerned that their work-life balance might be disrupted. A few working as consultants stated they had never been busier with work and therefore were not concerned about losing their jobs but were concerned about meeting deadlines. Two interviewees stated that having mentors helped them adjust to the new remote work environment but that they missed visiting facilities and clients. A new operator, on the other hand, candidly admitted: "I started just before the pandemic began and now I'm isolated from the other operators working 10-hour shifts. It's not what I signed up for but I appreciate the profession more than ever. We keep it [the water] running." Adaptability and dedication seemed to be a common theme in the young professional category and utility personnel, respectively.

### 3.3 Mid-Level Employees (10–15 Years)

Of the 42 interviewees, 23 were considered mid-level in their careers. For those on the municipal operations side, the sentiment was "business as usual." Most interviewees stated that they had no fears of losing their essential jobs but struggled to find supplies and parts (e.g., equipment), making completing tasks difficult at times. The challenges of finding PPE was also echoed, as well as the desire for more education on the virus, its effects, its prevalence in the workplace, and the usage of proper PPE. Supervisors also stated they were having greater difficulty in hiring skilled labor (mechanical and electrical), and some expressed concerns that the private sector "looked more attractive" to job seekers, especially because it enabled remote options.

### 3.4 Senior Employees (15+ Years)

Interviewees possessing the most experience had the most variable responses. Whereas some echoed a sentiment of "business as usual," others voiced concern about risks to employees' health. Two interviewees also mentioned retirement as an option. Like the mid-level interviewees, these individuals were most concerned with finding supplies and parts (e.g., machine parts and tools) and accessing PPE. Overall, these interviewees seemed practical in their approaches while remaining focused on the task at hand.

## 4.0 SKILLS AND COMPETENCIES FOR WATER PROFESSIONALS

Although the water sector employs a diverse workforce to meet its needs and the needs of the communities it serves, to ensure the workforce has the relevant training and experience necessary, the next sections will address the skills and competencies recommended during an outbreak.

## 4.1 Skills—Specific Learned Abilities Needed to Perform a Job Well

Skills are often listed on job postings because of their practicality and usefulness to recruiters and hiring managers. They help identify the types of abilities required by a position. Skills are learned and applied abilities that use one's knowledge effectively in execution or performance. However, skills can be vague and often require examples (e.g., observable behaviors). For example, what does it mean when someone says a candidate has excellent customer service skills? Does this mean that the candidate is good with people, shows patience and empathy, and/or is able to deescalate heated conversations while actively listens? Skills are frequently confused with competencies, but they are not the same thing. Skills make up competencies, and competencies account for the "how we can get the job done," which is covered in the next section.

## 4.2 Competencies—Behaviors That Lead to Job Success

Competencies constitute the knowledge, behaviors, attitudes, and even skills that lead to the ability to do something successfully or efficiently. The ability to make business decisions would be a competency. Competencies, like skills, help identify abilities. For example, let's return to the candidate with excellent customer service skills. A competency that demonstrates excellent customer skills could be evidenced by the candidate's ability to receive a 90% customer satisfaction rate over the last 2 years in an environment where the average customer satisfaction rating in that department was 80%. Competencies may include skills but not vice versa.

Academic institutions and institutes of professional learning often cite competency models as a basis for future graduates and/or employees. One popular model developed by Cheetham and Chivers (1996) can be applied to water professionals, as summarized here by Uhlenbrook and de Jong (2012):

- *Cognitive competence* includes possessing the knowledge required to perform a job and applying that knowledge—for example, the theoretical or technical knowledge of how a wastewater treatment process works.
- *Functional competence* includes the ability to perform job-specific tasks effectively—for example, writing and preparing budgets effectively.
- *Behavioral competence* includes the ability to personally adopt behaviors in a work-related environment, as appropriate—for example, controlling emotions or work-related environment, as appropriate.

- *Ethical competence* includes professional values and the ability to make appropriate judgements; these judgements might be based on rules, laws, or morals.

As Uhlenbrook and de Jong (2012) wrote, "Each core competence is made up of various constituents that all interact to produce specific outcomes" (2939). Although other competency models can be found online and in print, this particular model is holistic. In the water industry, these outcomes have consequences; they may affect public and/or rate payers' perceptions on the industry.

## 4.3 Job Descriptions, Competencies, and Checklists for Each Group

### *4.3.1 Field Crews (Water, Wastewater, and Collection Systems)*

The pandemic has accelerated the onset of the silver tsunami, resulting in retirements, even early retirements, or staff attrition. The section of the water sector workforce that has been extensively affected are field crews residing in the operations and maintenance departments of organizations. Retirements not only are causing an increased number of unfilled positions but also are contributing to extensive drain in institutional knowledge and experience. Although unfilled positions can be filled over time, the loss of institutional knowledge and experience is irreplaceable and directly affects the day-to-day operations and maintenance of water sector infrastructure. Due to a lack of effective salary adjustments—pay increases to keep up with inflation—water sector professionals, particularly field crews, are seeing better-paying opportunities in private sector organizations, such as Walmart or Amazon.

Although the water sector does a great job of defining the roles and responsibilities as well as knowledge, skills, and abilities necessary for field crews, the water industry should consider a comprehensive pay analysis that would try to mitigate field crew attrition to other industries. The sector should also consider strategies for succession planning that would ensure the recording and retention of institutional knowledge and experience and enable workforce business continuity. A list of competencies for water operators can be found at wioa.org.au (WIOA Australia, 2019) and is summarized below. The skills bolded are especially relevant during infectious disease outbreaks.

Levels certificate II, III, or IV in appropriate water industry operations require the following skills:

- operating water (drinking water, wastewater, and recycled water) treatment systems, processes, and/or networks;

- proficient water quality sample collection and field and laboratory-based analysis;
- capability to perform basic mechanical, and in some cases electrical and/or instrumentation, maintenance;
- applying quality management systems and procedures;
- using computers and process control systems;
- understanding water quality/environmental risk-management principles;
- understanding and applying health and safety processes and requirements; and
- pathogen awareness training.

Personal attributes include the following:

- being able to work unsupervised;
- completing tasks on time;
- having a logical approach to problem solving and trouble shooting;
- being able to work in a team;
- having well-developed communication skills; and
- being able to multitask.

A team lead or lead operator has additional responsibilities that require additional competencies:

- managing water treatment systems, processes, and/or networks;
- coordinating mechanical, electrical, and/or instrumentation maintenance;
- managing contractors;
- managing quality systems and procedures;
- understanding and applying health and safety processes and requirements;
- using computers and process control systems;
- applying of water quality risk-management principles;
- implementing actions to ensure compliance at an operational level and site-specific risk management to meet permit requirements;
- having a general understanding of legislative and/or regulatory rules;
- understanding health and safety systems; and
- pathogen awareness training.

Personal attributes include the following:

- good communications skills;
- coaching/mentoring skills;
- well-developed problem-solving and critical thinking ability;
- leadership and management skills;
- good planning and organizational skills (delegation skills);
- adaptability and out-of-the-box thinking; and
- basic mental health training.

### 4.3.2 Public Health Professionals

The core competencies for public health have been established by the Council on Education for Public Health (2023), the accrediting body for schools and programs of public health. In the context of water and wastewater professionals, there are multiple skills that translate across these professions. The most evident ones are listed below. For a full list of public health core competencies, please refer to the Council on Education for Public Health accreditation criteria at ceph.org (skills specifically related to infectious disease outbreaks are bolded).

- apply epidemiological methods to settings and situations in public health practice;
- select quantitative and qualitative data collection methods appropriate for a given public health context;
- analyze quantitative and qualitative data using biostatistics, informatics, computer-based programming, and software, as appropriate;
- interpret results of data analysis for public health research, policy, or practice;
- compare the organization, structure, and function of health care, public health, and regulatory systems across national and international settings;
- assess population needs, assets, and capacities that affect communities' health;
- apply awareness of cultural values and practices to the design, implementation, or critique of public health policies or programs;
- design a population-based policy, program, project, or intervention;
- explain basic principles and tools of budget and resource management;
- select methods to evaluate public health programs;
- discuss the policy-making process, including the roles of ethics and evidence;

- propose strategies to identify stakeholders and build coalitions and partnerships for influencing public health outcomes;
- advocate for political, social, or economic policies and programs that will improve health in diverse populations;
- evaluate policies for their impact on public health and health equity;
- apply leadership and/or management principles to address a relevant issue;
- select communication strategies for different audiences and sectors;
- communicate audience-appropriate (i.e., nonacademic, nonpeer audience) public health content, both in writing and through oral presentation;
- describe the importance of cultural competence in communicating public health content;
- integrate perspectives from other sectors and/or professions to promote and advance population health; and
- apply systems thinking tools to visually represent a public health issue in a format other than standard narrative.

### *4.3.3 Laboratory Technicians*

The core competencies for wastewater laboratories and laboratory professionals have been established by the NELAC Institute (TNI), a nonprofit organization whose mission is to foster the generation of environmental data of known and documented quality, and the Association of Boards of Certification (ABC), an association dedicated to fostering water and wastewater certification throughout North America, respectively. It should be noted that there are no federal competency/certification requirements for either a laboratory or an individual analyst; however, many states have developed their own local regulations, and analysts are able to become ABC certified on a voluntary basis. For more detailed information about TNI, please visit https://nelac-institute.org, and find detailed "need to now" lists for laboratory professionals at http://www.abccert.org/abc_certification_program/lab_analyst_certification.asp.

The following is based on a WEFTEC 2008 presentation by Jerry Parr from TNI titled "History and Future of Laboratory Accreditation" (Parr, 2008).

In 1978, the U.S. Environmental Protection Agency (U.S. EPA) initiated a laboratory certification program for laboratories involved in analyzing drinking water and delegated the authority for operation of the program to state agencies. Over the following years, many states expanded this program to include matrices like wastewater and biosolids. The result is the creation of a National Environmental Laboratory Accreditation Program (NELAP), now managed by TNI. Laboratory accreditation serves multiple purposes for different constituents. In general, NELAP accreditation attests to the competency of a laboratory for conducting environmental measurements.

For the public, NELAP accreditation promotes confidence that environmental data used to make policy decisions to protect public health and the environment are generated by laboratories with demonstrated competence. For data users, NELAP accreditation serves a consumer protection purpose. It provides assurance that the laboratory has been evaluated and has met accepted standards of competency established by and within the profession. For the profession, NELAP accreditation advances the field by promoting accepted standards of practice and advocating rigorous adherence to these standards. For government agencies, NELAP accreditation provides a basis to determine if environmental monitoring data are adequate for their intended use. For the laboratory, NELAP accreditation provides ongoing internal and external evaluations, demonstrates a commitment to continuous improvement, provides an effective mechanism for accountability, and enhances its reputation.

At the individual laboratory personnel level (taken from the ABC Need-to-Know Criteria for Wastewater Laboratory Analysts), the core competencies are clustered into the following job duties (ABC, 2003-2005):

- collect and preserve samples,
- prepare samples for analysis,
- analyze samples and interpret results,
- operate and maintain equipment and instruments,
- handle chemicals and wastes,
- implement quality assurance/quality control,
- manage laboratory, and
- laboratory safety.

A more extensive list of laboratory competencies includes but is not limited to the following:

- ability to determine appropriate sample location, type, and volume;
- familiarity with safety procedures for sample collection and preservation;
- familiarity with sample identification and labeling procedures;
- knowledge of sampler setup;

- knowledge of sampling techniques and equipment;
- familiarity with sample holding times, preservatives, and storage conditions;
- familiarity with chain of custody procedures;
- ability to identify, use, and clean common laboratory apparatus and glassware;
- ability to calibrate, operate, and maintain equipment/instruments;
- ability to prepare reagents;
- ability to store and handle chemicals;
- ability to weigh/measure accurately;
- familiarity with dilution techniques;
- knowledge of apparatus preparation;
- knowledge of interferences;
- knowledge of reagent purity;
- knowledge of laboratory pure water classification (types I, II, III) and standards;
- knowledge of, and ability to select approved analytical methods (local and U.S. EPA);
- knowledge of method limitations;
- ability to create and follow written standard operating procedures;
- familiarity with QA/QC practices;
- familiarity with permit requirements;
- familiarity with the sterilization process;
- knowledge of biology, chemistry, and basic math and statistics;
- knowledge of wastewater treatment processes, normal characteristics of wastewater, and wastewater pathogens;
- ability to accurately transcribe data and determine what needs to be recorded;
- ability to perform calculations, and evaluate, interpret, and summarize data;
- ability to recognize abnormal analytical results and determine appropriate corrective action;
- knowledge of customer service, communication, management, project management, and public relations principles;
- familiarity with material safety data sheets, chemical hygiene plans, chemical compatibility, storage limitations, disposal procedures, spill cleanup procedures, and hazard management plans;

- knowledge of compressed gas cylinder handling hazards, confined space characteristics, and fume hood operation; and
- knowledge of PPE, safety procedures, and emergency plans.

### 4.3.4 Health and Safety Professionals

One of the most recognized accrediting bodies for programs related to environmental health and safety is ABET. ABET is a nonprofit, nongovernmental organization that accredits college and university programs in the disciplines of applied science, computing, engineering, and engineering technology. ABET accredits more than 3100 programs at more than 660 colleges and universities in 23 countries (2020). Within ABET, the Applied Science Accreditation Commission covers the health-and-safety-related professions. The learning outcomes recommended from associate's to master's levels are outlined below.

- environmental, health, and safety fundamentals;
- physiological and/or toxicological interactions of physical, chemical, biological, and ergonomic agents, factors, and/or stressors with the human body;
- anticipation, identification, and evaluation of potentially hazardous agents, conditions, and practices;
- fundamental exposure assessment techniques (both qualitative and quantitative);
- environmental, health, and safety data interpretation including statistical and epidemiological principles;
- development of hazard control designs, methods, procedures, and programs;
- accident/incident investigation and analysis;
- industrial and construction safety;
- legal aspects of environmental, health, and safety practices;
- environmental, health and safety program management;
- hazardous materials/waste recognition, control, and remediation;
- air pollution fundamentals and control technologies;
- water pollution fundamentals and control technologies;
- environmental and occupational regulations, standards, and permitting processes;
- environmental sampling and measurement methodologies;

- industrial hygiene data interpretation including statistical and epidemiological principles;
- applicable business and managerial practices;
- anticipation recognition, evaluation, and control of hazards and exposures affecting people, property, and the environment;
- principles and practices of risk management and business justification;
- development of policies, procedures, and systems using research and evidence-based data;
- legal principles relevant to safety professions;
- root cause analysis, incident investigation, and acceptable risk causation and relative risk;
- safety data analysis, interpretation, and utilization of statistical and epidemiological principles; and trends of injuries, illnesses, and workplace incidents;
- principles and practices of continuous quality improvement and sustainable safety programs and management systems; and
- applications of adult learning theory and behavior change to safety and health training methodology.

Environmental, health, and safety programs are expected to provide breadth across the range of topics implied by the title. Thus, these curricular content areas are considered to be minimum requirements. Other areas may be added as dictated by the mission and educational program objectives of the specific program. Additionally, the extent to which each content area is developed and emphasized in a given program must also be consistent with the program's mission and objectives. Depending on the program, a given area may be addressed in a devoted course, a portion of a course, or an appropriate extracurricular experience.

### *4.3.5 Public Affairs Professionals*

The core competencies for professionals working in public affairs within a utility or municipality may vary but have been summarized here from the National Institutes of Health (NIH, 2023) as it pertains to communications related to matters of public health and safety. Although many utilities may not have a dedicated communications or public affairs department or individual on staff, the 10 suggested competencies (message delivery, communication, adaptability, advisory assistance, collaboration, innovation, interpersonal relations, planning and prioritization, data gathering and analysis, and external and organizational awareness) will allow employees to select the competencies most relevant to their positions, to include many of the following:

- evaluates and plans a communication plan;
- creates a variety of communications such as media releases, pamphlets, and fact sheets and evaluates their efficacy for the intended audience;
- creates communications materials tailored to specific audiences (e.g., based on different levels of education, interests, or points of view);
- clearly communicates ideas through writing and/or speaking;
- actively listens, asks questions, and summarizes information to verify understanding;
- uses skill in presenting information through stories, facts, analogies, and/or other techniques with the goal of tailoring communications to the intended audience;
- thinks and acts effectively and appropriately under pressure;
- identifies ways to incorporate new information or practices into existing communications plan;
- maintains flexibility and open to new ideas or information with the goal of inducing change;
- provides accurate and timely guidance for the best interest of the organization;
- may serve as lead or public information officer when dealing with the media;
- supports leadership and/or other staff by planning, preparing, and managing information;
- identifies and facilitates internal and/or external partners, stakeholders, alliances, and so on;
- works with others to share information and accomplish set goals and/or objectives;
- fosters an environment that emphasizes information sharing;
- fosters an environment that encourages creative and innovative thinking;
- seeks opportunities for growth and positive change;
- explores new ideas, methodologies, and/or alternatives to accomplish objectives;
- relates to people in an open, friendly, and professional manner;

- cooperates and works to gain support and commitment from others when performing tasks;
- fosters an environment conducive to open, transparent communications among all levels;
- assesses relative importance of tasks and prioritizes when appropriate;
- uses resources available to complete assignments;
- uses time efficiently while remaining committed to juggling multiple tasks and/or priorities;
- gathers and interprets data to identify issues or inaccuracies;
- meets with customers to understand their needs or concerns to ensure data, analytics, and reporting are used appropriately;
- organizes and develops data to respond to audits and data calls for an organization;
- demonstrates an understanding of the organization's functions and responsibilities;
- reviews external information and news to learn about political and social issues that might affect the organization; and
- identifies individuals who have the technical expertise to respond to inquiries on specific topics.

To review additional competencies and the complete list of key behaviors, please see the NIH website (NIH, 2023) and search for the Suggested Public Affairs (GS-1035) Competency Model.

## 5.0 OPPORTUNITIES FOR THE WATER SECTOR

Although the 2020 pandemic disrupted life as the world knew it in both negative and positive ways, it also benefited the water sector by bringing focus to opportunities for the sector moving forward.

### 5.1 Shape Public Perception in a Positive Manner

One of the best examples during the COVID-19 pandemic of an industry-shaping public perception was the attention medical personnel received in their communities. Social media campaigns, news reports, and banners posted along roadsides printed with slogans such as "Heroes work here" helped to raise an awareness of these essential employees on the frontlines.

Unlike hospitals or medical centers, water infrastructure and service providers often go unnoticed by the public. Even treatment facilities are usually set far back from roads, guarded by fences and/or tall hedges. Although security is a concern, a balance must be established between the industry, its infrastructure, and public awareness. Many WRRFs could have benefited greatly during the pandemic as an essential service in the fight against spreading disease. For instance, a few suggested public awareness campaigns could have included the following:

- how treated water is used for washing hands and helping to stop the spread of the virus;
- how wastewater treatment is designed to kill/inactivate pathogens and protect the community;
- the proper disposal of PPE (e.g., masks, gloves), as opposed to flushing PPE down a toilet;
- spreading awareness that alternatives and substitutes to toilet paper, like scented wipes, should not be flushed;
- reminders that water and wastewater employees, like medical personnel, are essential workers;
- misconceptions and/or concerns about how water and wastewater are treated—more specifically, the disinfection processes used by utilities to prevent disease outbreaks; and
- careers available in essential positions.

Without a positive public perception and support, the industry lacks the support it needs to raise the funds needed to repair infrastructure and/or hire skilled and competent employees. Furthermore, without a positive public perception, the industry and its associated workers will not receive the respect and recognition they deserve for keeping essential services operating. Furthermore, we can raise awareness of the essential services we provide while spotlighting the various careers that make it possible. Simply put, public awareness is an investment.

## 5.2 Promote Clear and Transparent Communications

Public relations and strategic communications build mutually beneficial relationships between organizations and the public. This is a growing and evolving field within the water sector where best practices are still being defined. Although not a new field, many utilities, especially smaller ones, do not have full-time staff dedicated or trained to communicate with the public. This presents many challenges because without public relations staff, the public may perceive utilities as unresponsive, lazy, or dishonest, or operating 24/7 without any issues, none of which are accurate.

Although hiring staff to manage public relations may be an additional cost to a utility, it is an operational cost that can benefit the utility in many ways:

- maintaining communications during normal operations and not just during emergencies;
- providing a positive presence in the communities they serve;
- educating the public on the importance of daily utility operations;
- communicating with the public through popular and appropriate channels such as social media;
- using multiple approaches to communicate with diverse audiences;
- creating a narrative of the utility (who, what, where, when, and why); and
- using staff trained to work with the news media.

Providing clear and transparent communications will also enable the industry to better position itself from bad press, public outcries, and misinformation. For more information on communications plan structure and creation, see Chapter 5.

## 5.3 Utility Participation in Research

In 2020, researchers and epidemiologists around the world made the case for surveillance of the SARS-CoV-2 virus in wastewater as an early detection method for COVID-19 spread in communities (called wastewater-based surveillance or WBS; see Chapter 7). Researchers needed wastewater samples to pursue WBS, and this prompted an overwhelming and positive response from utilities. Several fruitful collaborations between universities, utilities, government agencies, and state and local health officials quickly materialized, resulting in opportunities for utility participation in research. Even utilities that were not actively engaged prior to COVID-19 got involved in this national effort due to its public health value.

The constant discovery of new chemical and microbial contaminants and an ever-changing regulatory landscape are major drivers for utilities to get involved in research opportunities. Studies piloting new treatment technologies benefit both the utility and the manufacturer of the technology. The utility gets a sneak peek at how the technology works, to evaluate its maintenance demands, and the chance to make an informed decision on whether to implement it in full-scale. The manufacturer or proponent of the technology gets to pilot in a real-world scenario in which any shortcomings or issues can be identified and improved on. Evaluating new technologies or processes to reduce or remove contaminants is the most common form of utility participation in research and is a part of the continual improvement of treatment practices.

The other, less common form of research involvement for utilities is participating in forward-looking monitoring studies that investigate contaminants that either are not on the regulatory radar yet or are known to be present but their human health effects are uncertain (see Chapter 4). For example, the Stage 2 Disinfectants and Disinfection Byproducts Rule (Stage 2 DBPR) implemented in 2012 tightened compliance for two groups of disinfection byproducts (i.e., total trihalomethanes and haloacetic acids) based on the results of epidemiology and toxicology studies that demonstrated these contaminants were harmful to human health above a certain concentration threshold. This information on the health effects of exposure to specific levels of disinfection byproducts was not available during the initial rule making when Stage 1 DBPR was finalized in 1998. Decades-long studies conducted on water samples collected from participating utilities helped inform Stage 2 DBPR. Another example is the presence of antibiotic resistant bacteria and genes in treated wastewater. These contaminants have been studied for a couple of decades despite a lack of regulatory limits. That body of research is currently proving useful in states where a potable reuse regulatory framework is under development. To summarize, utilities that participate actively in forward-looking research are commendable for donating time and resources toward increasing the knowledge base of the water industry. They are willing participants despite recognizing that this information might not be useful or immediately actionable for their operations and management.

There are several ways utilities can engage in research. Partnerships with local high schools, colleges, and universities are symbiotic, as students gain work experience by getting involved in projects and the utility collects data through the efforts of these students. Some utilities offer volunteering/shadowing opportunities, while others offer paid internships. Research universities have historically entered into formal sponsored research agreements with utilities with defined terms stating the scope of work, deliverables, budget, timelines, and data-sharing protocols. Such contracts ensure timely completion of work within the budgeted expenses to provide well-documented technical reports with recommendations based on research findings. The utility selects projects that are of specific interest such as alpha or beta testing of new field products, laboratory equipment, or methods for improving water analysis or treatment. Graduate students can build their thesis around these research studies to obtain a master's or doctoral degree. Undergraduate students get mentored by professors, postdoctoral associates, or graduate students and gain valuable experience in planning and executing research projects.

Several for-profit and nonprofit organizations provide funding to advance water-related research. For example, the Water Research Foundation (WRF) publishes requests for proposals funding specific research areas in water treatment, conservation, source protection, innovation, distribution, and management. Successful WRF teams often include participating utilities, academics, consulting firms, manufacturers, commercial laboratories, and government and public health agencies. These teams garner support from utilities to provide in-kind contributions of staff time spent collecting and shipping water/wastewater samples, review results and periodic reports, and provide input based on years of water industry experience. Some teams propose customized projects tailored to the utility's specific challenges and might request cash contributions. For these studies to be successful and relevant, utility partners are essential because data collected from laboratory-scale experiments can only be validated by samples collected from the field. As such, utilities play an integral role in collecting data, which adds to the water industry knowledge and informs future regulations. This type of research participation is usually practiced by a small set of utilities; those that have the staff time and resources and have progressive thinking leadership.

To summarize, the recent COVID-19 pandemic changed that and introduced more utilities into the research arena as a result of WBS. The pandemic provided several opportunities for utilities to engage in collaborative research, which can be continued post-pandemic to solve existing issues and to investigate future challenges. These partnerships are mutually beneficial and provide valuable information to the water industry.

## 5.4 Create a Diverse, Equitable, and Inclusive Workforce

The water industry has begun to address the weaknesses within the occupation with regard to critical needs for increasing diversity, equity, and inclusivity in the water workforce. As the industry faces unprecedented challenges due to emerging pathogens, workforce shortages, climate-change-related weather crises, and infrastructure demands, the importance of bringing different and inclusive perspectives to the table becomes even more critical. By fully embracing inclusive practices, utilities can benefit from increased work productivity and competitive edge in the emerging markets primarily driven by technology and generational differences (Ewoh, 2013). These benefits should be the driver, beyond the legal requirements traditionally relied on, to move diversity and inclusion forward. The mission and values of an organization, in particular a utility, should reflect the prioritization of an inclusive, diverse workforce that is held to equitable standards.

## 5.5 Promote Trade Careers Within Schools—Universities Are Not the Only Path

High school students often choose career paths based on projected salaries, as opposed to the work environment, the benefits, or the satisfaction of helping their communities or the environment. By promoting trade careers to students in secondary grades (grades 6 to 12), the water sector can begin offering students career options they did not know existed before.

In an operational setting, a skilled workforce keeps the facility running and can even cut operating costs if employees possess specialized skillsets like welding, plumbing, or operating heavy machinery. In the past, many of the manual skills required to operate or maintain a utility are not typically offered to high school or university students. Today, with more high schools offering certificate programs to students in trending industries, like business or computers, the water industry simply cannot afford not to promote trade careers within the local schools. This is giving other industries an advantage to recruit before the water sector.

## 5.6 Identify and Highlight Advocates for the Industry

All employees of the water sector can be advocates for the industry at large. Some of the ways employees can do this include the following:

- developing a short elevator speech about what they do for a living and its importance to prepare for talking to those outside the industry;
- participating or volunteering in public outreach events;
- using a skill, hobby, or personal talent to design and/or create something for the industry that helps educate the public—examples could be using woodworking skills to build a tabletop display of a sewer system that can be used at outreach events or using photography skills to capture the sunsetting from different perspectives around a facility for display in a public lobby;
- participating on committees of professional associations that specifically target the public through education and/or outreach;
- mentoring students or interested candidates in the field; and
- posting about their work on social media in a positive manner.

The possibilities for how employees can advocate for the water sector are limitless but sometimes require the industry and its leadership to become less formulaic and more creative. More importantly, as essential as the industry is to providing safe and clean drinking water and sewer services, so is our duty to advocate for its place as an essential service in a modern world.

# 6.0 RESOURCES

## 6.1 Publicly Available

Associations of Boards of Certification. (2003–2005). *ABC Need-to-Know Criteria for Wastewater Laboratory Analysts.* http://www.abccert.org/pdf_docs/abc2005wwlabntk.pdf

ABET Applied and Natural Science Commission. (2020, December 30). *Criteria for accrediting applied and natural science programs.* https://www.abet.org/wp-content/uploads/2021/01/R001-21-22-ANSAC-Criteria.pdf

Bloom, N. (2020, June). *How working from home works out.* Institute for Economic Policy Research Policy Brief, 1–9. https://drive.google.com/file/d/15nB9nscTj38bIADQYLlQWlzUBNPeDl5_/view

Centers for Disease Control and Prevention. (2002). *Guidance for controlling potential risks to workers exposed to class B biosolids.* National Institutes for Occupational Safety and Health. https://www.cdc.gov/niosh/docs/2002-149/pdfs/2002-149.pdf

Council on Education for Public Health. (2023). FAQs: Competencies. https://ceph.org/constituents/schools/faqs/2016criteriafaq/competencies/

Government of Canada. (2011). The mental health continuum model. https://www.canada.ca/en/government/publicservice/wellness-inclusion-diversity-public-service/health-wellness-public-servants/disability-management/fundamentals-employers-responsibilities.html

National Institutes of Health. (2023). *Suggested public affairs (GS-1035) competency model.* https://hr.nih.gov/working-nih/competencies/occupation-specific/suggested-public-affairs-gs-1035-competency-model#:~:text=Suggested%20Public%20Affairs%20%28GS-1035%29%20Competency%20Model%20This%20suggested,technical%20competencies%2C%20but%20the%20general%20competencies%20will%20vary.

Parr, Jerry L. (2008). History and future of lab accreditation. Presented at WEFTEC 2008 October 20, 2008. https://nelac-institute.org/docs/pubs/NELAC%20history2008.pdf

Uhlenbrook, S., & de Jong, E. (2012). T-shaped competency profile for water professionals of the future. *Hydrology and Earth System Sciences Discussions, 9*, 2935–2957.

WIOA Australia. (2019). *Technical competency for water industry operations.* https://wioa.org.au/wp-content/uploads/2019/04/TechnicalCompetencyHandbook.pdf

## 6.2 Require Access

Adams-Prassl, A., Boneva, T., Golin, M., & Rauh, C. (2022). Work that can be done from home: Evidence on variation within and across occupations and industries. *Labour Economics, 74*, 102083.

Ahmadipour, M., Dehghan, M., Ahmadinejad, M., Jabarpour, M., Mangolian Shahrbabaki, P., & Ebrahimi Rigi, Z. (2022). Barriers to hand hygiene compliance in intensive care units during the COVID-19 pandemic: A qualitative study. *Frontiers in Public Health, 2763*.

Ammirato, S., Linzalone, R., & Felicetti, A. M. (2020). Knowledge management in pandemics. A critical literature review. *Knowledge Management Research & Practice, 19*(4), 415–426.

Bao, L., Li, T., Xia, X., Zhu, K., Li, H., & Yang, X. (2022). How does working from home affect developer productivity?—A case study of Baidu during the COVID-19 pandemic. *Science China Information Sciences, 65*, 1–15.

Bdeir, F., Hossain, L., & Crawford, J. (2013). Emerging coordination and knowledge transfer process during disease outbreak. *Knowledge Management Research & Practice, 11*(3), 241–254. https://doi.org/10.1057/kmrp.2012.1

Cheetham, G., & Chivers, G. (1996). Towards a holistic model of professional competence. *Journal of European Industrial Training, 20*, 20–30.

Ewoh, A. (2013). *Managing and valuing diversity: Challenges to public managers in the 21st century.* https://doi.org/10.1177/0091026013487048

Lonska, J., Mietule, I., Litavniece, L., Arbidane, I., Vanadzins, I., Matisane, L., & Paegle, L. (2021). Work–life balance of the employed population during the emergency situation of COVID-19 in Latvia. *Frontiers in Psychology, 12*.

Stoker, J. I., Garretsen, H., & Lammers, J. (2022). Leading and working from home in times of COVID-19: On the perceived changes in leadership behaviors. *Journal of Leadership & Organizational Studies, 29*(2), 208–218.

Water Environment Federation & Centers for Disease Control and Prevention. (2022). *Water Environment Federation–Centers for Disease Control and Prevention NWSS utilities community of practice 2022. The future of wastewater surveillance*. Water Environment Federation. www.nwbe.org

# Index

## Y



## Z

www.ingramcontent.com/pod-product-compliance
Lightning Source LLC
LaVergne TN
LVHW060557240526
839643LV00003B/32